W9-CRV-386

COLLOID–POLYMER INTERACTIONS

COLLOID–POLYMER INTERACTIONS

From Fundamentals to Practice

Edited by

Raymond S. Farinato
Cytec Industries
Stamford, CT

Paul L. Dubin
Department of Chemistry
Indiana University-Purdue University at Indianapolis
Indianapolis, IN

A WILEY-INTERSCIENCE PUBLICATION
JOHN WILEY & SONS, INC.
New York · Chichester · Weinheim · Brisbane · Singapore · Toronto

This book is printed on acid-free paper. ∞

Copyright © 1999 by John Wiley & Sons, Inc. All rights reserved.

Published simultaneously in Canada.

For ordering and customer service, call 1-800-CALL-WILEY.

Library of Congress Cataloging-in-Publication Data:

Colloid-polymer interactions : from fundamentals to practice / edited
   by Raymond S. Farinato, Paul L. Dubin.
      p.   cm.
      "A Wiley-Interscience publication."
      Includes index.
      ISBN 0-471-24316-7 (alk. paper)
      Adsorption. 2. Polymers. 3. Colloids   I. Farinato, Raymond S.
   II. Dubin, Paul L.
   OD547.C636   1999
   541.3'3--dc21   98-50702

Printed in the United States of America.

10 9 8 7 6 5 4 3 2 1

CONTRIBUTORS

GERALD F. BELDER, Philips Research Laboratories Eindhoven (WA04), Prof. Holstlaan 4, 5656 AA Eindhoven, The Netherlands

FRANK D. BLUM, Department of Chemistry, University of Missouri-Rolla, Rolla, Missouri 65409-0010

MARCEL R. BÖHMER, Philips Research Laboratories Eindhoven (WA04), Prof Holstlaan 4, 5656 AA Eindhoven, The Netherlands

PER M. CLAESSON, Laboratory for Chemical Surface Science, Department of Chemistry, Physical Chemistry, Royal Institute of Technology, SE-100, 44 Stockholm, Sweden, and Institute for Surface Chemistry, Box 5607, SE-114 86, Stockholm, Sweden

TERENCE COSGROVE, University of Bristol, School of Chemistry, Bristol BS8 ITS, United Kingdom

RAYMOND S. FARINATO, Cytec Industries, 1937 W. Main St., Stamford, Connecticut 06904

PETER C. GRIFFITHS, Department of Chemistry, University of Wales, Cardiff CF1 3 TB, United Kingdom

PATRICK HARTLEY, Advanced Mineral Products Special Research Centre, School of Chemistry, University of Melbourne, Victoria 3052, Australia

PETER HAWKINS, Cytec Industries, 1937 W. Main St., Stamford, Connecticut 06904

WILLEM HOOGSTEEN, Philips Research Laboratories Eindhoven (WAO4), Prof. Holstlaan 4, 5656 AA Eindhoven, The Netherlands

SUN-YI HUANG, Cytec Industries, 1937 W. Main St., Stamford, Connecticut 06904

STEPHEN M. KING, Large-Scale Structures Group, ISIS Facility, Rutherford Appleton Laboratory, Chilton, Didcot, Oxon OX I I 0QX, United Kingdom

ANDREW MILLING, IRC in Polymer Science, Department of Chemistry, University of Durham, Durham DH1 3LE, United Kingdom

M. MUTHUKUMAR, Polymer Science and Engineering Department and Materials Research Science and Engineering Center, University of Massachusetts, Amherst, Massachusetts 01003

v

H. DANIEL OU-YANG,   Department of Physics, Lehigh University, Bethlehem, Pennsylvania 18015

ROBERT H. PELTON,   McMaster Centre for Pulp and Paper Research, Department of Chemical Engineering, McMaster University, Hamilton, Ontario L8S 4L7, Canada

GUDRUN PETZOLD,   Institute of Polymer Research, Hohe Strasse 6, 01069 Dresden, Germany

MARIA M. SANTORE,   Department of Chemical Engineering, Lehigh University, Bethlehem, Pennsylvania 18015

JOSEPH B. SCHLENOFF,   Department of Chemistry and Center for Materials Research and Technology (MARTECH), Florida State University, Tallahassee, Florida 32306

ROBERT D. TILTON,   Department of Chemical Engineering, Colloids, Polymers and Surfaces Program, Carnegie Mellon University, Pittsburgh, Pennsylvania 15213

BRIAN VINCENT,   School of Chemistry, University of Bristol, Bristol BS8 ITS, United Kingdom

CONTENTS

Preface   ix
Paul L. Dubin and Raymond S. Farinato

I   Applied Technologies   1

  1 Polyelectrolyte-Assisted Dewatering   3
  Raymond S. Farinato, Sun-Yi Huang, and Peter Hawkins

  2 Polymer–Colloid Interactions in Pulp and Paper Manufacture   51
  Robert H. Pelton

  3 Dual-Addition Schemes   83
  Gudrun Petzold

  4 Role of Polymers in Particle Adhesion and Thin Particle Layers   101
  Marcel R. Böhmer, Willem Hoogsteen, and Gerald F. Belder

II   Fundamentals of Colloid–Polymer Interaction   125

  5 Diffusion-Controlled Phenomena in Adsorbed Polymer Dynamics   127
  Maria M. Santore

  6 Depletion-Induced Aggregation and Colloidal Phase Separation   147
  Andrew Milling and Brian Vincent

  7 Polyelectrolyte Adsorption: Theory and Simulation   175
  M. Muthukumar

  8 Small-Angle Neutron Methods in Polymer Adsorption Studies   193
  Terence Cosgrove, Stephen M. King, and Peter C. Griffiths

III   Methods for Investigating Polymer Adsorption   205

  9 Nuclear Magnetic Resonance of Surface Polymers   207
  Frank D. Blum

10 **Radiochemical Methods for Polymer Adsorption**   225
Joseph B. Schlenoff

11 **Measurement of Colloidal Interactions Using the Atomic
Force Microscope**   253
Patrick G. Hartley

12 **Surface Forces Apparatus: Studies of Polymers, Polyelectrolytes,
and Polyelectrolyte–Surfactant Mixtures at Interfaces**   287
Per M. Claesson

13 **Scanning Angle Reflectometry and Its Application to Polymer
Adsorption and Coadsorption with Surfactants**   331
Robert D. Tilton

14 **Total Internal Reflectance Fluorescence**   365
Maria M. Santore

15 **Design and Applications of Oscillating Optical Tweezers for Direct
Measurements of Colloidal Forces**   385
H. Daniel Ou-Yang

Index   407

PREFACE

Scientific or technological progress results from expansion and refinement of the physical picture that underlies research and development. This book attempts to fuel that sort of progress in the field of colloid–polymer interactions.

Work in this field is usually approached from one of three directions: theory, fundamental experiments, or applications. This book attempts to bridge the gaps among these approaches. In so doing, we first hope to put into clearer focus the models used to organize and rationalize observations. A second goal is to provide technologists and engineers with an appropriate introduction to the kind of fundamental information that can be derived from modem experimental techniques, often best applied to model systems. Third, we wish to present to the nonspecialist some of the practical technologies that are based on colloid–polymer interactions. The limitations of these technologies may indicate where fundamental understanding is incomplete. All of these goals may be seen as efforts toward bridging the ivory tower and the sewage plant.

The highly complex nature of colloid–polymer systems encountered in technology often impedes information transfer among the three realms of activity. This complexity can distract the technologist from the basic underlying physics and frighten away the fundamental researcher from practical and relevant systems. The choice of reasonable systems for study represents a compromise between direct relevancy and adequate characterization; but the selection of an appropriate system is often prerequisite to generating an important insight that bridges gaps among the three areas. This kind of advance can lead to a new way of looking at the usual collection of facts and trigger a cascade of understanding that invigorates the spirit and sometimes the profitability of the research/technology community. We hope to spark that kind of enthusiasm with this book.

The participants in this field are often unaware of each other's work: Engineers implement technology in "real" (commonly large-scale) situations, typically involving complex and only partially characterized materials, such as wastewater, wood pulp suspensions, preceramic dispersions, or paints; application specialists may carry out similar efforts but on a smaller and reproducible scale; synthetic chemists manipulate polymer structures; research experimental scientists apply sophisticated methods to model systems; and theoretical/computational chemists and physicists attempt to model these systems from basic principles. The corresponding enormous variety of approaches, objectives, and jargon frequently precludes communications among these disparate groups. Engineers and application specialists can find basic research efforts obscure and irrelevant, while those doing basic research often throw up their hands when confronted with the complex and uncontrollable qualities of the "real" situations. It is our belief nevertheless that much valuable cross-fertilization could take place if suitable efforts were made. These may require those closest to technological applications to abandon some imprecise (albeit comfortable) jargon, and those at the other end to provide qualitative assessments and physical pictures along with their rigorous (mathematical) conclusions. Suitable venues must be found for these efforts; this book we hope provides one.

Certainly another motivation for such an effort at the present time is the range of new experimental methods developed in the past decades. The accompanying molecular insights deserve a wider audience in technology-driven areas. Similarly, new theories of polyelectrolyte adsorption can assist progress in applications.

Each chapter begins in the form of a tutorial that acquaints the scientist or engineer with the basics and current state of affairs in a particular subject. The remainder of the chapter is typically devoted to an exposition of the authors' most recent contributions and concludes with indications of the likely directions for future progress.

Part I pertains to three technologies that are strongly based upon colloid–polymer interactions: wastewater treatment, papermaking, and the nano-engineering of colloidal particle layers. These chapters serve as examples of the accomplishments and challenges in several commercially important areas.

Part II introduces fundamental topics that provide some of the basis for understanding colloid–polymer interactions. These chapters discuss the dynamics of polymer transport to interfaces, models for the adsorption of polymer chains, and the role played by nonadsorbed polymers.

Part III focusses on modern experimental techniques and related recent findings. These results are answering questions about how polymers arrive at, reside at, and control the interfaces of colloidal particles with liquid media. Macroscopic phenomena that form the basis of large-scale applications rest squarely on the molecular processes being illuminated by these techniques.

No single volume could describe the full range of applications involving polymer–colloid interactions. For example, biological particles or biopolymers are not represented here. Similarly, no one text could describe the entire array of experimental methods that probe phenomena concerning macromolecules at interfaces. Our purpose, rather, is to assemble presentations that bridge theory and simulations, model systems, and technology within one volume. We hope that applications specialists will thereby gain a broader view of the available instrumental and conceptual tools while basic researchers will see more clearly the utility of their efforts in "the real world."

Paul L. Dubin

Raymond S. Farinato

COLLOID–POLYMER INTERACTIONS

PART I
Applied Technologies

1 Polyelectrolyte-Assisted Dewatering

RAYMOND S. FARINATO, SUN-YI HUANG, and PETER HAWKINS
Cytec Industries, 1937 W. Main St., Stamford, Connecticut 06904

1.1 INTRODUCTION

1.1.1 Solid–Liquid Separations

Solid–liquid separations are an integral part of many industrial and practical processes. The value of the individual phases are enhanced in this separation process. The focus of such a separation may be the reclamation of a valuable solid phase (e.g., ceramic processing, mineral recovery, paper production), a valuable soluble substance (e.g., pharmaceuticals from fermentation), or perhaps one of the most intrinsically valuable substances—clean water. In some cases the natural density difference between the solid and the liquid is sufficient to drive the separation under normal gravity; however, in many instances when the suspended solids are in the colloidal to mesoscopic size range, this separation process is thwarted by the natural stabilizing forces among the suspended particles. This results in either incomplete or intolerably slow separations. Such situations are commonplace in many technologies such as water treatment, mining, papermaking, ceramic processing, product recovery from bacterial fermentation, and erosion control. A diverse collection of equipment and technologies has been designed and developed to mechanically separate solids from aqueous liquids.[1,2] In a vast majority of situations, some method of conditioning the solids is required to improve the mechanical separation efficiency. When the liquid phase is predominantly water, the use of polyelectrolytes is often an important component in the solid–liquid separation process.

Colloid–Polymer Interactions: From Fundamentals to Practice, Edited by Raymond S. Farinato and Paul L. Dubin
ISBN 0-471-24316-7   © 1999 John Wiley & Sons, Inc.

3

1.1.2 Current State of Polyelectrolyte-Assisted Dewatering

In this chapter we will be dealing exclusively with the separation of suspended solids from aqueous phases, with a special emphasis on the use of synthetic water-soluble polymers for this purpose. Commercially significant use of synthetic polyelectrolytes for dewatering began in the 1960s in most industrialized nations. Usage is currently large and projected to grow on a worldwide basis.[3] Virtually any process for the separation of solids from aqueous liquids is a candidate for enhancement by water-soluble polymers if an economic advantage in the form of process throughput, solids capture, minimization of solids' moisture content, liquid purity, or environmental impact can be achieved. In the current state of affairs, a relatively small number of monomers are used to produce the polymers of significant economic value. During the same time period that industrial research and development (R&D) labs were developing these materials for specific technologies, researchers in both the academic and industrial communities were investigating the fundamental elements that needed to be understood in order to control the functioning of polyelectrolytes in dewatering applications. These elements, shown schematically in Figure 1.1, include the behavior of water-soluble polymers in solution and at interfaces; the mechanisms of polymer and particle transport, collision, and adsorption; the concepts of colloid destabilization and floc formation; the stresses on aggregated materials during mixing, settling, and compaction; and the mechanics of fluid flow in porous media. Often, polyelectrolyte-assisted dewatering is practiced under highly nonequilibrium conditions. A complete understanding thus also requires a knowledge of the dynamics of the above processes.

Wastewater treatment is an example that shows that solids control in municipal and industrial applications is not only critical to the effective treatment of the water but represents one of the highest costs of the treatment scheme. Improvements in removal of primary solids (those solids entering the treatment plant with the influent water) allow for better overall treatment by removing solids and organic load from the secondary treatment operation (often a biological stage[4,5]), by reducing costs in the form of reduced energy requirements in secondary treatment, and possibly by removing additional pollutants (e.g., phosphorus and metals). Improvements in the removal of secondary solids (those generated biologically in secondary treatment) result in cleaner final effluent water, thus contributing to cleaner water resources or allowing for a number of effluent water reuse options.[6] The use of polymers to improve settling or flotation rates and the level of compaction of the solids can have the effect of reducing equipment requirements and thus capital costs. Final disposal of solids from these operations is a huge undertaking involving transportation, incineration, reuse (e.g., land application, composting), and landfilling. Dramatically reducing the moisture content and weight of the solids to be handled in dewatering applications results in significant advantages in the form of reduced energy costs (evaporation of water in incinerators) and transportation costs. Improving particle capture efficiencies leads to a reduction of the amount of solids that must be recycled and retreated in the plant. In applications where the solids removed represent a

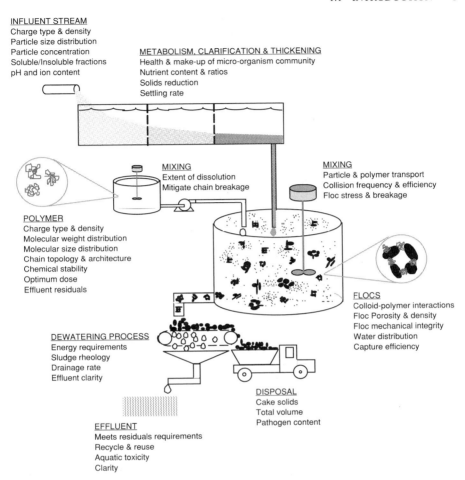

INFLUENT STREAM
Charge type & density
Particle size distribution
Particle concentration
Soluble/Insoluble fractions
pH and ion content

METABOLISM, CLARIFICATION & THICKENING
Health & make-up of micro-organism community
Nutrient content & ratios
Solids reduction
Settling rate

MIXING
Extent of dissolution
Mitigate chain breakage

MIXING
Particle & polymer transport
Collision frequency & efficiency
Floc stress & breakage

POLYMER
Charge type & density
Molecular weight distribution
Molecular size distribution
Chain topology & architecture
Chemical stability
Optimum dose
Effluent residuals

FLOCS
Colloid-polymer interactions
Floc Porosity & density
Floc mechanical integrity
Water distribution
Capture efficiency

DEWATERING PROCESS
Energy requirements
Sludge rheology
Drainage rate
Effluent clarity

DISPOSAL
Cake solids
Total volume
Pathogen content

EFFLUENT
Meets residuals requirements
Recycle & reuse
Aquatic toxicity
Clarity

FIGURE 1.1   Schematic of flocculation and dewatering process.

valuable product (e.g., pigments, sugar, and pharmaceutical intermediates), improved capture and throughput, removal of impurities, handling ease, and reduced drying requirements easily translate into worthwhile improvements. Much of this can be accomplished by using polyelectrolytes in the solid–liquid separation operation.[7]

The recent development of water-soluble polymer systems for use in soil conditioning and erosion control[8] demonstrates the current desire to mitigate environmental impact in agriculture. Polyacrylamide copolymers can be dosed to croplands to improve water quality, as well as nutrient and fertilizer retention. These polymers can also effectively reduce soil erosion in furrow irrigation. These attributes can result in significant improvements in the quality of nearby surface waters that are impacted by agricultural runoff. Such nonpoint sources of pollutants are currently regarded as the largest source of degraded water quality.

The use of polyelectrolytes in solid–liquid separations raises several concerns. These include their potential aquatic toxicity, their handling properties, the fate of the polymer and vehicle (e.g., emulsion) constituents, and the influence of their volatile organic contents on air quality. Although it is out of the scope of this discussion to address these issues in detail, suffice it to say that when handled properly and knowledgeably, polyelectrolytes provide a valuable tool for the continued improvement of aqueous liquid–solids separation applications.

This chapter is organized as follows. First, we discuss the physics of polyelectrolyte–colloid destabilization and floc formation. The important elements of polymer and particle transport, the adsorption event, and the state of the adsorbed polymer will be only mentioned, since these topics are covered in more depth in other chapters in this book. Next, we discuss the physics of suspension thickening and dewatering. A survey of commercially significant polyelectrolytes for dewatering applications will follow. Some examples of polyelectrolyte-assisted dewatering in wastewater treatment technology will illustrate useful commercial manifestations of the principles outlined. Several practical methods used for evaluating polymers for potential dewatering applications will then be discussed.

1.2   COLLOID DESTABILIZATION AND FLOC FORMATION

Fine particles in suspension often must be aggregated in order to improve their settling rate and dewaterability. This aggregation into flocs requires a destabilization of the particles, a means of bringing them together in the destabilized state, and adhesion between destabilized particles to occur. Once formed, the flocs must then adequately survive any disruptive forces during the solid–liquid separation process. The sizes of objects involved in the flocculation process are depicted in Figure 1.2.

1.2.1   Consequences of Colloid–Polymer Interactions

When interactions between colloids and polymers occur, the outcome is generally either stabilization, destabilization, or phase separation. In stabilization, the ubiquitous London–van der Waals forces of attraction are counterbalanced by the repulsive electrostatic and/or steric forces arising from either adsorbed or anchored polymer chains. Stabilization usually occurs at high surface coverage, and for polyelectrolytes, there is often an overcompensation of surface charges by adsorbed polymer charges (i.e., there is a reversal of the sign of the zeta potential). Destabilization by polyelectrolytes can result from charge neutralization, charge patch formation, or bridging. The second and third phenomena usually require low to intermediate amounts of surface coverage. Phase separation can occur when the attractive potential created between particles is weak enough (~ 2 to 5 kT) to still allow transport of the two phases on a reasonable time scale. Such attractive potentials can result via a depletion mechanism involving nonadsorbing polymers. In this chapter we

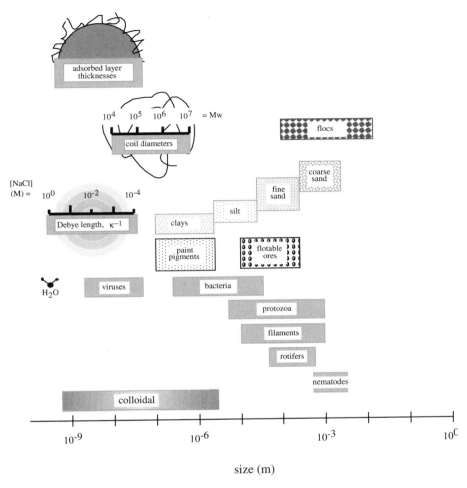

FIGURE 1.2 Size ranges for polyelectrolytes and substrates. Debye lengths are for 1:1 electrolyte; coil diameters are in 1 M NaCl for AMD-AETAC copolymers[62]; adsorbed layer thicknesses are in pure water and are estimates for high-molecular-weight polyelectrolytes[200,201]; particle sizes are from Refs. 117 and 202.

will be primarily concerned with colloid destabilization as a necessary condition for flocculation and dewatering.

Once destabilized, colloidal particles can form aggregates. Terms used to describe aggregation processes with different characteristics can have meanings that depend on context. While *coalescence* connotes the merging of smaller droplets into larger drops, *coagulation* and *flocculation* refer to the aggregation of solid (or at least mechanically rigid) particles. In the colloid literature,[9,10] *flocculation* refers to the formation of aggregates that can be readily redispersed, and *coagulation* to the formation of aggregates that cannot. In the dewatering literature, *coagulation* often implies destabilization by charge neutralization[11] (i.e., with electrolytes or low-molecular-weight polyelectrolytes), and *flocculation* often implies the involvement of a higher-molecular-weight polymer. The products of these aggregation mechanisms can range from compact to open, from large to small, and from strong to fragile.[12] The best form for the floc depends on the dewatering method used. For example, filtration requires strong porous flocs, sedimentation requires large flocs, and consolidation (reducing the amount of trapped liquid) requires dense flocs.[13]

1.2.2   Interactions in Particle–Polymer Mixtures

Predicting colloid stability forms a substantial enterprise in colloid science. The stability of colloidal systems depends on both thermodynamic and kinetic effects. The behavior of such systems is governed by both interparticle interactions and external (e.g., hydrodynamic) forces. The particle–particle interactions can be conceptually separated into equilibrium and dynamical contributions. Mathematical expressions for the equilibrium contributions are typically couched in terms of interaction potential energies. Potential energy diagrams serve to depict the equilibrium attractions or repulsions as a function of the separation distance between two particles. These static contributions are well described, for systems without polymers, by the DLVO (Derjaguin, Landau, Vervey, and Overbeek) theory discussed below. However, when polymers are present, additional concepts relating to the long-chain nature of the polymers are required to accurately describe even the equilibrium contributions to the interaction energies.

In systems that include either free, adsorbed, or anchored polymer chains, an additional component to the interparticle interaction must be included. These polymer-mediated forces fall into three categories. For nonadsorbing polymers, depletion forces (see Chapter 6) can exist due to unequal distributions of the polymers in the bulk and in the gap between two particles. This is a result of polymer conformational entropy loss in the gap.[14,15] However, these forces are typically not important in most dewatering applications. For polymers adsorbed or anchored to the particle surface at large surface coverage, a polymer-mediated force that can extend 10 to 100 nm will arise due to polymer–polymer interactions across the gap. These can be either repulsive or attractive depending on the solvent quality and the extent of the polymer excluded volume effect. Such forces, arising from the proximity and overlapping of polymer chains, have been traditionally referred to as steric forces. Description of such steric forces, in combination with the classical DLVO theory, comprise the extended DLVO theory. In the case of particle flocculation via high-molecular-weight polyelectrolyte adsorption, the steric and electrostatic contributions of the polymer to colloid stability are not easily separable. For adsorbed polymers at low to intermediate surface coverage, there exists the possibility of a single polymer chain bridging between two particles, creating a strong attractive force between the particles. Such bridging forces are central in the flocculation of colloidal particles by high-molecular-weight polyelectrolytes, especially at low polyelectrolyte charge densities.[16]

The conceptual scheme of the extended DLVO theory, in combination with the idea of polymer bridging, is strongly supported by direct measurement of the forces between polymer-bridged colloidal particles.[17–20] These measurements have been made possible by the elegant collection of contemporary methods for observing, confining, manipulating, and stressing colloids, macromolecules, and their combinations. The development of such methods constitutes a significant advance in colloid science.[21]

Dynamical contributions to the interparticle interactions are of two types. First, since the particles are moving in a fluid medium, a hydrodynamic interaction between particles takes place. That is, the flow fields transmit mechanical forces between the particles. This affects the kinetics of colloidal (e.g., diffusion, aggregation) and polyelectrolyte (e.g., collapse and rearrangement on a surface) processes, but not the equilibrium distributions of the particles. Second, as there are generally many internal degrees of freedom in systems of colloidal particles and polyelectrolytes (e.g., ion distributions, chain conformations), the different rates of equilibration of these internal degrees of freedom compared to the collision rate of the particles becomes important. Because of this coupling, nonequilibrium structures can predominate in many flocculation circumstances. This is especially true when the polymers and particles are mixed under turbulent conditions.

External forces on the system in dewatering applications are typically hydrodynamic in nature, although infrequently, electric and magnetic fields are sometimes present. Mixing effects are discussed below. The importance of externally imposed flow fields relative to thermally driven Brownian motion of the particles and polymers defines the orthokinetic (imposed > thermal) and perikinetic (thermal > imposed) regimes of behavior. In flocculation and dewatering processes in wastewater treatment, the crucial stage is almost always in the orthokinetic regime.

DLVO Theory   In fine-particle suspensions the equilibrium interaction forces between particles can be successfully understood as a superposition of three main components: London–van der Waals, electrostatic double layer, and forces mediated by polymer chains. While we conceptualize many other kinds of forces between matter, the importance of these three in colloid interactions is encapsulated in the extended DLVO theory,[22,23] which was developed independently by two groups in the 1940s: Derjaguin and Landau in the Soviet Union and Verwey and Overbeek in the Netherlands. The original DLVO theory demonstrated that the stability of charged colloidal systems was governed by a competition between the London–van der Waals forces of attraction and the electrostatic double-layer forces of repulsion. Both of these forces are long range in nature (extending to a few tens of nanometers), with well-defined analytical forms over the range of particle separations characteristic of most coagulation and flocculation phenomena. At short interparticle distances (<1 nm) there are a number of other forces that come into play such as solvent structural forces (due to differences between interfacial and bulk solvent), and finally the electron overlap repulsion (Born repulsion; Pauli exclusion). For the most part, it is not necessary to understand the details of these contributions in order to understand polyelectrolyte-assisted flocculation.

The combined effects of the attractive London–van der Waals interactions (V_{LvdW}) and the repulsive electrical double-layer interactions (V_{el}) are often summarized on a plot of total potential energy as a function of the particle–particle surface separation.[22] The potential energy of interaction between two particles controls the rate of coagulation or flocculation. The DLVO potential curve (Fig. 1.3) can have features (energy barriers, primary (A) and secondary (C) minima) that indicate coagulation and reversible flocculation. In the DLVO model, the energy barrier (B) must be overcome by sufficient impetus from either thermal ($k_B T$) or hydrodynamic forces for an irreversible association to occur. Barriers on the order of 10 to 15 $k_B T$ can prevent access to the primary minimum (A).[22]

An assessment of the energy barrier to coagulation can be made from the stability factor (W), which is determined experimentally as the ratio of the slow coagulation rate (with the barrier) to the rapid coagulation rate (without the barrier; i.e., diffusion controlled). For systems that obey the DLVO model, the barrier can be removed by adding sufficient electrolyte ($\geq$ critical coagulation concentration). Initial coagulation rates can be determined, for example, from turbidity measurements. The stability factor has been related to an integral of the total potential energy curve and, in an approximation, to the barrier height [$W \sim (1/a) \exp(V_{\text{max}}/k_B T)$] when the particles are moving only by Brownian diffusion.[24] The reciprocal of the stability ratio can also be thought of as the probability of particle–particle collision resulting in attachment.[25] Other parameters (H_{121}, ζ) of the interaction potential can also be estimated from simple coagulation kinetics data.[26] However, for destabilization mediated by polyelectrolytes, the same level of analytical detail is not available and the limits of the DLVO theory are usually exceeded.

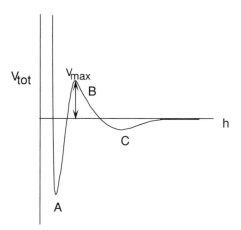

FIGURE 1.3 Schematic DLVO potential energy curve (with no polymer-mediated interactions) showing a (A) primary minimum associated with coagulation, (B) an energy barrier, and (C) a secondary minimum associated with reversible flocculation.

London–van der Waals Attractions   London–van der Waals (L–vdW) interactions are nearly always attractive ($V_{LvdW} < 0$) and are ubiquitous between all matter that supports electromagnetic fluctuations (i.e., that is polarizable). They are in large part responsible for the cohesion of condensed matter and can be calculated for simple geometries using the Hamaker theory. For example, the attractive potential between two identical spheres of radius a interacting at a short distance (surface separation = $h \ll a$) is given by[10]:

$$V_{LvdW}(h) = -\frac{a}{12}\frac{H_{121}}{h} \qquad (1.1)$$

A more complete collection of expressions for other geometries and separation ranges may be found in several sources.[23,27,28] The material properties of the system are described by the Hamaker coefficient (H_{121}), which depends on both the properties of the particles (subscript 1) and the medium (subscript 2). A more rigorous theory for computing L–vdW energies was developed by Lifshitz and colleagues.[29,30] In the Lifshitz theory, an effective Hamaker coefficient is computed from a knowledge of the frequency-dependent dielectric permittivities of the discrete and continuous phases. Such detailed spectroscopic data are rarely available over a wide enough frequency range for most materials, and usually the Hamaker coefficient is either determined empirically (e.g., from colloid stability measurements[31]) or estimated from tables of values for similar materials.[10,23,32,33] Some examples of the Hamaker coefficient are shown in Table 1.1. For example, the attractive energy between two 2-μm bacterial cells in water (a reasonable model for a waste-activated sludge) will exceed $k_B T$ when their separation is less than about 4-60 nm, depending on the cell composition. In general the L–vdW interaction energies are dictated by the situation and are not amenable to serious outside modification.

Electrostatic Repulsions   Many particles subjected to solid–liquid separations have a surface charge in water. The surface charges can arise from the ionization of surface groups (e.g., COOH → COO⁻ + H⁺), preferential adsorption of potential determining ions (e.g., Al^{+3}), the dissolution of ions from the surface (e.g., F⁻ from apatite), or from

TABLE 1.1   Representative Hamaker Coefficients

Material 1	Material 2	H_{121} (10^{-20} J)	$H_{121}/k_B T$	Reference
Water	Air	3.7	9	23
Quartz	Water	0.8	1.9	23
Mica	Water	2.1	5.1	23
Polystyrene	Water	1.0	2.4	23
Metals	Water	5–30	12–73	10
Oxides	Water	0.5–5	1–12	10
Protein	Water	1–5	2–12	33
Cell	Water	0.02–0.3	0.05–0.7	33

the adsorption of polyelectrolytes or charged surface-active species (ionic surfactants). Such charged particles organize a diffuse cloud of counterions and coions around them to form an electrical double layer, as depicted in the Gouy–Chapman model. Additionally, a tightly bound layer of counterions (Stern layer) is often included in this construct. When the electrical double layers for like-charged particles overlap, a repulsive interparticle potential results, which acts to stabilize the dispersion, and which must be overcome in all flocculation processes. Chemical modifications and additions to the system are most likely to alter the electrical double layer and are the chief route to affecting colloid stability.

The electrical double-layer interaction potential, $V_{el}(h)$, between two charged particles accompanied by their ion clouds can be calculated as a function of the surface separation (h; cm) from a knowledge of the electrical potential, $\Phi(z)$, around a single such particle; $\Phi(z)$ is determined from solution of the Poisson–Boltzmann equation. Various approximations to solving the Poisson–Boltzmann equation have been employed to yield analytical solutions for $\Phi(z)$. In the case of weak overlap of the double layers, an analytical expression for $V_{el}(h)$ can be derived that illustrates the important parameters in the electrostatic repulsion between two identical spheres of radius a.[10,23]

$$V_{el}(h) = \frac{64\pi a k_B T n(\infty)\Gamma_0^2}{\kappa^2} \exp(-\kappa h) \tag{1.2}$$

where $n(\infty)$ is the ion number concentration far from the particles, $1/\kappa$ (cm) is the Debye screening length, and Γ_0 is related to the surface potential (Φ_0)[23] and hence to the surface charge density (σ; C/cm^3):

$$\frac{1}{\kappa} = \left(\frac{\varepsilon_r \varepsilon_0 k_B T}{\Sigma_i (z_i e)^2 n_i(\infty)}\right)^{1/2} \tag{1.3}$$

where ε_r is the relative dielectric permittivity, and ε_0 is the permittivity in vacuum.

For a simple model of the electrical double layer consisting of a charged surface and a diffuse ion cloud (i.e., no strongly bound surface counterions),

$$\Gamma_0 = \tanh\left(\frac{ze\Phi_0}{k_B T}\right) \approx \frac{1}{4}\left(\frac{ze}{k_B T}\right)\Phi_0 \quad \text{(for small } \Phi_0) \tag{1.4}$$

$$\sigma = (8\, k_B T n(\infty)\varepsilon_r \varepsilon_0)^{1/2} \sinh\left(\frac{ze\Phi_0}{k_B T}\right) \tag{1.5}$$

These relations serve to indicate that the important parameters that control V_{el} are the surface potential and the Debye screening length. The actual situation in practical dewatering applications is somewhat more difficult to model due to complex particle geometries and topologies, heterogeneous surface chemistries with complex equilibria and kinetics, and hydrodynamic interactions among the particles. For example, the effective V_{el} will be greater for rougher surfaces than for smooth surfaces.[31,34]

The surface potential, σ, can be difficult to determine. A good practical approximation is to use the zeta potential (ζ, mv) instead. This is the effective potential at a point in the vicinity of the shear plane of a moving particle. It can be estimated from the electrophoretic mobility of the particle and is commonly used as an index of the electrical potential of the double layer when there are no adsorbed polymers. Adsorbed or anchored polymers shift the shear plane away from the particle surface, making it difficult to identify exactly where in the diffuse layer the zeta potential is located. Nonetheless, the zeta potential can be a useful particle characterization parameter.[35] Several electrokinetic methods are available for determining ζ, or in some cases an index related to ζ. They all require a model of the electrical double layer in order to calculate ζ from the measured quantities.[31] These methods include electrophoresis [particle motion in a direct current (DC) electric field], electrophoretic light scattering (Doppler shifting of scattered light due to particle motion in a DC electric field), streaming potential or current (voltage difference or current generated by electrolyte motion passed a stationary surface), electroosmosis (fluid motion passed a stationary surface due to applied DC electric field), sedimentation potential (voltage difference generated by a particle falling through an electrolyte), diffusiophoresis (particle migration in solute gradient), acoustoelectrophoresis (acoustic wave generated by pulsed electric-field-induced particle motion), and electroacoustophoresis (voltage generated by acoustic pulse-induced particle motion).

Another useful index related to the electrical state of the particle surface is the zero point of charge (ZPC) or isoelectric point (IEP). These are measures of the solution conditions required to neutralize the surface charge ($\sigma \cong 0 \Rightarrow \zeta = 0$) as measured by any of the above techniques. In aqueous solution this is typically an acid–base titration, and the ZPC is reported as the pH at which one obtains a null electrokinetic effect. However, other titrants such as charged surfactants and polyelectrolytes can also be used, whereby the results are reported as an equivalent charge per weight of particle. Examples of ZPCs for common materials are shown in Table 1.2.

When neutralization of the surface charge is the dominant destabilization mechanism, coagulation and flocculation typically are the most rapid near the ZPC. However, for flocculation mediated by high-molecular-weight polyelectrolytes, the kinetics are often fastest near, but not at, the ZPC.[36–38] One explanation for this may be gleaned from recent simulations of flocculating spherical particles with adsorbing polymer chains performed by Dickson and Euston.[39] They showed that a soft repulsive potential between the particles produced a more open and deformable floc, more reminiscent of the kinds of structures seen in bridging flocculation experiments. These kinds of structures might be expected to drain better than those formed from the aggregation of hard spheres, which would be more reminiscent of electrostatic patch

TABLE 1.2   Zero Point of Charge (ZPC) for Materials in Aqueous Media[10]

Material	ZPC (pH)
Quartz (SiO_2)	2–3.7
Cassiterite (SnO_2)	4.5
Rutile (TiO_2)	6
Hydroxyapatite ($Ca_5(PO_4)_3OH$)	7
Hematite (Fe_2O_3); natural	4.8
Hematite (Fe_2O_3); synthetic	8.6
Corundum (Al_2O_3)	9.0
Calcite ($CaCO_3$)	9.5
Magnesia (MgO)	12

neutralization results. A small, but nonzero zeta potential (i.e., not at ZPC) would translate into a soft repulsive potential between the particles.

In summary thus far, from the original DLVO theory (no polymers) we learn that the stability of charged colloids is determined by their interparticle potentials, which are a sum of an attraction contribution due to London–van der Waals interactions (V_{vdW}), and a repulsion contribution due to electrical double-layer interactions (V_{el}). From simple models the important scaling relations are $V_{vdW} \sim a\,H_{121}/h$ and $V_{el} \sim a\,\zeta^2\,f(\kappa h)$, illustrating the decay lengths (h dependencies) for these potentials. From the point of view of flocculation with polyelectrolytes, however, this picture is incomplete.

Polymer-Mediated Interactions   The long-chain nature of polymers must be taken into account when modeling their effects on the forces between particles. In the case of polyelectrolyte-assisted dewatering, the dissolution, transport, and adsorption of the polymer to the particle surface are necessary preludes to the operation of these forces (Fig. 1.4). As these are all dynamic processes, their time scales relative to particle collision rates are often quite crucial in flocculation applications. This necessarily means that mixing and the rate of polymer addition are important variables as well as the polymer properties.[40]

The transport of polymer to the vicinity of the particle surface in flocculation applications is most commonly accomplished with mechanical mixing (orthokinetic regime). Collisions of polymers and particles must be sufficiently frequent and successful at producing attachment. The high-affinity binding isotherms of most polyelectrolytes on oppositely charged surfaces suggest that ionic polymer segments can greatly increase the attachment probability. This effect is mitigated with increasing ionic strength. In spite of the fact that polyelectrolytes bind quite strongly to particles of the opposite charge, the adsorption process in solution is often neither strongly exothermic nor driven by enthalpy. For example, microcalorimetric measurements of a poly(acrylate) adsorbing on $CaCO_3$ have shown that the adsorption enthalpy is weakly endothermic; only about +2 kJ/mol.[41] It was postulated that the driving force for adsorption in this case was entropy gain (release of bound ions and

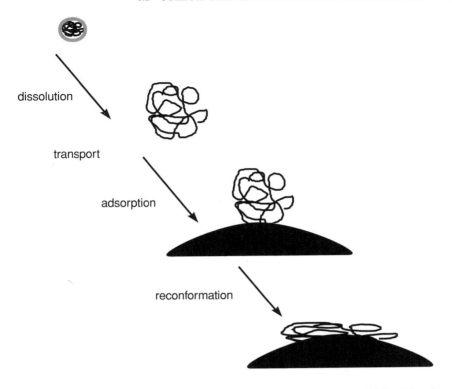

FIGURE 1.4   Polyelectrolyte dissolution, adsorption, and rearrangement at a solid–liquid interface.

solvent). It is the multiple attachments of a polymer chain to the particle surface that results in its kinetically irreversible binding.

An understanding of the adsorbed state of the polymer, including its dynamics, is central to modeling the polymer-mediated interactions. The details of attached polymer segment density distributions (trains, loops, and tails) and their influence on particle interactions has been a field of intensive study where much progress has been made both theoretically[42] and experimentally.[17,43–49] The situation is more complex for polyelectrolytes due to their long-ranged Coulombic interactions. Here the combined steric and electrostatic interactions must be taken into account and the theory is less developed.

The dynamics of adsorbed polymer rearrangements can have profound effects.[40,50] In the incipient stage of polyelectrolyte adsorption, the chain has a conformation similar to its solution conformation. However, soon thereafter, the polyelectrolyte will tend to flatten on the surface to an extent related to its charge density (Fig. 1.4). The outcome depends on the balance of interactions among the polymer segments, surface sites, and solvent. In the few examples studied,[50–52] the time for this collapse of a high-molecular-weight polyelectrolyte onto the surface is on the order of seconds to minutes, which can be long compared to the collision times in agitated systems of particles and polymers. During such a nonequilibrium event the opportunities for polymer bridging can be significantly enhanced (Fig. 1.5). Once the polymers become affixed, we need to understand the interactions of particles with adsorbed chains.

It is helpful to conceptualize two general regimes for adsorbed polymer interactions. At high surface coverage, polymer–polymer interactions are important. These steric interactions are usually, but not always, repulsive in good solvents and form the basis of the polymeric stabilization of colloidal dispersions.[14,53] At lower surface coverage polymer–particle interactions are more important. They result in two key effects for understanding dewatering applications: polymer bridging and electrostatic patch interactions. Also, in both regimes, adsorbed polyelectrolytes will neutralize some fraction of the surface charge and will thereby alter the electrostatic repulsive forces.

In polymer bridging (Fig. 1.5) the same polymer chain is simultaneously adsorbed onto more than one particle. Due to multisegment binding to surface sites, the attachment is essentially irreversible on the time scale of the flocculation process, and this provides a particle–particle binding strength that is commensurate with some multiple of the polymer backbone chain strength (dissociation energy ~6×10^{-16} J/bond => breakage force ~1 to 10 nN).[54,63] As such, the resulting flocs can be quite strong, but once broken they tend not to reform.[55,63] In cases where bridging is important it is often found that the optimum flocculation rate occurs when the average surface charge on the particles has been reduced to a small, but nonzero, value of the same sign as the original charge.[56] It turns out this yields the best opportunity for polyelectrolyte bridging. Flocculation by bridging tends to dominate with high-molecular-weight polymers having a low charge density[57,58] and improves with polymer molecular weight. In some cases, polyelectrolyte bridging can occur when the polymer and particle have the same sign of charge. This requires the presence of a low-molecular-weight counterion that binds strongly both to the particle surface and

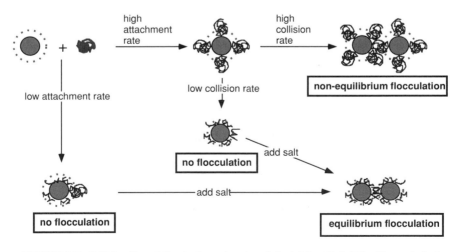

FIGURE 1.5   Bridging flocculation in charged systems (adapted from Ref. 203 with permission).

the polyelectrolyte. Some examples of this are the Ca^{2+} promotion of poly(acrylamide-*co*-acrylate) bridging in clay suspensions,[59] and the use of alum or polyaluminum chloride (PAC) prior to flocculating mineral refuse,[60] cellulose slurries, or raw waters[61] with anionic polyelectrolytes.

Electrostatic patch interactions occur between two particles having surface regions with differences in local charge densities. When two patches of opposite sign on different particles interact, the effect is an attraction. These electrostatic patches arise from a nonuniform distribution of adsorbed polyelectrolyte, wherein local (surface + near-surface regions) charge neutralization has occurred to varying degrees. Local overcompensation of surface charge can occur if the spacing of the charged groups on the polymer is less than that on the particle surface or if the conformational freedom of the adsorbing chain does not allow all charged groups on the polymer to reach the surface charges. The optimum flocculation in situations where a charge patch mechanism dominates tends to occur at the ZPC[62] and is nearly independent of polymer molecular weight. The binding strength of flocs with electrostatic patch interactions are thought to be greater than those with simple charge neutralization,[63] wherein only the remnant L–vdW forces are producing particle attractions.

Steric (polymer–polymer) forces derive from the free energy of interaction between adsorbed or anchored polymers. They are relevant in polyelectrolyte-assisted flocculation insofar as they can exist under overdosed conditions and are to be avoided. The presence of even nonadsorbing polymers can affect the stability of colloidal dispersions. Depletion forces arise from an osmotic pressure difference between the bulk fluid and the fluid between two particles. Polymers are restricted from narrow gaps. This increases the bulk fluid osmotic pressure and contributes an attractive force between the colloidal particles. Typically these effects are less important in polyelectrolyte-assisted dewatering because of the very strong adsorption of the polyelectrolyte onto the particle surface.

In summary to this point, the dominant concepts in the case of polyelectrolyte-mediated destabilization and flocculation of colloidal particles are surface charge neutralization (ion exchange), charge patch formation (attached but unneutralized charge), and polymer bridging.

1.2.3 Floc Formation Process

Collision, Attachment, and Breakage   The process of intimately mixing polymers and particles is known as conditioning. During this process the polymer and particles must become interdispersed and be driven to interact with one another in order to promote both destabilization and floc formation. The details of the aggregation process determine the physical characteristics of the floc, which in turn determine the rheology of the suspension,[64] the structure and mechanical properties of the consolidated sludge, and the dewatering properties of the system. Because of the nonequilibrium nature of many of the subprocesses in floc formation, the essential elements in describing floc formation must include mixing dynamics and polymer addition rates, in addition to the traditional descriptors of the polymer, solvent, and particles.

The details of the aggregation process are affected by several factors that influence the balance between floc growth, breakage, and rearrangement. In the case of floc growth in polyelectrolyte-assisted dewatering, a polymer must collide with and stick to a particle; then a polymer-decorated particle must collide with and stick to another particle, and so on. Floc growth rates are therefore functions of the collision rate (J_S) and the collision efficiency (α_0 = orthokinetic sticking probability). The initial rate of floc formation is given by[65] $J = \alpha_0 J_S$. An expression for the collision rate, J_S, was first derived by Smoluchowski[66] in 1917 for particles with no hydrodynamic or interparticle interactions. The exact form of J_S depends on the driving force for the collision: Brownian motion, laminar shear, turbulent shear, turbulent inertia, or differential settling.[54,67,68] In wastewater treatment particle collisions are usually driven by fluid motion (orthokinetic). Under conditions of turbulent shear, the collision rate term ($J_S \Rightarrow J_T$) for N_j particles/cm^3 of radius r_j hitting a reference particle of radius r_i is[69]

$$J_T = \left(\frac{8\pi e_0}{15\nu}\right)^{1/2} N_j(r_i + r_j)^3 \qquad (1.6)$$

To obtain the rate of total contacts per unit volume of suspension, multiply the above expression by N_i. The energy dissipation/mass (e_0) and kinematic viscosity (ν) are often combined to calculate an effective velocity gradient,[70] $G = (2/15)^{1/2} (e_0/\nu)^{1/2}$. Estimates of the numerical prefactor vary but are generally of $O(1)$. Using this approach, Saffman and Turner[70] estimated the effective shear rate in jar-to-jar mixing to be ≈ 60 s^{-1}. For a stirred tank configuration an estimate of G is given by $(P/\mu V)^{1/2}$, where P is the power input (a function of impeller speed, size and geometry, and suspension density), μ is the suspension viscosity, and V the volume of fluid.[71,72] Glasgow and Liu[68] estimated that for a Phipps–Bird impeller stirring at 30 rpm in an octagonal box (see Glasgow and Liu[68] for geometry), $G \approx 30$ s^{-1} in the vicinity of the vortex. The velocity gradients in several configurations of jar test equipment common to flocculation testing have also been determined.[73,74,85]

Methods for determining flocculation rates measure either the rate of reduction of total particle number concentration, $N(t)$, or the rate of increase of the average particle size; these involve a variety of particle analysis techniques based on scattering (turbidity, single- or multiangle intensity,[25,75] photon correlation spectroscopy,[25,76] intensity fluctuations[55,77]), microscopy,[78,79] and electrical properties (Coulter counter[79,80]). Less direct methods utilize settling rates and rheological measurements. In the most sophisticated experiments on flocculation rates, the time evolution of the particle size distribution has been tracked. Single-particle optical sizing (SPOS) techniques[81] have been developed that allow this determination for initially narrowly dispersed (in size) systems.

Analytical solutions of the differential equations for floc growth rate are available for simple situations. For example, in the early stages of the orthokinetic flocculation of an initially monodisperse suspension of particles of number concentration N_0, and constant volume fraction ϕ, the time-dependent number concentration $N(t)$ is given by[82]:

$$\frac{N(t)}{N_0} = \exp\left[\frac{-4\alpha_0\,\phi Gt}{\pi}\right] \tag{1.7}$$

The expression for the growth of the average particle size with time is of a similar form except for a sign reversal in the argument of the exponential.[83] The time scale for aggregation under these conditions is $\tau_0 \sim 1/(\alpha_0\,\phi G)$. The dimensionless parameter $G{\cdot}t$ characterizes the opportunity for particle contact, and is often used, with mixed success, in scaling flocculation data under different mixing conditions.[84,85] Complications arise since floc breakage can also occur during agitation.

Most of the colloid and polymer chemistry is contained in the parameter α_0, which is the orthokinetic collision efficiency (probability) to produce aggregation. It is affected by the particle interaction potentials, the amount of adsorbed polymer, its charge and conformation on the surface, and the hydrodynamic forces. A full computation of α_0 including all these effects under typical turbulent conditions found in water treatment applications is still lacking, although some progress has been made.[86] However, expressions relevant for particle coagulation by small ions are available,[25,87] wherein the added complexity of the polymer dynamics is absent. Examples of this are the aggregation laws for coagulation under conditions of: (1) no interparticle energy barrier (i.e., only L–vdW and hydrodynamic forces $\Rightarrow$ fast coagulation $\Leftrightarrow$ diffusion-limited aggregation) and (2) a DLVO-type energy barrier (i.e., add electrical double-layer forces to above $\Rightarrow$ slow coagulation $\Leftrightarrow$ reaction-limited aggregation). A full accounting of polymers in this picture has yet to be made.

Because of the dynamical nature of several important microprocesses (transport, collision, adsorbed polymer, and floc rearrangements), the history of how polymers and particles are brought together (addition profiles and mixing conditions) can be influential.[85,88] Variation of flocculation behavior with mixing technique can especially occur with bridging flocculation.[89] Improvements in flocculation efficiency have also been witnessed upon adjusting the polymer addition scheme.[68,90–93] This can be used to balance the polymer adsorption rate relative to the floc growth rate. The tailoring of addition rates and shear history represent a significant area for improving the quality, consistency and economics of dewatering operations. Process control via these variables would require feedback sensors capable of reporting the state of flocculation. Some progress in this direction has been seen.[94–99]

In agitated suspensions the floc formation process is concurrent with floc breakage[100–102] and floc rearrangement processes. Floc breakage can occur due to hydrodynamic stresses or collision events. Whereas the formation rate $\sim\alpha_0\,\phi G$, the breakage rate of a floc of diameter d_f has been modeled[100] as being $\sim d_f G^y$, where $y \approx$ 1.6. Numerical solution of a flocculation–fragmentation equation yielded a steady-state floc size distribution that was self-preserving (i.e., independent of shear rate when normalized by average floc size) and approximately log-normal.[100,101] Quantitative prediction of the formation (k_A) and breakage (k_{Br}) rate constants would require modeling the collision efficiency (α_0) and floc strength in terms of the particle interactions and hydrodynamic forces. Most often, an empirical assessment of the rate constants are made by comparing observations of flocculation rates and steady-state properties with the numerical solutions. The ratio k_{Br}/k_A is a measure of the shear sensitivity of the flocs.[102] These empirical models do not account for internal floc rearrangement due to labile linkages.

To summarize, there are several potential effects of adsorbed polyelectrolyte on the flocculation rate. The adsorbed polymer increases the effective collision radius of the particle, which increases J_S.[80,103,104] At the correct dose, the adsorbed polyelectrolyte also reduces the interparticle repulsion (charge neutralization) and provides polymer-mediated attractive forces (charge patch and bridging), both of which increase α_0.[105] Third, polymer bound to more than one particle surface acts as a binder for the floc.[105] This does not improve the flocculation rate, but it does reduce the breakage rate, with the overall effect of improving flocculation.

Floc Structure and Properties   The physical structure, deformability, and integrity of the flocs are important in settling and dewatering operations. For example, dewatering by filtration requires porous flocs, whereas sedimentation is best carried out with larger flocs, and consolidation with denser flocs.[106] The floc properties are a result of the combined effects of formation, breakage, and rearrangement. While this convolution of effects can be obfuscating, in some simple circumstances the physical properties of the resulting structures can be used to infer something about the aggregation physics.[106–108] For example, more compact flocs can imply interparticle attachments that were labile under the existing hydrodynamic forces. Less labile linkages might result in more open flocs.

Direct observation and calculation support the idea of a hierarchy of floc structures.[109,110] That is, one can identify general aggregate classes with different physical characteristics (e.g., density, porosity, and strength): primary particles $\Rightarrow$ flocculi $\Rightarrow$ flocs $\Rightarrow$ floc aggregates. The intra-aggregate connections are best modeled as being elastic[111,112] and breakable. This means that mixing history (strength and duration) will affect the final floc structure. The structure of the flocculi carry the signature of the early stages after destabilization. The shear history during this period is especially important. Its effect has been studied for both charge-destabilized systems and systems in which polymer bridging was important.[88,113,114] For example, in a wastewater sludge flocculated with a high-molecular-weight cationic polymer, it was found that turbulence had to prevail for ~ 1 s followed by a sustained low shear period in order to obtain the best results.[88]

In some cases there are multiple types of building blocks for floc formation. For example, waste-activated sludges typically contain exocellular biopolymers and single bacteria loosely associated with a more rigid structure of flocs consisting of fibers, filamentous bacteria and bacterial colonies.[115–117] The natural hydrophilic biopolymers can also act as a sink for flocculation polymers.[118]

The statistical nature of the flocculation process often leads us to using average properties as metrics for floc structure, such as the floc density or porosity. The value of these parameters strongly influence the drainage properties of the individual flocs as well as the consolidated phases formed in dewatering applications. Floc densities are often estimated from microscopic observations, settling rates, terminal settling velocities, hydraulic conductivity,[119] or buoyant density measurements (isopycnic centrifugation).[120,121] Empirical power law relations for the floc density (ρ_f) are of the form[54,109] $(\rho_f - \rho_s) \sim d_f^{-k}$, where ρ_s is the density of the liquid phase, and d_f is the floc size. The exponent k (>0) depends on the details of the aggregation and particle rearrangement processes. Note that the floc density decreases with size (d_f). Such power law density–size relationships arise naturally in the description of fractal (fragmented and irregular) objects.[122,123] Numerical simulations of aggregate structure are often compared with observations via the exponent k, which is related to the fractal dimension, D_F, of the aggregate ($k = 3$-D_F in 3-space). Since the specific aggregation physics used in the model affects D_F, some inferences about flocculation mechanism can be developed.[54,109]

Also of great importance are the mechanical properties of the floc—its strength and deformability. During mixing and transport of the flocs, they should not fragment into or shed small particles that degrade the quality of the separated liquid. If the flocs do break, then reformation kinetics can be important.[95] Rearrangements of the particles within a floc can alter its density and hence settling characteristics. These rearrangements can be driven by hydrodynamic forces or adsorbed polymer conformational changes, and are resisted by the mechanical properties of the interparticle attachments.

The total energy required for the deformation and breakage processes consists of network (distributed stresses), creep, and viscous contributions.[124] The stresses on a floc come from the fluid and from collisions. Simple models for the hydrodynamic stresses in laminar and turbulent flow are available.[125–127] For example, turbulent stresses depend on the floc size (d_f) relative to eddy scale (Kolmogoroff length scale, λ_0). For agglomerates larger than λ_0, the hydrodynamic force on a floc under turbulent conditions is $F_H \sim \rho_s e_0^{2/3} d_f^{8/3}$. The eddy scale is estimated from[127] $\lambda_0 = [\mu^3/(\rho_s^2 e_0)]^{1/4}$, where μ and ρ_s are the fluid viscosity and density, and e_0 is the energy dissipation rate. Note that breakage would be more likely for larger flocs.

A simple model for the binding force (F_B) in a porous aggregate of diameter d and porosity ε, composed of primary particles of diameter x adhering to each other with a force F, yields[127]:

$$F_B = \frac{\pi d^2 (1 - \varepsilon) F}{4 \varepsilon x^2} \tag{1.8}$$

Since flocs are often self-similar (fractals), $(1 - \varepsilon) \Rightarrow (x/d)$. In other words, as the flocs become larger, they become more porous. This simple model only serves to indicate important variables. Measurements of the critical force for floc rupture often indicate a different scaling law. For example, Glasgow and Hsu[128] determined this rupture force by directly observing the process in a turbulent jet for kaolin flocculated with an anionic polyacrylamide at pH 7.2. They found $F_B = 24.8d^{2.45}$ (F_B, dynes; d, cm). Yeung and Pelton[129] also developed a direct way of studying the strength and breakup of flocs through a micromechanics approach. The effort to accumulate sufficient data with adequate statistics through any of these direct measurements can be substantial.[54] It is not surprising that a number of less direct metrics of floc strength have found more use. Determining the effect of postconditioning mixing or centrifugal force and duration on any of a number of performance parameters (e.g., drainage or settling rates, floc size,[130] filtrate clarity, flocculation index[95]) has been tried. For example, floc strength can be evaluated from the slope of a plot of capillary suction time (CST) versus stirring time in a given reactor at a fixed stirring rate.[58,102] The CST is related to the drainage properties of the flocs.

In summary, polymers act to destabilize the suspended colloids, increase the collision efficiency, and behave as a binder for the floc. The physical form of the floc must be engineered with a mind to the solid–liquid separation process. This engineering is accomplished via a manipulation of the key parameters that govern floc formation, breakup, and rearrangement. These include the interparticle forces, collision rates, floc mechanical properties, and the dynamics of the subprocesses. Optimal conditioning of the sludge is mainly accomplished through the correct choice of destabilizing and flocculating agents, addition schemes, and mixing conditions.

1.3   CLARIFICATION, THICKENING, AND DEWATERING

Solids consolidation is a necessary step in the solid–liquid separation process (Fig. 1.6). Several different categories of water in sludges can be distinguished[131–133]: free, capillary (physicomechanically immobilized in floc), adsorbed (physicochemically bound), and chemically bound (intracellular). The water removed in clarification, thickening, and dewatering operations is mainly the free and capillary water. In practice, several different unit operations in the solid–liquid separation process can be identified,[7] each typically requiring different procedures and equipment.

The influent raw waste stream and mixed liquors from secondary and tertiary treatment might have from 0.01 to 0.1 wt/vol % solids content. The first stage in concentrating the solid phase is clarification. Many types and configurations of clarifiers have been developed over the years to allow gravitational separation in a continuous process. Clarifier design and operation has become a science unto itself. Generally, these devices are part of the liquid treatment process and are designed to remove relatively small amounts of solids from large quantities of liquid. The simplest designs include rectangular and circular clarifiers. Rectangular clarifiers can have either co-current or countercurrent takeoff of effluent water and settled solids. Circular clarifiers can be modified for different combinations of either peripheral or center feed and effluent takeoff and may include major modifications such as squircles, slurry recirculation, and sludge blanket control. There are also some advanced design settling clarifiers that include lamella clarifiers and tube settlers. In some circumstances, solids are separated by flotation, wherein small air

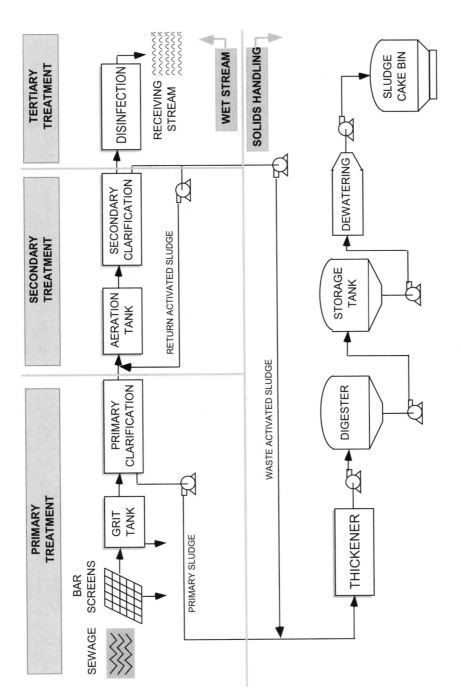

FIGURE 1.6   Flow diagram of a typical wastewater treatment process.

bubbles add buoyancy to the solid particles that are then removed from the surface of the fluid. After the clarification stage, the solids content might reach 1.5 to 2.5 wt/vol %.

Next the suspended solids are consolidated into a zone of increasingly higher solids in a process referred to as thickening. Thickening methods include gravity settling, rotary screens, gravity belts, elutriation (which also washes the sludge), dissolved air flotation (DAF) and centrifugation (disk, solid-bowl decanter style, or basket).[132] Much of the same equipment used for clarification can also be used in thickening, but with a feed stream more concentrated in solids. The solids concentrations after gravity thickening can range from only 3 to 5% for extended aeration sludges to 5 to 12% for primary sludges to 6 to 14% for digested sludges to 10 to 15% for heat-treated activated sludges.[132]

Dewatering is most typically used prior to final recovery or disposal of solids. In the dewatering stage, sufficient water is removed so that the sludge behaves like a solid. For example, beyond about 20 wt % solids the sludge is no longer pumpable but can be shoveled and trucked.[132] Methods include the use of lagoons, drying beds (water drains through a sand bed allowing solids to air dry above the bed), coil filters, rotary vacuum filters (vacuum-generated pressure differential across a filter cloth), pressure filtration, filter presses (roll, screw, plate and frame/chamber, belt, capillary belt), freeze–thawing, and centrifugation. Probably the two most common pieces of dewatering equipment are belt filter presses[134] (BFPs) and centrifuges.[135,136] Belt filter presses can be found in a number of configurations that usually combine a gravity drainage zone, a wedge zone (solids are initially compressed between the two belts of filter cloth), and a pressure zone. Pressure zone configurations can vary widely but are generally found in two main categories: S-roll type where the belt/solids/belt sandwich is stretched around rollers in an "S" pattern (as roller diameter decreases, pressure on the solids increases); and P-roll type, where rollers are squeezed against the belt/solids/belt sandwich with hydraulic pressure provided by an ancillary hydraulic system. Centrifuges can be found as batch operation types (basket, disk stack, and disk bowl) and continuous operation types (solid-bowl decanter style). After dewatering, the solids have been pushed to 12 to 45 wt/vol %. Further reductions in moisture content typically require heating or incineration.

1.3.1 Settling

In gravity settling one can identify three general regimes: free settling (little interparticle interaction), hindered settling (significant interparticle interaction), and compression zone settling (stresses distributed over a particle network). Once flocs have become large enough they will begin to settle at a rate that depends on their size and density relative to the fluid medium. In the absence of electrostatic forces, the terminal settling velocity, v_T, of an isolated nonporous spherical particle of diameter d and density ρ_p in a fluid of density ρ_s and viscosity μ is given by[54]:

$$v_T(0) = \frac{d^2 g(\rho_p - \rho_s)}{18 \, \mu} \tag{1.9}$$

where g is the gravitational acceleration. For a 1-μm clay particle, v_T is about 1 μm/s. Flocs are typically porous, with this porosity increasing with floc diameter. For example, Li and Ganczarczyk[137] found that nearly all activated sludge flocs > 60 μm were permeable during gravity settling. Such porosity can greatly impact the floc settling properties. Observations on the gravity settling of isolated porous flocs were fitted according to $v_T \sim d_f^m$, where $m \approx 0.67$ for digested sludge and $m \approx 0.51$ to 0.66 for raw-water coagulated flocs.[138]

As the suspension of flocs consolidates during settling, the individual flocs will interact and the mean settling velocity will decrease according to[139]:

$$\overline{v_T(\phi)} = \frac{v_T(0)(1 - \phi)^2}{r(\phi)} \tag{1.10}$$

where $r(\phi)$, the hindered settling factor, is a function of the local solids volume fraction (ϕ).

Flocculated slurries are typically faster settling and result in larger settled volumes than nonflocculated (dispersed) slurries. Above a critical slurry solids either system can form an infinite network with solidlike rheological properties (i.e., it will support a stress). Flocculated slurries typically solidify at lower solids concentrations than do dispersed slurries. The percolation limit for network formation depends on floc structure and thus the mechanism of aggregation. A rheological model for flocculated suspensions with a network structure has been developed by Landman and White,[139] and was successfully used by them to model several conventional solid–liquid separation processes. The key parameters in their constitutive model were the compressive yield stress of the network, $P_y(\phi)$, and the hindered settling factor, $r(\phi)$. The yield stress is an implicit function of the interparticle binding forces and the shear history. The hindered settling factor accounts for hydrodynamic interactions; $P_y(\phi)$ can be determined empirically from centrifugation or filtration experiments and $r(\phi)$ from sedimentation or filtration experiments.[139] The link between $P_y(\phi)$ and interparticle forces has not yet been put into mathematical form, however, the semiempirical constitutive relation allows for a realistic modeling of sedimentation, gravitational thickening, and filtration processes.

1.3.2 Dewatering

Dewatering of the thickened sludge results in formation of a sludge cake. Prior to cake formation, drainage of the suspension can be modeled as plug flow of the suspension. Upon cake formation, fluid drag forces in the cake become important. That is, the cake becomes a filtration medium during the dewatering step. Filtration models are essentially porous media flow models (e.g., Darcy equation[140–142]) with some concern for pressure-dependent media properties. The general form of porous media flow models relevant at this stage is[143]

$$Q = \left(\frac{K}{\mu}\right) f\left(\frac{d}{L}\right) L^2 P \tag{1.11}$$

where Q is the volumetric fluid flux relative to the medium, $f(d/L)$ is a function describing the porous medium geometry, d and L are characteristic length scales of the particles (flocs) and the medium, K is the permeability coefficient, μ is the fluid viscosity, and P is the fluid pressure gradient. Cake filtration properties are often expressed using a normalized parameter called the specific resistance to filtration (SRF; cm/g):[144–146]

$$SRF = K_K S_A \left(\frac{1-\varepsilon}{\varepsilon^3 \rho}\right) \tag{1.12}$$

where K_K is the Kozeny constant (a permeability), S_A is the specific surface area of the particles (surface area/volume = $6/d$) of diameter d and density ρ, and ε is the porosity of the filter cake. In practice, the SRF is determined from filtration rate curves (e.g., Buchner funnel test).[7,102,143] Typical filtration rates for sewage sludges may be found in Coackley[147] and Gale.[148] For compressible cakes, the SRF is pressure dependent. Typically SRF $\sim P^s$, where s is a coefficient of compressibility.[7,140,149] For typical sewage sludges $0.6 \leq s \leq 0.9$.

A convenient measure of sludge filtration properties is the capillary suction time.[102,141] In the CST measurement a small sample of the sludge is confined in an open-ended tube positioned on top of a piece of filter paper or ceramic. The water in the sludge drains through the filter cake into the paper under gravity and capillary action. The rate of filtrate wicking through the paper is measured using an interval timer triggered by the expanding wetted perimeter passing electrical contact points. Note that the CST does not separate permeability and compressibility and is nonlinearly related to SRF.[148] However, in some instances CST is a better performance indicator than the SRF.[7,141]

At some stage in the dewatering, the liquid is being replaced by a gas, and capillary forces must be overcome. Semiempirical relations for the capillary pressure curve allow for a modeling of the dewatering kinetics in this stage.[140,150]

Efforts at modeling cake formation and filtration dewatering of suspensions from first principles have been attempted. The important parameters include the permeability of the filter plus deposited particles and the stiffness of the solids matrix. These are both functions of the solidosity of the cake, and the stiffness is also a function of the interparticle potential. Models for which the particles interact via simple electrical double-layer potentials have been reasonably successful.[151] Extensions to flocculated particles interacting via polymer-mediated forces would be useful in many dewatering scenarios.

1.4 CHEMISTRY OF POLYELECTROLYTES USED IN DEWATERING

The chemical and physical properties of polyelectrolytes govern their effectiveness in dewatering applications. The primary parameters include the usual descriptors: molecular weight and size (distributions), charge type and density, chain architecture and topology, and finally, chemistry. The allowable chemistry is influenced by one's ability to achieve the necessary polymer properties and to maintain them under the conditions of use; all at a reasonable price. In some instances the chemical moiety that binds with the particle surface is critical (e.g., red mud flocculation using hydroxamated polyacrylamides). However, in many instances there is some leeway with the chemical structure, and it is the polyelectrolyte character that is most important. Natural selection of flocculants in the solid–liquid separation markets has culled several dominant chemistries from the possible set of polyelectrolytes.[152] These will be discussed below. Also important is the physical form of the concentrated product (dry powder, dispersion, emulsion), as this determines the resources necessary to dissolve the polymer and condition it with the substrate.

1.4.1 General Description

Water-soluble polymers (WSPs) generally can be divided into two main categories: ionic (polyelectrolytes) and nonionic. The large number of functional groups summarized in Table 1.3 can impart water solubility to a polymer chain. For nonionic WSPs, hydration results from an interaction with a polar site in the functional group (e.g., —OH, —O—, or —$CONH_2$). Typical examples are poly(acrylamide) [1], poly(ethyleneoxide) [2], and poly(vinylalcohol) [3].

$$\left(\!\!-CH_2\!-\!\underset{\underset{CONH_2}{|}}{CH}\!-\!\!\right)_n \qquad \left(\!\!-CH_2\!-\!CH_2\!-\!O\!-\!\!\right)_n \qquad \left(\!\!-CH_2\!-\!\underset{\underset{OH}{|}}{CH}\!-\!\!\right)_n$$

[1]   [2]   [3]

Polyelectrolytes comprise the most useful materials for industrial dewatering applications. The macroions possess formal cationic, anionic, or both (polyampholytes) types of charges on the polymer. Associated counterions ensure an overall charge neutrality. The ionic nature of the polymer chains dramatically affects their dimension in solution (polyelectrolyte effect), with intramolecular charge repulsions tending to increase the solvated coil size. The charge-bearing moieties in the great majority of polycationics are nitrogen based, whereas in polyanionics, oxygen is usually involved. Several important structures whose macroions remain fully ionized over large pH ranges, due to either the quaternary nitrogen or sulfonate groups, are shown below in [4] to [8].

A number of important water-soluble polymers contain functional groups that are ionized over only a narrow pH range. Polymers containing weakly acidic —COOH groups are examples [9, 10]. A polymer containing a weak base (e.g., —NH—) can

TABLE 1.3   Functional Groups in Water-Soluble Polymers

—OH	—COO⁻ M⁺	—NH₃⁺
—O—	—OSO₂⁻ M⁺	—⁺NH₂R
—COOH	—OPO₃²⁻ M²⁺	—⁺NHR₂
—NH₂	—CS₂⁻ M⁺	—⁺NR₃
—NHR		=⁺NH₂

$$—COO^{-}\ M^{+}\qquad —NH_3^{+}$$
$$—OSO_2^{-}\ M^{+}\qquad —^{+}NH_2R$$
$$—OPO_3^{2-}\ M^{2+}\qquad —^{+}NHR_2$$
$$—CS_2^{-}\ M^{+}\qquad —^{+}NR_3$$

Chemical structures including: —N< , NH—C(=O)—NH₂ (urea), NH—C(=NH)—NH₂ (guanidine), triazine ring (H₂N, N, NH, NH₂), quaternary ammonium zwitterions ⁺NR₂(CH₂)ₙCO₂⁻ and ⁺NR₂(CH₂)ₙSO₃⁻, —N⁻—N⁺—, —CH₂·N⁺(O⁻)(O), =⁺NH, ≡⁺NH, imidazolium ring, piperazinium N—N⁺—, morpholinium O—N⁺—, —⁺SR₂, —⁺PR₃

Polymer structures:

[4] $-\!(CH_2\!-\!C(R))_n-$; C=O ; O ; (CH₂)₂ ; CH₃—N⁺—CH₃ ; CH₃ ; Cl⁻ ; R = H, CH₃

[5] $-\!(CH_2\!-\!C(R))_n-$; C=O ; NH ; (CH₂)₃ ; CH₃—N⁺—CH₂ ; CH₃ ; Cl⁻ ; R = H, CH₃

[6] $-\!(CH_2\!-\!CH\!-\!CH\!-\!CH_2)_n-$; N⁺ Cl⁻ ; R₁, R₂ ; R₁ = R₂ = CH₃ ; R₁ = R₂ = CH₃CH₂

[7] $-\!(CH_2\!-\!CH)_n-$; C=O ; NH ; CH₃—C—CH₃ ; CH₂ ; SO₃⁻ Na⁺

[8] $-\!(CH_2\!-\!CH)_n-$; OSO₃⁻ Na⁺

be protonated to a polycation in an acidic medium, but it remains an undissociated polybase in alkaline media [11–13].

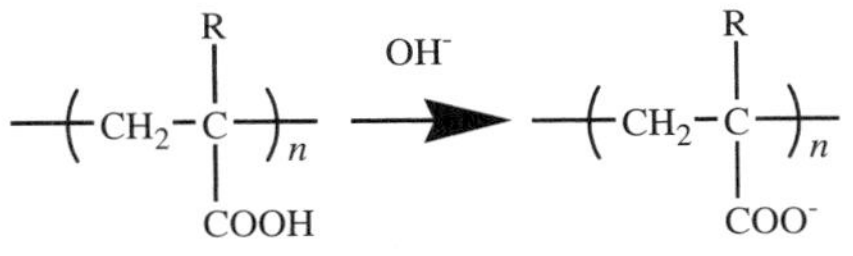

$$\left(CH_2-\underset{\underset{COOH}{|}}{\overset{\overset{R}{|}}{C}}\right)_n \xrightarrow{OH^-} \left(CH_2-\underset{\underset{COO^-}{|}}{\overset{\overset{R}{|}}{C}}\right)_n \qquad \left(\underset{\underset{HOOC}{|}}{CH}-\underset{\underset{COOH}{|}}{CH}\right)_n$$

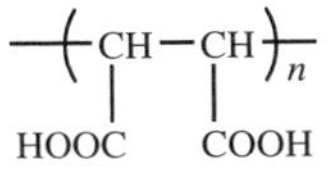

R = H, CH_3

[9]   [10]

$$\left(CH_2-CH_2-NH\right)_n \xrightarrow{H^+} \qquad \left(CH_2-\underset{\underset{NH_2}{|}}{CH}\right)_n \qquad \left(CH_2-\underset{|}{CH}\right)_n$$

$$\left(CH_2-CH_2-NH_2{}^+\right)_n$$

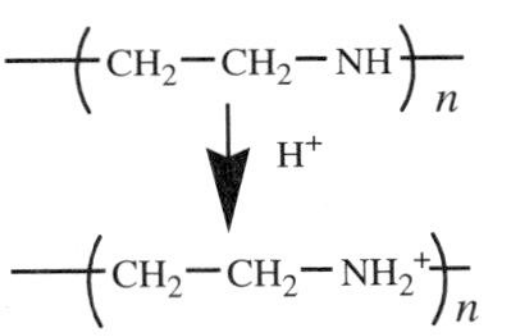

[11] 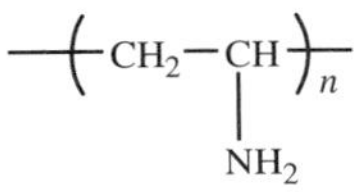   [12] 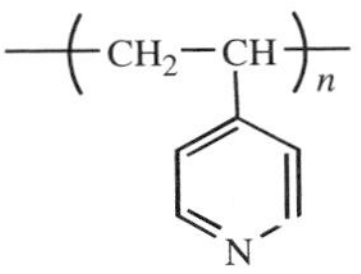   [13]

Ampholytic monomers such as aminimide or betaine contain both positive and negative charges in the same monomer. Their polymers find no use in dewatering due to the unstable nature of polyaminimides; and the poor solubility of polybetaines in water, which results in hydrogel formation. Other approaches to preparing polyampholytes are the copolymerization of anionic with cationic monomers[153] and the postpolymerization derivatization of a monocharged polyelectrolyte to yield a polyampholyte.

Ecological reasons generally preclude the use of phosphorous-containing polyelectrolytes for wastewater treatment. However, these polymers are important in the areas of corrosion inhibition, deposit control, and chelation. Nitrilo-tris-methylene phosphonic acid [14] is a well-known example.

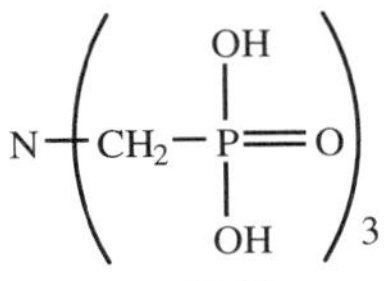

$$N\left(CH_2-\underset{\underset{OH}{\overset{\overset{OH}{|}}{|}}}{P}=O\right)_3$$

[14]

There are also high-molecular-weight cationic water-soluble polymers possessing the cyclic amindine moiety[154–157] and other heterocyclic quaternary structures.[158] Their pendant groups possess very high cationicities. Representative examples are polyvinylimidazolines [15], polypyrimidine salts [16], and poly(morpholinium methylvinylacrylate) quaternary salts [17].

$$\left(\!\!CH_2\!-\!CH\right)_{\!n}$$

R—N N$^+$—R$_3$

R$_1$ R$_2$

(CH)$_m$

R = R$_1$ = R$_2$ = H, CH$_3$
R$_3$ = H, alkyl, benzyl
m = 0, 1
X$^-$ = halide

[15]

$$\left(\!\!CH_2\!-\!CH\right)_{\!n}$$

N

N$^+$ X$^-$

R

R = H, CH$_3$
X$^-$ = halide

[16]

$$\left(\!\!CH_2\!-\!C\right)_{\!n}$$

R$_1$

C=O

O

CH$_2$

X$^-$ N$^+$—R$_2$

R$_1$ = R$_2$ = H, CH$_3$
X$^-$ = halide

[17]

1.4.2   Synthetic Methods

High-molecular-weight water-soluble polymers are prepared by step-growth polymerization, whereas low- to medium-molecular-weight polymers are usually prepared by polycondensation. Polymer chain topology can be made nonlinear (i.e., branched) by copolymerization with multifunctional comonomers. Structural modification of preformed polymers has also been proven to be commercially and economically feasible. The most commercially successful polymers for dewatering applications have been prepared from functional monomers via free radical polymerization.

Water-soluble azo compounds (1.13), persulfates (1.14) and red-ox systems [(1.15) and (1.16)] are the most commonly used for free radical initiation:

$$
\begin{array}{c}
\text{H}_2\text{N} \\
 \\
\text{HN}
\end{array}
\!\!C\!\!-\!\!\underset{\text{CH}_3}{\overset{\text{CH}_3}{\text{C}}}\!\!-\!\!N\!=\!N\!\!-\!\!\underset{\text{CH}_3}{\overset{\text{CH}_3}{\text{C}}}\!\!-\!\!C\!\!\begin{array}{c}\text{NH}_2\\\text{NH}\end{array}\cdot 2\text{ HCl} \xrightarrow{\;\Delta T\;} 2\ \cdot\underset{\text{CH}_3}{\overset{\text{CH}_3}{\text{C}}}\!\!-\!\!C\!\!\begin{array}{c}\text{NH}_3{}^+\\\text{NH}\end{array}\text{Cl}^- + \text{N}_2 \tag{1.13}
$$

$$ \text{S}_2\text{O}_8{}^{2-} \xrightarrow{\;\Delta T\;} 2\ \text{SO}_4^-\cdot \tag{1.14} $$

$$ \overset{\text{CH}_3}{\underset{\text{CH}_3}{\text{C}}}\!\!-\!\text{OOH} \;+\; \text{Fe}^{2+} \longrightarrow \text{Fe}^{3+} + \text{OH}^- + \overset{\text{CH}_3}{\underset{\text{CH}_3}{\text{C}}}\!\!-\!\text{O}\cdot \tag{1.15} $$

R: C(CH$_3$)$_3$ > Cl > CH(CH$_3$)$_2$ > CH$_3$(CH$_2$)$_9$CHCH$_3$ > CH$_3$ > CH$_3$(CH$_2$)$_2$CHCH$_3$ > H

$$ \text{S}_2\text{O}_8{}^{2-} + \text{HSO}_3^- \longrightarrow \text{SO}_4^-\cdot + \text{HSO}_3\cdot + \text{SO}_4{}^{2-} \tag{1.16} $$

Unique features of a red-ox polymerizations include a very short induction period and a relatively low energy of activation; 10 to 20 kcal/mol compared to ~30 kcal/mol for thermal initiation. Red-ox initiated polymerizations are easily carried out at low temperatures, thus minimizing side reactions.

Chain growth involves three competing processes: initiation, propagation, and termination. The instantaneous rate of polymerization (R_p) and the instantaneous number-average degree of polymerization ($\overline{DP_n}$) under steady-state conditions are[159]

$$R_p = (R_t)^{1/2}(k_p/k_t^{1/2})\,[\text{M}] = k_p(fk_d/k_t)^{1/2}\,[\text{IN}]^{1/2}\,[\text{M}] \qquad (1.17)$$

$$\frac{1}{DP_n} = \frac{k_t}{k_p^2}\frac{R_p}{[M]^2} + \frac{k_{fi}}{k_p}\frac{[\text{IN}]}{[\text{M}]} + \frac{k_{fm}}{k_p} \qquad (1.18)$$

where R_t is the chain termination rate; k_p, k_t, and k_d are the chain propagation, termination, and catalyst decomposition rate constants; f is the radical decomposition efficiency; [M] and [IN] are the instantaneous concentrations of monomer and initiator; k_{fm} is the transfer-to-monomer rate; and k_{fi} is the transfer-to-initiator rate constant. If the monomer concentration and initiation rate $[R_i = (fk_d/k_t)[\text{IN}]^{1/2}]$ are held constant, then the instantaneous rate of polymerization depends on $k_p/k_t^{1/2}$. Monomers having large k_p and small k_t values have faster rates of polymerization and greater molecular weights. Most frequently, the molecular weights obtained are lower than predicted from these parameters because of premature termination of the propagating free radical by transfer to monomer, solvent, or initiator.

Acrylamide Polymerization   Acrylamide, although nonionic itself, is one of the most widely used comonomers in the preparation of polyelectrolytes for dewatering applications. Acrylamide has an unusually high value of the ratio of $k_p/k_t^{1/2}$ (~4.2). This indicates acrylamide's ability to polymerize to very high molecular weights. Values of k_p and k_t are 1.72×10^4 and 1.63×10^7 L mol^{-1}s^{-1} at 25°C and pH 1.[160] These values were found to vary over the pH range 5.5 to 13,[161,162] however, $k_p/k_t^{1/2}$ was found to be practically constant at all pHs. Shawki et al.[163] observed a moderately high chain transfer constant to persulfate initiator, but it appeared to be zero to both monomer and polymer in water. In comparison, the structurally similar methacrylamide monomer possesses a lower value of $k_p/k_t^{1/2}$ (~0.2)[164]; k_p and k_t were 0.08×10^4 and 1.65×10^7 L mol^{-1}s^{-1}. The lower rate and degree of polymerization with this monomer is further influenced by the stability of the allylic radical formed by hydrogen abstraction from the methacrylamide monomer. The rate of acrylamide polymerization was found to be proportional to [IN]$^{1/2}$ [Eq. (1.17)] and to the 1.2 to 1.5 power of acrylamide monomer concentration. Termination of the active polymer was mainly by disproportionation.[160]

The initiation system affects the overall energy of activation (E_a) for the polymerization of acrylamide. Persulfates typically result in E_a ~70 kJ/mol,[165] persulfate/mercaptoethanol ~134 kJ/mol,[166] and hydrogen peroxide/hydroxylamine ~27 kJ/mol.[167] Problems encountered in the polymerization of acrylamide include dissipating the approximately 83 kJ/mol heat evolved (which can result in a rapid temperature rise), processing the resulting viscoelastic solutions, and depletion of the initiator before complete conversion. In industrial processes using an adiabatic gel polymerization, the evolved heat is effectively utilized to reduce residual monomer. A wide variety of polymerization methods have been utilized, such as photochemical, radiation-induced, red-ox, thermal, electro-initiated, ultrasonic, and others.[168]

Chemical Modification of Polyacrylamide The amide functionality of polyacrylamide is relatively chemically inert, however, several postpolymerization derivatization strategies are commonly used. Polyacrylamide (PAM) can be hydrolyzed to produce copolymers containing acrylamide and acrylate by reacting PAM with sodium hydroxide at elevated temperatures. Polyacrylamide can also be reacted with formaldehyde and sodium bisulfite to produce sulfonate polymers. Cationic derivatives can be obtained by reacting the amide group with formaldehyde and a secondary amine, to form a Mannich base. Current examples of these approaches are summarized below.

Cationic Carbamoyl Polymers The polymerization of polyacrylamides in inverse microemulsions was a significant advance in the state of the art.[169,170] This approach can overcome the inherent settling problems of conventional inverse emulsions and even low-settling formulated emulsions. Water-in-oil polyacrylamide microemulsions tend to have very small droplet sizes, for example, less than ~100 nm, and are characterized by low bulk viscosities, thermodynamic stability, and optical clarity. Because of the smaller droplet size, a higher surfactant level than regular inverse emulsions is necessary. Preparations of the monomer microemulsion are clear and stable with either a discrete or bicontinuous aqueous phase depending on the combination of oil, surfactants, and aqueous monomer solution. In such cases no homogenization is required. Recently, microemulsion technology has been used to develop a new synthetic route to a functionalized polyacrylamide. A polyacrylamide microemulsion[171] has been reacted with formaldehyde and dimethylamine to form a Mannich carbamoyl polymer inside the aqueous droplets, each containing only a few polyacrylamide molecules[172] [see Eq. (1.19)]. Subsequent reaction of the Mannich polymer and an alkylating agent yielded a quaternary ammonium polymer.[172,173]

$$
-\!\!\left(CH_2\text{--}CH\right)_{\overline{n}} + (CH_3)_2NH + HCHO \xrightarrow{K_{eq}} -\!\!\left(CH_2\text{--}CH\right)_{\overline{n}} + H_2O
$$

with pendant $C{=}O$, NH_2 on left and $C{=}O$, NH, $CH_2N(CH_3)_2$ on right.

$$
\xrightarrow{CH_3Cl} -\!\!\left(CH_2\text{--}CH\right)_{\overline{n}} + H_2O \qquad (1.19)
$$

with pendant $C{=}O$, NH, $CH_2N^+(CH_3)_3\ Cl$.

Products based on this scheme have been commercially successful as high-performance cationic organic flocculants for municipal and industrial sludge dewatering.[173–175]

Formation of the Mannich derivative is a second-order process. The reaction between the rapidly formed dimethylaminoethanol (from formaldehyde and dimethylamine) and polyacrylamide was studied by carbon-13 nuclear magnetic resonance (^{13}C-NMR).[176] The Mannich reaction was reversible and pH dependent. The equilibrium constant has been estimated at 7.9×10^3 at pH 10,[176] indicating a very fast base-catalyzed condensation mechanism. Decreasing the pH of the medium decreased the rate of substitution. At pH 5, the Mannich substitution was very slow. The effect of pH on a model compound, formed from equimolar quantities of dimethylamine and formaldehyde, and half mole of cyclohexanone, has been studied.[177] The rate of reaction at 25°C was 45-fold faster at pH 10.6 than at pH 5.05.

Solution Mannich polyacrylamides, prepared using the older technology, are inexpensive and are sold in high volume as aqueous solutions containing only 4 to 6 wt % polymer due to their high bulk viscosities and propensity to crosslink upon standing. Their shelf life can be improved by the addition of formaldehyde scavengers such as dicyandiamide and urea. Their applications in sludge dewatering are usually carried out with the substrate pH < 7. Low-charge quaternized Mannich solution polyacrylamides are also sold as aqueous solutions of approximately 3 wt % polymer or less, again due to their high solution viscosities. However, there is virtually no pH dependence of their efficacy in dewatering applications.

Sulfomethylation Unlike the base-catalyzed Mannich reaction, the reaction of polyacrylamide with formaldehyde and sodium bisulfite at alkaline pH and elevated temperature to produce sulfomethylated polyacrylamide has not been realized. A recent study[178] suggested that the ammonia obtained from the hydrolysis of polyacrylamide reacted with sodium formaldehyde bisulfite to produce hexamethylenetetraamine, methanesulfonates, and other species. Upon lowering the reaction pH to acidic conditions, under high pressure and temperature, the sulfomethylation can take place smoothly.

$$\left(CH_2\text{-}CH\right)_{n} + NaHSO_3 + HCHO \xrightarrow{\ \ H^+\ \ } \left(CH_2\text{-}CH\right)_{n} + H_2O \qquad (1.20)$$

with pendant $C{=}O$, NH_2 on the left, and pendant $C{=}O$, NH, $CH_2SO_3^-\ Na^+$ on the right.

Ultra-High-Molecular-Weight Hydrolyzed Polyacrylamide Poly(acrylamide-*co*-sodium acrylate) and poly(acrylamide-*co*-ammonium acrylate) are flocculants of considerable importance. They can be prepared directly by the acid- or base-catalyzed hydrolysis of polyacrylamide. The rate of alkaline hydrolysis proceeds rapidly at first and then decreases with increasing conversion. This is due to the growing content of neighboring carboxylate groups next to the amide groups, which leads to more and more electrostatic repulsion of the approaching OH^- reactants. It has been shown[179] from ^{13}C-NMR data that a nearly random sequence of carboxylates resulted for hydrolyzed polyacrylamide made under alkaline conditions, whereas hydrolysis at low pH resulted in blocks of carboxylate groups. A process for preparing ultra-high-molecular-weight (above 30×10^6) hydrolyzed polyacrylamides in a stable inverse emulsion has been patented.[180] The resulting random copolymer had an intrinsic viscosity of 32 dL/g in 1 N NaCl.

Cationic Acrylamide Copolymers The manufacture of high-molecular-weight cationic polymers for dewatering applications has been largely dominated by the copolymerization of acrylamide (AMD) and cationic comonomers. These comonomers are generally either amino-derivatives of (meth)acrylic acid esters or (meth)acrylamides or diallyldimethylammonium chloride. The best ratio of AMD to quaternary ammonium monomer depends on cost and performance parameters of the applications. Low to medium charge copolymers are typically used for paper waste applications, while medium to high charge copolymers are used for sludge dewatering. For most applications in wastewater treatment, the cationic charge is in the range of 10 to 60 mol %. The molecular weights for flocculants are usually >5 × 10^6 g/mol. Commercially important cationic comonomers, along with their reactivity ratios with acrylamide, are shown in Table 1.4.

Copolymers of acrylamide and AETAC (see Table 1.4 for abbreviations) have become preferred due to their favorable reactivity ratios, which in turn results in a uniform sequence distribution and lessens compositional drift. However, these esters are very susceptible to base-catalyzed hydrolysis.[181] In the manufacture and application of these products, great care is needed to avoid this hydrolysis, for example, by maintaining acidic pH conditions.

The rate of hydrolysis of AETAC copolymers increases with pH and is composition dependent. For a copolymer composition of 43 mol % AETAC, it was determined[181] that the ester loss was lowest at pH 2 to 3. Above pH 3, ester loss increased dramatically with pH (Fig. 1.7). The rate of isolated ester hydrolysis was first order in $[OH^-]$. If there were neighboring acrylamides (AMD) in the chain, then there was a second-order dependence of ester disappearance on $[OH^-]$. This indicated that the imidization reaction was first order in $[OH^-]$ over the pH region studied.[181] The percentage of esters cleaved increased as the number of AETAC groups with neighboring AMD groups increased. Esters with no neighboring AMD, such as those found in homopolymer of AETAC, were very difficult to hydrolyze (Fig. 1.8). The chemistry of ester cleavage and imidization with neighboring amide groups is described below:

poly(AMD–AETAC)   (1.21)

TABLE 1.4 Acrylamide Monomer (M_1) Reactivity Ratios

M_1 / M_2	r_1	r_2	Initiator[a]	Temp. (°C)	Reference
Cationic comonomers (M_2)[b]					
AMD/AETAC	0.61	0.47	TBHP/MBS	40	181
AMD/AETAC	0.64	0.48	KPS	40	195
AMD/MAETAC	0.24	2.47	TBHP/MBS	40	181
AMD/MAETAC	0.25	1.71	KPS	40	195
AMD/DMAPAA	1.1	0.47	KPS	40	195
AMD/DMAPMA	0.47	0.96	KPS	40	195
AMD/MAPTAC	0.57	1.13	KPS	40	195
AMD/DADMAC	6.4–7.54	0.05–0.58	APS, ACV	20–60	195–197
Anionic Comonomers (M_2)[b]					
AMD/AA	0.25–0.95	0.3–0.95	KPS	30	191, 198
AMD/AA	0.89	0.92	AIBN	45	192
AMD/MAA	2.8–0.39	0.2–0.51	KPS	30	198
AMD/NaAMPS	0.98	0.49	APS	30	199

[a]TBHP, *tert*-butylhydroperoxide; KPS, potassium persulfate; ACV, azocyanovaleric acid; MBS, sodium meta-bisulfite; APS, ammonium persulfate; AIBN, 2,2-azobisisobutyronitrile.

[b]AETAC, acryloyloxyethyltrimethylammonium chloride: $CH_2=CHCO_2(CH_2)_2N^+(CH_3)_3Cl^-$
MAETAC, methacryloyloxyethyltrimethylammonium chloride: $CH_2=C(CH_3)CO_2(CH_2)_2N^+(CH_3)_3Cl^-$
DMAPAA, dimethylaminopropylacrylamide: $CH_2=CHCONH(CH_2)_3N(CH_3)_2$
MAPTAC, methacrylamidopropyltrimethylammonium chloride: $CH_2=C(CH_3)CONH(CH_2)_3N^+(CH_3)_3Cl^-$
DMAPMA, dimethylaminopropylmethacrylamide: $CH_2=C(CH_3)CONH(CH_2)_3N(CH_3)_2$
DADMAC, diallyldimethylammonium chloride: $(CH_2=CHCH_2)N^+(CH_3)_2Cl^-$
AA, acrylic acid: $CH_2=CHCOOH$
MAA, methacrylic acid: $CH_2=C(CH_3)CH_2COOH$
Na AMPS, sodium 2-acrylamido-2-methylpropanesulfonate: $CH_2=CHCONHC(CH_3)_2SO_3^-Na^+$.

Cationic copolymers derived from amides such as MAPTAC or APTAC (acrylamidopropyl trimethylammonium chloride) are considered to have quite random comonomer distributions, are more resistant to hydrolysis, and offer a wider range of application pH. However, cationic ester monomers are less expensive than the cationic amide monomers.

The relatively inexpensive water-soluble monomer diallyldimethylammonium chloride (DADMAC) is synthesized commercially[182] by reacting allylchloride, dimethylamine, and sodium hydroxide. The DADMAC is usually not isolated, but 60 to 70% solutions are purified and used for polymerizations. DADMAC is a nonconjugated diene monomer that can form highly charged cationic linear polymers under radical initiation.[183,184] This polymerization is known as an inter-intra-cyclopolymerization. The initiator radical attacks a terminal carbon atom on one allyl group. The radical thus formed attacks the internal carbon on the other allyl group in the same molecule to form a five-membered pyrrolidinium ring with a cis to trans ratio of 6 to 1,[185] and <2% pendant double bond. The less stable primary radical is the predominate propagation species.

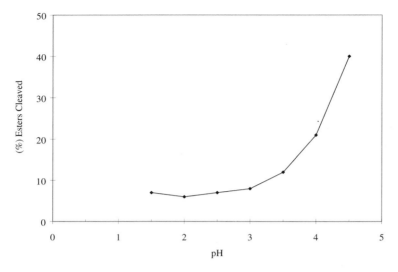

FIGURE 1.7   Plot of (%) esters cleaved versus pH for an AMD/AETAC (57/43 mol %) copolymer heated for 9 h at 90°C.

The kinetics of DADMAC polymerization has been studied extensively. The rate of cyclopolymerization in aqueous solution, when persulfate was used as an initiator,[186] was $R_p = k \, [S_2O_82-]^{0.8} \, [DADMAC]^{2.9}$. The unusually high exponents are mainly due to complicated initiation reactions. Even at the high monomer concentrations typical of commercial syntheses (>1.5 mol/L), high rates of

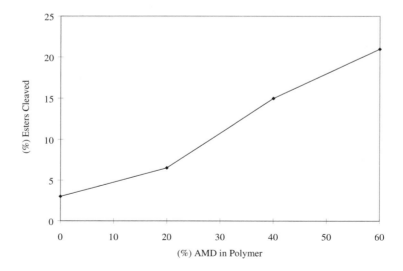

FIGURE 1.8   Effect of AMD content on ester stability at pH 4 heated for 9 h at 90°C.

propagation resulted, and the [DADMAC] exponent remained 2.9. This deviation from the usual kinetic scheme can be explained mainly by the formation of dimeric DADMAC, due to π-π interactions. NaCl can also enhance both the rates of initiation and propagation. A study of a DADMAC polymerization in an inverse emulsion[187] stabilized with sodium di-2-ethylhexyl sulfosuccinate (AOT) and sorbitan monooleate (SMO), and initiated with the oil-soluble initiator 2,2'-azo bis(2,4-dimethyl valeronitrile) (ADVN) in a stirred-tank reactor, yielded a rate of polymerization, $R_p = k$ [ADVN]$^{0.4}$ [AOT]$^{0.5}$ [SMO]$^{-0.4}$ [DADMAC]^3. The influences of ionic strength and partitioning effects were considered to be the causes of the third-order dependence on monomer concentration. The negative order of [SMO] may have been due to the fact that SMO acts as a radical scavenger. The molecular weights of poly(DADMAC)s are not usually as high as cationic acrylic polymers due to the chain-transfer constant of allylic radicals. Molecular weights of 5×10^5 for the homopolymer are achievable, and applications are limited to potable water and color removal. Production of very high molecular weight poly(AMD-co-DADMAC)s with a high cationic charge and a high polymer solids is difficult. Problems include a vast difference in reactivity ratios, the high chain-transfer constant of allylic radicals, and crosslinking reactions due to the small amount of pendent double bonds present. Production of more linear copolymers at high molecular weights can be done by the staged addition of AMD and initiator to the copolymerizing mixture. After monomer addition is complete, a chain-transfer agent can then be added and the batch heated to high temperature to complete the conversion.

Polyamines Quaternary polyamines are manufactured by ring-opening polyadditions and polycondensations. In both cases, low-molecular-weight branched, crosslinked, or thermosetting resins are often obtained. For example, methylamine can be reacted with epichlorohydrin to form poly(2-hydroxypropyl-1-*N*-methylammonium chloride), which is a commercial flocculant.[188] However, the charge on the secondary amine along the polymer backbone varies with the pH of the application. To overcome this problem, quaternized polyamines are used. Dimethylamine can be reacted with epichlorohydrin to form poly[2-hydroxypropyl-*N*,*N*-dimethyl ammonium chloride].[189] The molecular weight of this polymer can be increased substantially with a small amount of a polyfunctional amine or ammonia.[190] Methods have been developed to reduce the residual toxic epoxy compounds.[189] Quaternized polyamines are excellent low-cost flocculants for raw-water clarification, filter aids in treating coal washings and sludge, and color removal.

Anionic Acrylamide Copolymers Similar to cationic acrylamide copolymers, anionic acrylamide copolymers such as poly(acrylamide-*co*-sodium acrylate), poly(acrylamide-*co*-ammonium acrylate), and poly(acrylamide-*co*-sodium 2-acrylamido-2-methylpropanesulfonate) are of considerable practical importance. They are mainly prepared by copolymerization of acrylamide (AMD) with sodium acrylate, ammonium acrylate, or sodium 2-acrylamido-2-methylpropanesulfonate (AMPS). Comonomer reactivity ratios of acrylamide and commercially important anionic monomers are also shown in Table 1.4.

AMD can be copolymerized with acrylic acid (AA) or acrylic acid salts by radical polymerization in solution, inverse emulsion,[191] or microemulsion.[192] Reactivity ratios vary with pH. At low pH, the polymer formed at low conversion from acrylamide and acrylic acid system will contain a higher proportion of acrylic acid units than the feed composition.[191] But at higher pH, the copolymer first formed will be richer in acrylamide than in acrylic acid units. At an intermediate pH range the compositional drift can be mitigated (i.e., $r_1 \sim r_2$), and the azeotropic copolymer composition can be formed.[192]

AMD/AMPS copolymers have been studied extensively. Their ability to maintain charge at low pH and their tolerance to high divalent cation concentration are important features. They are used as flocculants for phosphate slimes, uranium leach residues, and coal refuse.

1.5   APPLICATIONS

A vast and varied market exists for solids–aqueous liquid separation enhanced by water-soluble polyelectrolytes. This market is expanding daily both in size and in the variety of applications. Major portions of this commercial opportunity include water treating, papermaking, mineral processing, and enhanced oil recovery. The example of water treating is discussed below.

The water-treating industry is typically subdivided into municipal and industrial sectors, which are then further subdivided into applications. Municipal water treating includes treatment of sewage, which may have domestic and industrial components, and potable water. Water-soluble polymers are used in potable water treatment to aid in the removal of suspended solids (clay, silt, and microorganisms such as bacteria and viruses) via clarification and/or filtration. Municipal sewage treatment offers a number of opportunities for the use of enhanced solid–liquid separation processes. These treatment plants typically treat very large flows and volumes of influent solids on a continual basis by a biological treatment process (which generates even larger quantities of solids). Thus optimum solids handling is critical for efficient and cost-effective operation. Solids handling is a major factor affecting equipment requirements (and thus capital costs) in all plant processes, solids final disposal options, and most importantly final water quality. Application points for polyelectrolytes in these plants include (Fig. 1.6): primary clarification (to enhance removal of influent solids, organic loading, phosphorus, and metals); secondary and tertiary clarification (to enhance solids removal from treated effluent); solids thickening (to reduce the volume of solids to be handled by downstream processes such as digestion and dewatering); effluent filtration and dewatering (for preparation of solids for final disposal). In the dewatering step, solids must be captured to the maximum possible extent, as the water discharged from this equipment is typically returned to the head of the plant. Significant solids in this stream account for rapidly rising treatment costs, as these solids add load to the plant, which may be unaccounted for in the design.

Industrial water treating consists of waste applications and in-process applications. Industrial wastewater treatment generally is similar to that of domestic wastewater as described above, with some notable additional requirements. Typically, industrial wastewater contains a more varied mixture of contaminants that subsequently require treatment strategies, equipment and microbiological populations vastly different from those seen in treating domestic wastewater. The advantages of polyelectrolyte-enhanced solid–liquid separation discussed above are vital to these industrial wastewater treatment plants. These treatment plants constitute a financial overhead and thus have a significant direct effect on the profitability of the business. Therefore, unlike municipal plants that are built using federal grants or loan guarantees and operated with tax or rate-payer generated revenue, survival of these industrial entities literally depends on the wastewater treatment plants being built, staffed, maintained, and operated as cost efficiently as possible. In many cases, use of polyelectrolytes to improve the efficiency of all aspects of solids handling is effective and inexpensive in light of the alternatives. Larger equipment needed to compensate for inefficient solids handling is costly from the standpoint of capital monies as well as footprint considerations (i.e., available real estate). Also, in many cases, the solids removed from these systems for disposal are designated as hazardous waste, and thus minimizing the entrained water can have an astronomical cost impact on the operation.

Industrial in-process (nonwaste) applications vary widely and, as discussed above, continue to develop as needs to improve economics and quality are affecting all industries. The following list, though not exhaustive and containing some nondewatering applications, will provide a flavor for the variety of industrial in-process applications using water-soluble polyelectrolytes:

1. Raw and recovered water used for heating, cooling, locomotion, processing, formulation, etc.
2. Fiberglass mat production
3. Enhanced oil recovery; including drag reduction, viscosity modifiers, drilling mud treatment, fracturing fluids, cements, effluent reclamation, dispersants, and antiscalants
4. Color removal; especially in the textile industry
5. Sugar processing
6. Soil conditioning and erosion control
7. Fat–oil–grease (FOG) removal
8. Animal feed recovery from waste (rendering, etc.)
9. Pharmaceuticals; solids removal from fermentation broths
10. Brick making
11. Papermaking "on-machine" applications including retention aids, wet and dry strength agents, and sizing
12. Papermaking "off-machine" applications including recovery of fiber from wastewater for fuel or reuse, de-inking, and secondary waste solids handling

13. Mineral processing (primarily anionic polyelectrolytes) applications including improved recovery and processing of coal, "red mud" (Bayer process for aluminum recovery), metal sulfides, magnetite, copper, zinc, gold, uranium, potash, etc.

1.6   TESTING AND EVALUATION METHODS

Consider the multitude of substrates, applications, generating processes, and performance requirements; then factor in the wide range of water-soluble polyelectrolytes, varying in product form, chemistry, charge type and density, molecular weight, and architecture; and the magnitude of the task of selecting the proper polymer for a particular solid–liquid separation becomes apparent. However, being armed with some basic knowledge of the substrates, applications, generating processes, and the principles of flocculation science will allow one to narrow the selection group.

In wastewater solids dewatering, solids from different sources or treatment processes will often have different requirements with respect to charge demand, polymer type, molecular weight and dose to yield the best capture, and dry solids content. Examples of different sludges include primary paper solids, primary municipal solids, raw blends of primary and secondary solids, anaerobically digested sludge blends, waste-activated sludges (WAS) and their variations (extended aeration WAS, whole aerated sludges, aerobically digested WAS, etc.), and ATAD (autothermal thermophilic aerobic digestion) sludges. Knowledge of the substrate, along with a customer's application and performance criteria, will allow one to narrow the choices of products to consider for laboratory testing. Some of the substrate properties that can be used in guiding this decision include the solids content (total, total dissolved, total suspended, and volatile), pH, temperature, capillary suction time (CST), conductivity, and particle size distribution. The compatibility of the polymer chemistry with the sludge must also be considered, since some chemistries perform better under conditions of high salt content, pH extremes, and so forth.

Bench-scale testing to determine the optimum polyelectrolyte for a particular application is best preceded by a thorough evaluation of both the end user's requirements and the actual application conditions. For example, under most circumstances it would be senseless to recommend an expensive polymer program that would yield superlative performance if the end user required only a minimum performance level at a minimum cost. Conversely, a low-cost polymer program with baseline performance might not be advisable when polymer costs were minuscule compared to the economic impact of improved solids capture or dehydration. It is also usually advantageous to undertake a walk-through of the full-scale application in order to evaluate operating conditions, including characteristics of the water source (e.g., quality), equipment used for polymer makeup, dilution water availability and quality, pump capacities, tankage for polymer solution aging, polymer addition points to the substrate, conditioning mixing intensity and time, required throughput, shear history of the formed floc, and the like.

Laboratory test strategies should imitate the essentials of the full-scale application conditions (e.g., settling, flotation, or filtration) as much as possible. Observations made on the flocculated substrate might include floc size, drainage, settling or rise rate, CST, streaming current, floc strength, solids compaction, turbidity of separated water, cake solids, release from filter, dose window (dose range over which optimum performance is achieved), and of course optimum polymer dose. Usually only a subset of these parameters are relevant and are used. Polymer dose must be evaluated on a consistent basis, since different polymer forms have radically different levels of active polymer (typical activities: dry powders > 80%, emulsions ~25 to 50%, concentrated solutions ~2 to 10%).

Some consideration should be given regarding the preparation of the polymer solutions for testing. When makeup water composition is an issue, lab polymer solutions should be made up in a water similar to that used full scale. This can be important since many end users employ process water or treated effluent that may yield a polymer solution that performs differently than a solution made in deionized or city water. The makeup polymer concentration should be chosen to fulfill solution property requirements (e.g., proper inversion of emulsions and manageable viscosity), and to allow the optimum dose of polymer to be delivered with existing pump capacities.

Consideration should also be given to the agitation available for polymer makeup, polymer solution age time, mixing conditions during polymer conditioning with the substrate and subsequent shearing of the flocs. Mixing of the polymer solution and substrate can be achieved in a number of ways in the lab depending upon the mixing intensity required and the availability of equipment and facilities. "Boxing," or pouring from container to container, is probably the simplest method for field work, and it imparts a low level of shear. "Snap-shaking" in a sealed container is also easy in the field, and it imparts significantly more shear than boxing. A more flexible conditioning method for polymer testing is to use a variable-speed mixer that allows for a reproducible and adjustable mixing intensity and time. The most appropriate mixing method depends on the application and can be best judged from correlations between on-machine and lab testing results.

Basic bench-scale testing methods[193] for settling, clarification, and gravity thickening applications, as well as for preliminary evaluation in flotation applications, include multijar or gang stirrer testing. Usually in these methods several containers of the same substrate receive, nearly simultaneously, varying polymer doses using a multiple-paddle, variable-speed mixer. A typical settling or clarification evaluation might include comparisons of the settling rate, floc size, polymer dose, solids compaction, and supernatant clarity. For flotation applications, after selecting the best performing material(s) and dose(s), further work may be done to evaluate the solids rise rate and subnatant clarity. An example of performing this evaluation would be to to fill a 2-L cylinder to the 1-L level with substrate and mix in the proper polymer dose. Water and dissolved air would then be added to the cylinder using a commercial garden sprayer. The final volume should reflect the full-scale recycle rate.

Dewatering testing[193,194] can be more involved due to the requirements for dry cakes, capture, and floc shear stability, under conditions of varying feed solids concentration and the multitude of different full-scale equipment types. In these tests, one typically evaluates conditioning variables using metrics such as floc size, drainage rate (e.g., time to filter, gravity drainage rate, or specific resistance to filtration), CST, streaming current, floc strength, turbidity, total suspended solids (TSS) of separated water, cake solids, release from filter, and optimal polymer dose (relative to weight of dry solids). Often a relevant subset of these measures are employed. Cake solids can be evaluated for their dewaterability in the lab using any of a number of methods that simulate the full-scale equipment type and extent of dewatering. These methods include vacuum filtration (with rubber dam) and a Crown Press to simulate BFP applications; a Tetra Press to simulate BFP, screw press, and centrifuge applications (varying pressure and cycle according to the application); and laboratory centrifuges to simulate BFP and centrifuge applications (varying applied G force and time according to the application). Plate and frame or chamber filter presses can be evaluated in the lab by generating a very small floc size (like coffee grounds) and determining the lowest dose that both minimizes the CST and maintains it over the greatest dose range. Rotary drum vacuum filter applications are typically evaluated using the filter leaf test.

1.7   CONCLUSIONS, UNSOLVED ISSUES, AND AVENUES FOR PROGRESS

Polyelectrolytes can be used with great effectiveness to destabilize colloidal suspensions, increase their particle collision efficiencies and adhere together the resulting flocs. The dominant concepts in polyelectrolyte-mediated destabilization and flocculation are surface charge neutralization, charge patch formation, and polymer bridging. The physical form of the floc must be engineered with a mind to the solid–liquid separation process used and the desired characteristics of the dewatered sludge. This engineering is accomplished via manipulation of the key parameters that govern floc formation, breakup, and rearrangement. These include the interparticle forces, collision rates, floc mechanical properties, and the relative dynamics of the subprocesses. Optimal conditioning of the sludge is mainly accomplished through the correct choice of destabilizing and flocculating agents, addition rates, and mixing conditions for the prevailing solution conditions.

While the DLVO theory has unparalleled success in modeling the destabilization of colloids with low-molecular-weight substances, a quantitative incorporation of polyelectrolyte-mediated forces (e.g., bridging and charge patch interaction) into models of the flocculation and dewatering processes remains a challenge. Two key parameters requiring expression are the orthokinetic collision efficiency factor, α_0, and the compressive yield strength of the floc network, $P_y(\phi)$.

Quantitative determinations of the kinetics of the dynamic subprocesses, such as adsorbed polymer rearrangement and collision rate-dependent attachments, are a necessary step in developing in a more complete picture of their influence on floc structure and strength. There are no models for the internal rearrangement dynamics (especially compaction) of flocs undergoing agitation.

The amount of polymer required for efficient flocculation is often considerably less than needed for complete surface coverage, and can be affected by the mixing history. Insight into the distribution of the polymer in flocs of this sort might lead to a better approach to wastewater conditioning that would minimize polyelectrolyte usage.

Agitation effects on the conditioning of sludges with polyelectrolytes do not always scale with the contact opportunity parameter $(G \cdot t)$. Presumably this occurs because of floc breakage processes, whose physics are not captured by this scaling parameter. A better scaling law under such conditions is required.

Robust methods for on-line flocculation monitoring could be used as part of a feedback control loop for adjusting polymer dosing and mixing conditions. This would allow for an optimization of the dewatering operation.

REFERENCES

1. L.A. Adorjan, *NATO ASI Ser., Ser. E* **117**, 339 (1986).

2. T.A. Wolfe and P.E. Malmrose, *Proc.–Annu. Conf., Am. Water Works Assoc.* 19 (1992).

3. J. Goin, Water Soluble Polymers, CEH Marketing Research Report 852.0000 D-E, SRI International, August 1991.

4. W.W. Eckenfelder and J.L. Musterman, in W.W. Eckenfelder and P. Grau, eds., *Activated Sludge Process Design and Control: Theory and Practice*, Water Quality Management Library, Vol. 1, Technomic, Lancaster, PA, 1992, Chap. 4, pp. 127–266.

5. M. Hsu and T.E. Wilson, in W.W. Eckenfelder and P. Grau, eds., *Activated Sludge Process Design and Control: Theory and Practice*, Water Quality Management Library, Vol. 1, Technomic, Lancaster, PA, 1992, Chap. 2, pp. 37–68.

6. G. Culp, G. Wesner, R. Williams, and M.V. Hughes, Jr., *Wastewater Reuse and Recycling Technology*, Noyes Data Corp, Park Ridge, NJ, 1980.

7. W.L.K. Schwoyer, in W.L.K. Schwoyer, ed., *Sludge Dewatering*, CRC, Boca Raton, FL, 1981, Chap. 6, pp. 159–209.

8. R.D. Lentz and R.E. Sojka, *Soil Sci.* **158**, 274 (1994).

9. M.J. Vold and R.D. Vold, *Colloid and Interface Chemistry*, Addison-Wesley, Reading, MA, 1983.

10. S. Ross and I.D. Morrison, *Colloidal Systems and Interfaces*, Wiley, New York, 1988.

11. G.R. Rose and M.R. St. John, in H.F. Mark, N.M. Bikales, C.G. Overberger, G.M. Menges, eds., J.I. Kroschwitz, series ed., *Encyclopedia of Polymer Science and Engineering*, 2nd ed., Wiley, New York, 1987, Vol. 7, pp. 211–232.

12. H.H. Hahn, *Prog. Colloid Polym. Sci.* **95**, 153 (1994).

13. B.M. Moudgil, S. Behland, and V. Mehta, in B.M. Moudgil and P. Somasundaran, eds., *Dispersion Aggregation*, Proc. Eng. Found. Conf. 1992, Engineering Foundation, New York, 1994, pp. 419–426.

14. D.H. Napper, *Polymeric Stabilization of Colloidal Dispersions*, Academic, New York, 1983.

15. B. Vincent, J. Edwards, S. Emmett, and A. Jones, *Colloids Surf.* **18**, 261 (1986).

16. E. Dickinson and L. Eriksson, *Adv. Colloid Interface Sci.* **34**, 1 (1991).

17. P.M. Claesson, T. Ederth, V. Bergeronand, and M.W. Rutland, *Adv. Colloid Interface Sci.* **67**, 119 (1996).

18. T.G.M. van de Ven, P. Warszynski, X. Wu, and T. Dabros, *Langmuir* **10**, 3046 (1994).

19. M. Kamiti and T.G.M. van de Ven, *Macromolecules* **29**, 1191 (1996).

20. G.J.C. Braithwaite, A. Howe, and P.F. Luckham, *Langmuir* **12**(17), 4224 (1996).

21. V.A. Parsegian and E.A. Evans, *Curr. Opin. Colloid Interface Sci.* **1**, 53 (1996).

22. J. Lyklema, in B.M. Moudgil and B.J. Scheiner, eds., *Flocculation and Dewatering*, Proc. Eng. Found. Conf., 1988, Engineering Foundation, New York, 1989, pp. 1–20.

23. D.F. Evans and H. Wennerström, *The Colloidal Domain: Where Physics, Chemistry, Biology, and Technology Meet*, VCH, New York, 1994.

24. D.F. Evans and H. Wennerström, Ref. 23, 1994, p. 347.

25. H. Holthoff, S.U. Egelhaaf, M. Borkovec, P. Schurtenberger, and H. Sticher, *Langmuir* **12**, 5541 (1996).

26. H. Reerink and J.Th.G. Overbeek, *Discuss. Faraday Soc.* **18**, 74 (1954).

27. J. Mahanty and B.W. Ninham, *Dispersion Forces*, Academic, New York, 1976.

28. J. Israelachvili, *Intermolecular and Surface Forces*, 2nd ed., Academic, San Diego, 1991.

29. E.M. Lifshitz, *Sov. Phys.–JETP* [English trans.] **2**, 73 (1956).

30. I.E. Dzyaloshinskii, E.M. Lifschitz, and L.P. Pitaevskii, *Adv. Phys.* **10**, 165 (1961).

31. R. Hidalgo-Alvarez, A. Martin, A. Fernandez, D. Bastos, F. Martinez, and F.J. de las Nieves, *Adv. Colloid Interface Sci.* **67**, 1 (1996).

32. D.B. Hough and L.R. White, *Adv. Colloid Interface Sci.* **14**, 3 (1980).

33. J. Visser, *Adv. Colloid Interface Sci.* **3**, 331 (1972).

34. R.S. Chow and K. Takamura, *J. Colloid Interface Sci.* **125**, 226 (1988).

35. P. Coackley and F. Wilson, *Filtr. Separ.* **8**, 61, 64 (1971).

36. M. Abu-Orf, Proc. Water Environ. Fed. Conf., 66th, Anaheim, CA, Oct. 3–7, 1993, Water Environment Federation, Alexandria, VA, 1993, paper AC93-035-004, pp. 201–211.

37. S. Dentel and M. Abu-Orf, Proc. Water Environ. Fed. Conf., 67th, Chicago, IL, Oct. 15–19, 1994, Water Environment Federation, Alexandria, VA, 1994, paper AC943805, pp. 541–552.

38. Y. Nakamura, K.-I. Kameyama, C. Igarashi, K. Tanaka, T. Kitamura, and K. Fujita, Proc. Water Environ. Fed. Conf., 67th, Chicago, IL, Oct. 15–19, 1994, Water Environment Federation, Alexandria, VA, 1994, paper AC943804, pp. 533–540.

39. E. Dickinson and S.R. Euston, *Colloids Surf.* **62**, 231 (1992).

40. J. Gregory, in R.A. Williams and N.C. De Jaeger, eds., *Advances in Measurement and Control of Colloidal Processes*, Butterworth-Heinemann, Oxford, UK, 1991, pp. 3–17.

41. C. Geffroy, J. Persello, A. Foissy, B. Cabane, and F. Tournilhac, *Rev. Inst. Fr. Pet.* **52**, 183 (1997).

42. J.M.H.M. Scheutjens and G.J. Fleer, *Macromolecules* **18**, 1882 (1985); *J. Colloid Interface Sci.* **111**, 504 (1986); in Th.F. Tadros, ed., *The Effect of Polymers on Dispersion Properties*, Academic, London, UK, pp. 145–168.

43. K. Barnett, T. Cosgrove, T.L. Crowley, Th.F. Tadros, and B. Vincent, in Th.F. Tadros, ed., *The Effect of Polymers on Dispersion Properties*, Academic, London, UK, pp. 183–197.

44. K.G. Barnett, T. Cosgrove, B. Vincent, M. Cohen-Stuart, and D.S. Sissons, *Macromolecules* **14**, 1018 (1981).

45. T. Cosgrove and K.G. Barnett, *J. Magn. Reson.* **43**, 15 (1981).

46. K.G. Barnett, T. Cosgrove, B. Vincent, A.N. Burgess, T.L. Crowley, T. King, J.D. Turner, and T.F. Tadros, *Polymer* **22**, 283 (1981).

47. D. Cebula, R.K. Thomas, N.M. Harris, J. Tabony, and J.W. White, *Faraday Discuss.* **65**, 76 (1978).

48. X. Wu and T.G.M. van de Ven, *J. Colloid Interface Sci.* **183**, 388 (1996).

49. A. Takahashi and M. Kawaguchi, *Adv. Poly. Sci.* **46**, 1 (1982).

50. E.G.M. Pelssers, M.A. Cohen Stuart, and G.J. Fleer, *J. Chem. Soc. Farady Trans.* **86**, 1355 (1990).

51. M.A. Cohen Stuart and H. Tamai, *Macromolecules* **21**, 1863 (1988).

52. M.A. Cohen Stuart and H. Tamai, *Colloids Surfaces* **31**, 265 (1988).

53. B. Vincent, *Adv. Colloid Interface Sci.* **4**, 527 (1974).

54. S.X. Liu and L.A. Glasgow, *Adv. Transp. Processes* **9**, 103 (1993).

55. W. Ditter, J. Eisenlauer, and D. Horn, in T.F. Tadros, ed., *The Effect of Polymers on Dispersion Properties, [Proc. Int. Symp.]*, 1981, Academic, London, UK, 1982, pp. 323–342.

56. A.I. Cole and P.C. Singer, *J. Environ. Engr.* **111**(4), 501 (1985).

57. G. Durand-Piana, F. Lafuma, and R. Audebert, *J. Colloid Interface Sci.* **119**, 474 (1987).

58. L. Eriksson and B. Alm, *Water Sci. Technol.* **28**, 203 (1993).

59. P.A. Rey, in B.M. Moudgil and B.J. Scheiner, eds., *Flocculation and Dewatering*, Proc. Eng. Found. Conf., 1988, Engineering Foundation, New York, 1989, pp. 195–214.

60. F.F. Peng and P. Di, *J. Colloid Interface Sci.* **164**, 229 (1994).

61. N. Narkis, B. Ghattas, M. Rebhun, and A.J. Rubin, *Water Supply* **9**(1), 37 (1991).

62. F. Mabire, R. Audebert, and C. Quivoron, *J. Colloid Interface Sci.* **97**(1), 120 (1984).

63. J. Gregory, in C.A. Finch, ed., *Industrial Water Soluble Polymers*, Royal Society of Chemistry, Cambridge, UK, 1996, pp. 62–75.

64. Y. Otsubo, *Heterog. Chem. Rev.* **3**, 327 (1996).

65. T.G.M. van de Ven, *Colloidal Hydrodynamics*, Academic, London, 1989, p. 386.

66. M. Smoluchowski, *Z. Physik. Chem. (Leipzig)* **92**, 9 (1917).

67. H.J. Pearson, I.A. Valioulis, and E.J. List, *J. Fluid Mech.* **143**, 367 (1984).

68. L.A. Glasgow and S.X. Liu, *Chem. Eng. Commun.* **132**, 223 (1995).

69. Y. Adachi, M.A. Cohen Stuart, and R. Fokkink, *J. Colloid Interface Sci.* **167**, 346 (1994).

70. P.G. Saffman and J.S. Turner, *J. Fluid Mech.* **1**, 16 (1956).

71. T.R. Camp and P.C. Stein, *J. Boston Soc. Civil Engr.* **30**, 219 (1943).

72. S.J. Peng and R.A. Williams, *J. Colloid Interface Sci.* **166**, 321 (1994).

73. R.J. Lai, H.E. Hudson, Jr., and J.E. Singley, *J. Am. Water Works Assoc.* **67**, 553 (1975).

74. K.J. Ives, in J. Gregory, ed., *Solid-Liquid Separations*, Ellis Horwood, Chichester, UK, 1984, Chap. 15, pp. 196–217.

75. W.D. Young and D.C. Prieve, *Langmuir* **7**, 2887 (1991).

76. T.M. Herrington and B.R. Midmore, *J. Chem. Soc., Faraday Trans. 1* **85**(10), 3529 (1989).

77. L. Wagberg, *Svensk Papper.* **6**, R48 (1985).

78. P.T. Spicer and S.E. Pratsinis, *Water Res.* **30**, 1049 (1996).

79. X. Li and B.E. Logan, *Environ. Sci. Technol.* **31**, 1237 (1997).

80. Y. Adachi and T. Matsumoto, *Colloids Surf., A: Physicochem. Engr. Aspects* **113**, 229 (1996).

81. E.G.M. Pelssers, Dissertation Agric. Univ. Wageningen (1988).

82. J. Gregory, *Adv. Colloid Interface Sci.* **17**, 149 (1982).

83. R. Hogg, in B.M. Moudgil and B.J. Scheiner, eds., *Flocculation and Dewatering*, Proc. Eng. Found. Conf., 1988, Engineering Foundation, New York, 1989, pp. 143–151.

84. J.T. Novak, J.F. Prendeville, and J.H. Sherrard, *J. Environ. Eng.* **114**, 190 (1988).

85. C.P. Werle, J.T. Novak, W.R. Knocke, and J.H. Sherrard, *J. Environ. Eng.* **110**, 919 (1984).

86. F. Gruy and H. Saint-Raymond, *J. Colloid Interface Sci.* **185**, 281 (1997).

87. T.G.M. van de Ven and S.G. Mason, *Colloid Polymer Sci.* **255**, 468 (1977).

88. S.J. Langer and R. Klute, *Water Sci. Technol.* **28**, 233 (1993).

89. E. Killmann and J. Eisenlauer, in T.F. Tadros, ed., *The Effect of Polymers on Dispersion Properties, [Proc. Int. Symp.]*, 1981, Academic, London, UK, 1982, pp. 221–244.

90. L.A. Glasgow and Y.H. Kim, *Water Air and Soil Pollution* **47**, 153 (1989).

91. R.H. Cumming, P.M. Robinson, and G.F. Martin, *Bioseparation* **6**, 17 (1996).

92. S. Baran, A.A. Baran, and D. Gregory, *Colloid J.* **58**(1), 9 (1996).

93. P.M. Robinson, G.F. Martin, and R.H. Cumming, *Bioseparation* **4**, 247 (1994).

94. J. Gregory and T.O. Kayode, in B.M. Moudgil and B.J. Scheiner, eds., *Flocculation and Dewatering*, Proc. Eng. Found. Conf., 1988, Engineering Foundation, New York, 1989, pp. 645–655.

95. J. Eisenlauer and D. Horn, in J. Gregory, ed., *Solid-Liquid Separations*, Ellis Horwood, Chichester, UK, 1984, Chap. 14, pp. 183–195.

96. N. Böhm and W.M. Kulicke, *Colloid Polym. Sci.* **275**, 73 (1997).

97. H.W. Campbell and P.J. Crescuolo, *Water Sci. Tech.* **21**, 1309 (1989).

98. S.K. Dentel, K.M. Wehners, and M.M. Abu-Orf, in R. Klute and H.H. Hahn, eds., *Chemical Water and Wastewater Treatment III; Proc. 6th Gothenburg Symp.*, Springer, Berlin, Heidelberg, 1994, pp. 373–381.

99. A. Bartelt, D. Horn, W. Geiger, and G. Kern, *Prog. Colloid Polym. Sci.* **95**, 95 (1994).

100. P.T. Spicer and S.E. Pratsinis, *AIChE J.* **42**, 1612 (1996).

101. P.T. Spicer, S.E. Pratsinis, M.D. Trennepohl, and G.H.M. Meesters, *Ind. Eng. Chem. Res.* **35**, 3074 (1996).

102. L.H. Mikkelsen, A.K. Gotfredsen, M.L. Agerbaek, P.H. Nielsen, and K. Keiding, *Water Sci. Technol.* **34**, 449 (1996).

103. T.G.M. van de Ven, Ref. 65, 1989, p. 391.

104. M.J. Vold and R.D. Vold, Ref. 9, 1983, p. 281.

105. T.G.M. van de Ven, Ref. 65, 1989, pp. 369, 390.

106. B.M. Moudgil and T.V. Vasudevan, in B.M. Moudgil and B.J. Scheiner, eds., *Flocculation and Dewatering*, Proc. Eng. Found. Conf., 1988, Engineering Foundation, New York, 1989, pp. 167–178.

107. L. Eriksson, B. Alm, and L. Aldén, in B.M. Moudgil and B.J. Scheiner, eds., *Flocculation and Dewatering*, Proc. Eng. Found. Conf., 1988, Engineering Foundation, New York, 1989, pp. 179–193.

108. J.B. Farrow and L.J. Warren, in B.M. Moudgil and B.J. Scheiner, eds., *Flocculation and Dewatering*, Proc. Eng. Found. Conf., 1988, Engineering Foundation, New York, 1989, pp. 153–165.

109. N. Tambo and R.J. Francois, in A. Amirtharajah, M.M. Clark, M. Trussell, and R.R. Rhodes, eds., *Mixing Coagulation Flocculation*, 1991, Am. Water Works Assoc., Denver, 1991, Chap. 7, pp. 256–281.

110. F. Jorand, F. Zartarian, F. Thomas, J.C. Block, J.Y. Bottero, G. Villemin, V. Urbain, and J. Manem, *Water Res.* **29**(7), 1639 (1995).

111. B.A. Firth and R.J. Hunter, *J. Colloid Interface Sci.* **57**(2), 248 (1976).

112. T.G. van de Ven and R.J. Hunter, *Rheol. Acta* **16**, 534 (1977).

113. L. Eriksson and B. Alm, *Chemical Water and Wastewater Treatment II; Proc. 5th Gothenburg Symp.*, Springer, Berlin, Heidelberg, 1992, pp. 19–32.

114. L.A. Glasgow and S.X. Liu, *Environ. Tech.* **16**, 915 (1995).

115. K. Keiding and P.H. Nielsen, *Water Res.* **31**(7), 1665 (1997).

116. T.L. Poxon and J.L. Darby, *Water Res.* **31**, 749 (1997).

117. Task Force on Wastewater Biology, M.H. Gerardi and F.L. Horsfall III, chairpeople, *Wastewater Biology: The Microlife*, Water Environment Federation, Alexandria, VA, 1994.

118. L. Eriksson, *Water Sci. Tech.* **19**, 859 (1987).

119. S.X. Liu and L.A. Glasgow, *Separations Tech.* **5**, 139 (1995).

120. D.H. Li and J. Ganczarcyzk, *Crit. Rev. Environ. Control* **17**, 53 (1986).

121. W.R. Knocke, C.M. Dishman, and G.F. Miller, *Water Env. Res.* **65**, 735 (1993).

122. B.B. Mandelbrot, *The Fractal Geometry of Nature*, W.H. Freeman, New York, 1983.

123. M.Y. Lin, R. Klein, H.M. Lindsay, D.A. Weitz, R.C. Ball, and P. Meakin, *J. Colloid Interface Sci.* **137**(1), 263 (1990).

124. A.S. Michaels and J.C. Bolger, *Ind. Engr. Chem. Fund.* **1**, 153 (1962).

125. K. Muhle and K. Domasch, in H.H. Hahn and R. Klute, eds., *Chemical Water and Wastewater Treatment*, Springer, Berlin, 1990, pp. 105–115.

126. K. Mühle, in B. Dobiás, ed., *Coagulation and Flocculation, Theory and Applications*, Surfactant Series, Vol. 47, Dekker, New York, 1993, Chap. 8, pp. 355–390.

127. R. Hogg, in B.M. Moudgil and P. Somasundaran, eds., *Dispersion Aggregation*, Proc. Eng. Found. Conf., 1992, Engineering Foundation, New York, 1994, pp. 21–31.

128. L.A. Glasgow and J.P. Hsu, *A. I. Ch. E. J.* **28**(5), 779 (1982).

129. A.K.C. Yeung and R. Pelton, *J. Colloid Interface Sci.* **184**, 579 (1996).

130. M. Charon-Charles and J.-P. Gozlan, *Chem. Engr. Sci.* **51**(20), 4649 (1996).

131. M. Smollen, *Water Sci. Tech.* **22**(12), 153 (1990).

132. N.W. Schmidtke, *Conf. Proc.–Res. Program Abatement Munic. Pollut. Provis. Can.-Ont. Agreement Great Lakes Water Qual.* **6**, 173 (1978).

133. K.R. Tsang and P.A. Vesilind, *Water Sci. Tech.* **22**(12), 135 (1990).

134. M. Hashimoto and M. Hiraoka, *Water Sci. Tech.* **22**(12), 143 (1990).

135. E.A. Retter and R. Schilp, *Filtration Sep.* **June**, 387 (1994).

136. P.L. LaMontagne, *Mid. Atl. Ind. Waste Conf., [Proc.]* **10**, 93 (1978).

137. D. Li and J.J. Ganczarczyk, *Water Environ. Res.* **64**(3), 236 (1992).

138. J. Namer and J.J. Ganczarczyk, *Water Res.* **27**, 1285 (1993).

139. K.A. Landman and L.R. White, *Adv Colloid Interface Sci.* **51**, 175 (1994).

140. A.J. Carleton and A.G. Salway, *Filtr. Sep.* **30**(7), 641 (1993).

141. J.R. Christensen, P.B. Sorensen, G.L. Christensen, and J.A. Hansen, *J. Envr. Eng.* **119**(1), 159 (1993).

142. G. Stroh, in B. Dobiás, ed., *Coagulation and Flocculation, Theory and Applications*, Surfactant Series, Vol. 47, Dekker, New York, 1993, Chap. 14, pp. 653–695.

143. R.I. Dick and R.O. Ball, *CRC Crit. Rev. Environ. Control* **10**, 269 (1980).

144. T. Poxon, Proc. WEF Conf., Dallas, TX, 1996, pp. 167–178.

145. G.L. Christensen and R.I. Dick, *J. Env. Engr.* **111**(3), 258 (1985); **111**(3), 243 (1985).

146. R.S. Gale, *Water Pollut. Control* **66**, 622 (1967).

147. P. Coackley, *J. Inst. Public Health Eng.* **64**(8), 275 (1965).

148. R.S. Gale, *Filtr. Separ.* **8**, 531 (1971).

149. D.L. Ford, *Water Resour. Symp.* **3**, 341 (1970).

150. R.J. Wakeman, in L. Svarovsky, ed., *Solid-Liq. Sep. (2nd ed.)*, 1981, Butterworths, London, UK, pp. 452–471.

151. M.A. Koenders and R.J. Wakeman, *Chem. Eng. Res. Des.* **75**, 309 (1997).

152. N. Vorchheimer, in W.L.K. Schwoyer, ed., *Polyelectrolytes Water Wastewater Treat.*, CRC, Boca Raton, FL, 1981, pp. 1–45; S.-Y. Huang and D.W. Lipp, in J.E. Salamone, ed., *Polymeric Materials Encyclopedia*, CRC, Boca Raton, FL, 1996, Vol. 4, p. 2427.

153. J.-M. Corpart and F. Candau, *Macromol.* **26**, 1333 (1993).

154. U.S. Pat. 3,300,406 (1967), F.X. Pollio (to Rohm and Hass Co.).

155. U.S. Pat. 4,00,627 (1977), H.P. Panzer and M.N. O'Conner; U.S. Pat. 4,007,200 (1977), L.J. Baccei (to American Cyanamid Co.).

156. U.S. Pats. 4,137,416 (1979) and 4,137,415 (1979), H.P. Panzer and K.U. Acholona (to American Cyanamid Co.).

157. S. Machida, M. Araki, and K. Matsuo, *J. Polym. Sci.* **12**, 325 (1968).

158. U.S. Pat. 3,198,762 (1965), A. Maeder and O. Albrecht (to Ciba, Ltd.).

159. J. Brandrup and E.H. Immergut, *Polymer Handbook,* 3rd ed., Wiley, New York, 1989.

160. W.M. Thomas and D.W. Wang, in J.I. Kroschwitz, ed., *Encyclopedia of Polymer Science and Engineering,* 2nd ed., Wiley, New York, Vol. 1, 1985, p. 182.

161. D.J. Currie, F.S. Dainton, and W.S. Watt, *Polym.* **6**, 451 (1965).

162. V.F. Gromov, N.I. Galperina, T.O. Osmanov, P.M. Khomikovskii, and A.D. Abkin, *Eur. Polym. J.* **16**, 529 (1980).

163. S.M. Shawki and A.E. Hamielec, *J. Appl. Polym. Sci.* **23**, 3341 (1979).

164. F.S. Dainton and W.D. Sisley, *Trans. Faraday Soc.* **59**, 1369 (1963).

165. C. Walling, *Free Radicals in Solution*, Wiley, New York, 1957.

166. M.M. Hussain, S.N. Mishra, and A. Gupta, *Makromol. Chem.* **177**, 41 (1978).

167. S. Das, I.K. Kar, and S.R. Palit, *J. Indian Chem. Soc.* **51**, 393 (1974).

168. W.M. Kulicke, R. Kniewske, and R. Klein, *J. Prog. Polym. Sci.* **8**, 373 (1982).

169. U.S. Pat. 4,021,364 (1977), P. Speiser and G. Birrenbach (to Forsch. Switzerland).

170. F. Candau, Y.S. Leong, and R.M. Fitch, *J. Polym. Sci., Polym. Chem.* **23**, 193 (1985).

171. U.S. Pat. 5,545,688(1996), Sun-Yi Huang (to Cytec Technology Corp.).

172. U.S. Pat. 4,956,399 (1990), J.J. Kozakiewicz and S.-Y. Huang (to American Cyanamid Co.).

173. U.S. Pat. 5,037,881 (1991), J.J. Kozakiewicz and S.-Y. Huang (to American Cyanamid Co.).

174. U.S. Pat. 5,132,023 (1991), J.J. Kozakiewicz and S.-Y. Huang (to American Cyanamid Co.).

175. U.S. Pat. 5,627,260 (1997), S.-Y. Huang, A. Leone-Bay, J.M. Schmitt, and P.S. Waterman (to Cytec Technology Corp.).

176. C.J. McDonald and R.H. Beaver, *Macromolecules* **12**, 203 (1979).

177. T.F. Cummings and J.R. Shelton, *J. Org. Chem.* **25**, 419 (1960).

178. U.S. Pat. 4,762,894 (1988), D.W. Fong and D.J. Kowalski (to Nalco Chem. Co.).

179. H. Kheradmand, J. Francois, and V. Plazanet, *Polymer* **29**, 860 (1988).

180. U.S. Pat. 5,286,806 (1994), R.E. Neff and R.G. Ryles (to Cytec Industries, Inc.).

181. D.R. Draney, S.-Y. Huang, J.J. Kozakiewicz, and D.W. Lipp, *Polymer Preprints* **31**(2), 500 (1990).

182. Y. Negi, S. Harada, and O. Ishizuka, *J. Polym. Sci., Chem.:* Part A-1, **5**, 1951 (1967).

183. G.B. Butler and F.L. Ingley, *J. Am. Chem. Soc.* **71**, 3120 (1949).

184. G.B. Butler, *Acc. Chem. Res.* **15**, 370 (1982).

185. J.E. Lancaster, L. Baccei, and H.P. Panzer, *J. Polym. Sci., Polym. Lett. Ed.* **14**, 549 (1976).

186. M. Hahn and W. Jaeger, *Angew. Makromol. Chem.* **198**, 165 (1992).

187. P.C. Huang and K.H. Reichert, *Angew. Makromol. Chem.* **162**, 19 (1988).

188. A.T. Coscia, *Kirk-Othmer Encyclopedia of Polymer Science and Chemical Technology*, Wiley, New York, 1969, Vol. 10, p. 616.

189. U.S. Pat. 3,725,312 (1973), H.P. Panzer and R. Rabinowitz (American Cyanamid Co.).

190. U.S. Pat. 4,319,020 (1982), A.T Coscia, R.F. Tarvin, and D.F. Bardoliwalla (American Cyanamid Co.).

191. U.S. Pat. 4,439,332 (1984). S. Frank, A.T. Coscia, and A.J. Frisque (American Cyanamid Co.).

192. F. Candau, Z. Zekhnini, and F. Heatley, *Macromolecules* **19**, 1895 (1981).

193. S.K. Dentel, M.M. Abu-Orf, and N.J. Griskowitz, *Guidance Manual for Polymer Selection in Wastewater Treatment Plants*, Water Environment Research Foundation, Project 91-ISP-5, 1993.

194. J. Novak, W. Knocke, W. Burgos, and P. Schuler, *Water Sci. Tech.* **28**(1), 11 (1993).

195. H. Tanaka, *J. Polym. Sci., Polym. Chem. Ed.* **24**, 29 (1986).

196. W. Baade, D. Hunkeler, and A.E. Hamielec, *J. Appl. Polym. Sci.* **38**, 185 (1989).

197. Ch. Wandrey and W. Jaeger, *Acta Polym.* **36**, 100 (1985).

198. K. Plochocka, *J. Macromol. Sci., Rev. Macromol. Chem.* **C20**(1), 67 (1981).

199. C.L. McCormick and G.S. Chen, *J. Polym. Sci., Polym. Chem. Ed.* **20**, 817 (1982).

200. T.K. Wang and R. Audebert, *J. Colloid Interface Sci.* **121**(1), 32 (1988).

201. R. Aksberg, M. Einarson, J. Berg, and L. Ödberg, *Langmuir* **7**, 43 (1991).

202. R.C. Weast, M.J. Astle, and W.H. Beyer, eds., *CRC Handbook of Chemistry and Physics*, 67th ed., CRC, Boca Raton, FL, 1987, p. F-231.

203. E.G.M. Pelssers, M.A. Cohen Stuart, and G.J. Fleer, *Coll. Surf.* **38**, 15 (1989).

2 Polymer–Colloid Interactions in Pulp and Paper Manufacture

ROBERT H. PELTON

McMaster Centre for Pulp and Paper Research, Department of Chemical Engineering, McMaster University, Hamilton, Ontario L8S 4L7 Canada

Water-soluble polymers are widely used for the manufacture of pulp, paper, and paperboard. The strength, moisture resistance, oil resistance, and printing properties of paper are routinely enhanced by polymer treatment. Furthermore, polymers are used to improve pulp and paper manufacturing processes. The objective of this chapter is to survey the application of water-soluble polymers in the pulp and paper industry. Since this is rather a broad area that has been covered in many books and review articles, the emphasis will be on recent developments.

2.1 SOME BACKGROUND IN PULP AND PAPER SCIENCE AND TECHNOLOGY

Wood is a composite material containing long, slender cellulose fibers embedded in lignin, a polyphenolic, network polymer responsible for the brown color in wood. The chemical structure of cellulose is very well defined, whereas lignin displays no level of stereoregularity. That is, unlike nearly every other biologically derived organic material, lignin does not have a well-defined repeat structure nor does it display optical activity; instead it is present in wood as an irregular, crosslinked polymer network. The chemical composition of wood depends on the type of tree and its geographic location; however, cellulose accounts for about 50% of the mass of wood and lignin for approximately 30%. Other important components include hemicellulose that sits at the interface between crystalline cellulose and lignin-rich regions, and resin and fatty acids that are hydrophobic and sticky, thus tending to form

Colloid–Polymer Interactions: From Fundamentals to Practice, Edited by Raymond S. Farinato and Paul L. Dubin
ISBN 0-471-24316-7   © 1999 John Wiley & Sons, Inc.

troublesome "pitch" deposits.[1] In an aqueous pulp suspension the pitch is present as very dilute emulsion droplets that can accumulate on the paper machine as macroscopic sticky deposits which in turn cause paper breaks or give flaws in the paper sheet.

Paper and paperboard are manufactured from dilute aqueous suspensions of wood fibers. There are two main types of pulping process that convert wood to papermaking fibers: mechanical pulping and chemical pulping. From the perspective of the application of water-soluble polymers, the fibers resulting from these two pulping processes are very different. The key features are summarized below.

2.1.1 Mechanical Pulping

In mechanical pulping, wood chips, softened by steam and possibly chemical pretreatment, are forced between refiner plates where the chips are broken down into individual fibers. This process is popular because the resulting pulp contains more than 90% of the original wood materials and thus is an efficient use of the forest resource. The most distinctive feature of mechanical pulps is that most of the lignin originally present in the wood remains in the pulp. Therefore, mechanical pulps produce light brown papers, of which newsprint is a good example.

Mechanical pulps are characterized by a broad distribution of particle sizes and shapes. The normal industry practice is to define a fines fraction, which is the mass fraction of material that will pass through a 76-μm (200-mesh) screen. The particle size distribution of the fines can be measured by image analysis, and the average longest linear dimension is between 20 and 30 μm, which roughly corresponds to the width of a fiber.[2] Visual inspection reveals two types of fines: One type appears as flat sheets that have peeled off the fiber walls. The other type of fines are chunky, containing lignin and possibly resin and fatty acids.[3] Because fines are small, they have a large specific surface area, which, in turn, means that the fines have a larger capacity to adsorb water-soluble polymers. Early electrokinetic studies suggested that the surface charge characteristics of fines were similar to those of the parent fibers. However, recent work by Wågberg and Björklund[4] has revealed that the specific charge content of fines, both surface and total, is more than an order of magnitude greater than that of the parent fibers.

Another important feature of mechanical pulps is that all the water-soluble polymers and salts originally present in the wood are present in the pulp suspension because mechanical pulps are usually not washed. These dissolved and colloidal substances (DCS) are a mixture of dissolved hydrophilic polymers, mainly carbohydrates with bonded lignin fragments, dispersed resin and fatty acids (pitch) and dispersed fiber fragments.[5–7] In terms of polymer applications, the DCS have two significant effects. First, the soluble anionic components consume cationic polymers by forming polyelectrolyte complexes; processes such as peroxide bleaching which introduce more charged DCS tend to exacerbate this effect.[8] Second, some of the DCS components will adsorb onto hydrophobic surfaces such as polystyrene latex and render the particles more colloidally stable, presumably by electro-steric stabilization.[8–10]

2.1.2 Chemical Pulping

The objective of chemical pulping is to decompose and remove the lignin in wood chips to release pure cellulose fibers. This is usually done by the kraft pulping process, which treats the wood chips with a mixture of concentrated aqueous sodium hydroxide and sodium sulfide. After pulping and washing, the kraft pulp fibers contain about 2% residual lignin that colors the fibers. Unbleached or partially bleached kraft pulp is used to make the paper layers on corrugated boxes and paper grocery bags. The residual lignin in kraft pulp is removed by a multistage bleaching process to give bright white pulps that are used to make white office papers and a variety of other white paper products. More details about the kraft pulping process are given in Section 2.4.1.

Chemical pulps contain less fines than do mechanical pulps, which, in turn, leads to a lower capacity to adsorb polymers. Chemical pulps also have a much lower level of dissolved and colloidal substances (DCS) because the pulps are washed during the pulping and bleaching process.

2.1.3 Papermaking Fibers

To understand the mechanisms by which water-soluble polymers function in pulp and paper applications, it is important to appreciate the detailed structure of pulp fibers. What follows is a very brief summary; for a more detailed discussion see the excellent recent review by Wågberg and Annergren.[11] Fibers are about 5 mm long and about 20 μm wide with a hollow core called a lumen. The fiber wall is a complicated structure with a total thickness of about 2 μm. The detailed characteristics depend upon the wood type and the pulping process. Nevertheless, some general features can be described.

Cellulose in a fiber wall forms elementary fibrils that are about 35 Å in diameter and aggregate together to form microfibrils. The fibrils are present as sheets of parallel fibrils with different layers oriented relative to the fiber's longitudinal axis. The crystalline cellulose fibrils are embedded in a matrix of lignin and hemicellulose to give the structure shown in Figure 2.1.

Mechanical and chemical pulping delaminates the fiber wall to form a porous structure capable of absorbing more than twice its mass in water.[13] This is illustrated in Figure 2.2.[14] There have been many publications about the characterization of the pore structure distribution in fiber walls.[11] The most common approach is the solute exclusion method in which dextran or some other well-defined, nonadsorbing, water-soluble polymer is mixed with the pulp. The amount of water associated with the fiber wall that is not accessible to the dextran can be calculated from the mass of added dextran and the concentration of the dextran in the water

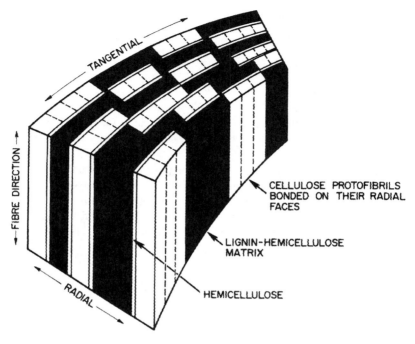

CELLULOSE PROTOFIBRILS
BONDED ON THEIR RADIAL
FACES

LIGNIN-HEMICELLULOSE
MATRIX

HEMICELLULOSE

FIGURE 2.1 Schematic representation of the wood fiber wall, after Kerr and Goring.[12]

exterior to the fibers. A pore size distribution can be constructed by repeating this experiment with a series of narrow molecular weight dextran samples of known size.[13] The average pore diameter is approximately 10 nm and the specific surface area is roughly 150 m^2/g. The porous nature of cellulose fiber walls complicates the interpretation of polymer adsorption experiments. Low-molecular-weight copolymers can access a much greater surface area than can high-molecular-weight polymers.[15]

Polymers and colloids first interact with the exterior surface of fibers. However, little is known about the detailed structure of the exterior surface. Optical microscopy reveals roughness at all visible size scales, and it seems reasonable to assume that this roughness persists down to molecular dimensions.[16] In most practical situations the fiber surface bears an adsorbed layer of hydrophilic polymer, either originating from the wood (i.e., hemicellulose or soluble lignin fragments) or from materials added to the process (starch, synthetic resins, etc.).

Wood fibers bear electrical charges that can influence the interaction of fibers with water-soluble polymers. Carboxylic and sulfonate groups, the latter originating from the pulping chemistry, are the most common charge types. Charge contents range from 10 to 100 microequivalents per gram.[11] The surface charge is reflected in zeta potentials, usually less than −50 mV, which can be measured by streaming potential experiments.

In summary, wood pulp fibers are rough, porous, complicated surfaces exhibiting behavior characteristic of both a hydrogel and a microporous solid.[16] With respect to

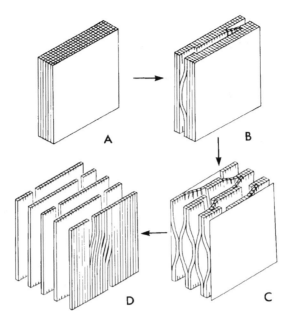

FIGURE 2.2   Illustration of the beating-induced delamination of the fiber wall that gives fibers the high capacity to retain water, from Scallan.[14]

the interaction of pulp with water-soluble polymers, the fines fraction of a pulp suspension presents the largest surface area and the most anionic charges to an absorbing polymer.

2.1.4   Papermaking Process

Paper is made by the rapid filtration of a dilute aqueous suspension of wood pulp fibers. The mass concentration of fibers, fines, and filler material delivered to the headbox of a modern paper machine is between 0.5 and 1%. The pulp in the headbox is exposed to hydrodynamic forces designed to disperse the fibers that have a natural tendency to aggregate into what papermakers call flocs. From the headbox the pulp is delivered onto one, or between two, moving plastic screens, called papermaking wires. The pulp supported on or between the wire(s) then travels over foils and vacuum boxes that remove most of the water to produce a weak sheet of paper containing about 60% water. The wet paper sheet is pressed between felt belts to mechanically remove as much water as possible. Then the sheet is dried by passing over steam heated drums. The dried paper is usually calendered (i.e., pressed between metal rolls) for improved surface properties. In some cases the paper is treated with a size press or coater to apply chemicals and/or coatings.

A modern paper machine runs at speeds as high as 4500 feet per minute (fpm)[17] so water removal on the wire is very rapid. Colloidal material dispersed in the pulp suspension, including fillers, fines, and pitch, is often not trapped (retained) during water removal because the pore sizes in the forming paper sheet are large relative to colloidal dimensions.[18] To retain colloidal material, papermakers routinely add water-soluble polymers (retention aids) to the pulp suspension before the headbox. Retention aids are flocculants that promote deposition of the colloidal material onto the surfaces of fibers and fines, which, in turn, are retained in the paper sheet during the dewatering on the wire.

2.2   POLYMER APPLICATIONS IN PULP AND PAPER—AN OVERVIEW

2.2.1   Polymers for Dry Strength

Water-soluble polymers are used to improve both the processes and the final product properties in the pulp and paper industry. This section summarizes the use of polymers to improve paper strength. The factors influencing paper strength are illustrated by the following equation[19]:

$$\frac{1}{T} \propto \frac{1}{F} + \frac{1}{B}$$

where T is the tensile strength of paper, F is the fiber strength, and B is the bond strength between fibers. The bond strength is in turn the product of the bond strength per unit surface area (specific bond strength) multiplied by the bonded area. Strength-enhancing polymers are believed to increase both the specific bond strength and the bonded area.[20,21] The latter effect will occur if the polymer can fill in the voids in the contact zone between two rough surfaces.

The paper industry divides strength-enhancing polymers into two categories: dry strength and wet strength resins. Dry strength resins are used in many paper and paperboard products to give stronger physical properties. Wet strength resins are used to extend the functionality of paper products into applications involving water where normal paper would disintegrate. Domestic paper towels are a good example. From a chemical perspective, most dry strength resins form physical bonds between fibers, whereas wet strength resins bear reactive chemical groups that can form covalent bonds with the paper structure.

Wet and dry strength resins are usually added to the dilute pulp suspension. Good retention is very important because unretained polymer is a waste of valuable material and can contribute to pollution. The usual approach to obtaining good retention is to design the strength-enhancing polymer to contain positively charged cationic groups. The cationic polymers have a natural tendency to adsorb onto the negatively charged surfaces of fibers and fines.

Starch derivatives are one of the most important dry strength-enhancing polymers.[22] The starch content of dried paper typically varies from 0.3 to 2% by mass. Before use, starch must be "cooked" to hydrate the starch granules to yield an aqueous dispersion of microgel particles and soluble starch molecules. In most cases papermaking starch has been chemically modified. Nonionic and anionic starch tend not to adsorb onto fibers and so are poorly retained in the papermaking process. On the other hand, starch that has been modified to contain cationic groups has a strong tendency to adsorb onto negatively charged pulp fibers and fines.[4,23]

Cationic copolymers of acrylamide, usually of intermediate molecular weight, are also used as dry strength resins. These resins can also give temporary wet strength improvement by the incorporation of reactive glyoxal groups.[24]

2.2.2   Polymers for Wet Strength

Water-borne reactive polymers are employed as wet strength resins. Chan[25] has extensively reviewed these in a recent monograph. Wet strength resins are added to the pulp as solutions or colloidal dispersions that adsorb onto the fibers and fines before the paper is formed. Chemical reactions between the resins in the paper, usually some form of condensation, occur at elevated temperatures during paper drying.

2.2.3   Polymers for Fines and Filler Retention

Polymeric retention aids were first developed for fine paper manufacture to help retain colloidal clay and titanium dioxide. For many years, the technology of choice was the use of high-molecular-weight cationic copolymers of acrylamide. Polymers having charge contents of less than 30 mol % ensured that they adsorbed in an expanded configuration that gave bridging flocculation.[26] These polymers give good colloidal flocculation but have two disadvantages. The first is that the polymers also induce fiber flocculation which, in turn, gives nonuniform paper. The second is that if the polymer-induced bonds between filler particles and fibers surfaces are broken by hydrodynamic forces, the bonds appear not to reform.

More recently, the single component retention aid systems have been replaced by microparticle systems. In this technology two components are added sequentially; the first, a high-molecular-weight flocculant, the second a colloidal particle. Often pulp suspensions are exposed to strong hydrodynamic forces between the addition of the first and second component. Examples of microparticle systems are cationic starch plus colloidal silica, and cationic polyacrylamide plus bentonite clay. In both cases, the polymeric component will adsorb onto the colloidal component. One of the major advantages of microparticle systems is that both good colloidal retention and good paper formation can be achieved at the same time. The explanation is that the fiber flocs are dispersed by hydrodynamic forces and that any colloidal flocs broken by these forces can reform.

2.2.4   Fundamental Aspects

Upon addition to an aqueous pulp suspension, water-soluble polymers either adsorb onto an interface or form complexes with other water-soluble species. Subsequently the adsorbed polymer or polymer–polymer complex may induce the aggregation of small particles or the deposition of small particles onto fiber surfaces.

2.3   POLYMER ADSORPTION ONTO WOOD PULP FIBERS

The adsorption of polymers at the solid-solution interface has been the subject of many experimental and theoretical investigations. This subject has been extensively covered in a book by Fleer et al.[27] as well as in other chapters in this book. Most academic studies of polymer adsorption involve well-defined, linear polymers interacting with flat, clean surfaces to give equilibrium configurations. By contrast, wood fibers have porous, gel-like surfaces, and adsorbing polymers must perform a function in time scales much smaller than that required to attain equilibrium. Many of the polymer adsorption studies with wood pulps come from Lindström, Ödberg, Wågberg, and co-workers at the Swedish Pulp and Paper Institute (STFI) in Stockholm; for a good review see Ödberg et al.[15]

Electrostatic interaction between positively charged water-soluble polymers and negatively charged wood pulp fibers is the major driving force for adsorption. For low-molecular-weight polymers, adsorption is stoichiometric, meaning the number of adsorbed cationic polymer groups equals the number of accessible charges on the fibers and fines. This is essentially an ion-exchange process, where adsorption results in the release of the counterions of the fibers (cations) and the counterions of the cationic polymers (anions). By contrast the capacity of fibers and fines to adsorb high-molecular-weight polymers is less than the stoichiometric charge balance.[28] Wågberg et al.[29] studied the adsorption dynamics by measuring the release of counterions when a cationic polymer segment formed a salt with a fiber carboxyl group. They showed that high-molecular-weight cationic copolymers adsorb within the first 2 s of being exposed to pulp fibers. However, about 60 s were required to release most of the counterions.

Polymers adsorbed onto cellulose fibers tend not to desorb when washed with water. However, they can undergo exchange with polymers in solution or be displaced by multivalent ions such as lanthanum.[30] Unlike the behavior of polymers on smooth surfaces, however, much less exchange will occur if the adsorbed polymers have an opportunity to penetrate the pores, a process that occurs over a time scale of weeks.[15]

Papermaking is a challenging application for water-soluble flocculants because the pulp suspension is subjected to very high hydrodynamic forces that tend to break down the polymer and destroy the bonds between fillers and fibers formed by the flocculants. The shear rate in a headbox can be as high as 10^5 s^{-1}.[31] One of the consequences of this situation is that care must be taken in the evaluation of retention aids in laboratory tests. Britt and Unbehend[32] developed an apparatus called the dynamic drainage jar (DDJ) (see Fig. 2.3) that is widely used to compare different retention aids as flocculants for papermaking suspensions. In a modification of the classical flocculation jar, a high-speed mechanically driven propeller generates hydrodynamic forces that challenge the ability of flocculants to induced the deposition of colloids onto fibers and to maintain these structures. Water and unretained fines and fillers are drained through a coarse screen at the bottom of the jar, whereas small particles deposited on the fibers remain in the jar. The amount of deposition is calculated from the solids content of the drained water.

The DDJ has proved to be an invaluable tool for screening potential flocculants for papermaking applications. However, the amount of material retained in the DDJ is a reflection of both particle deposition onto the fibers and particle removal by hydrodynamic forces. It is impossible to separate these two factors in a DDJ experiment. There have been a few publications focused on finding the fundamental strength of adhesion between particles and surfaces in the presence of aqueous polymers. The usual approach is to deposit spherical, well-defined particles onto a surface such as a regenerated cellulose film, and then use hydrodynamic forces to remove the particles. If the flow field and the particle shapes are well defined, it is

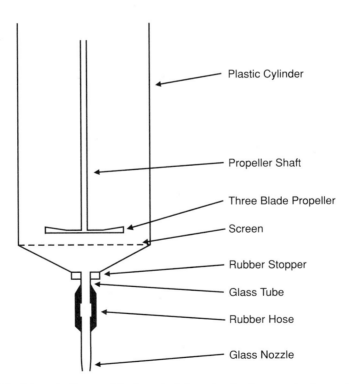

FIGURE 2.3 Schematic diagram of the dynamic drainage jar. A pulp suspension is poured into the top and drained through the screen. Unretained fines and fillers will be collected with the water, whereas fibers will remain in the jar.

possible to calculate strengths of adhesion. A few approaches have been used to expose deposited particles to well-defined laminar flow: Hubbe[33] used a concentric cylinder device; Pelton and Allen[34] used flow through a capillary tube; and van de Ven[35] used stagnation point flow. These techniques are very labor intensive because many measurements are required to obtain good statistics.

The basic adhesion studies have yielded some general conclusions. The force required to detach colloidal or near colloidal-sized particles is in the order of nano Newtons (nN). The strength of adhesion increases with the molecular weight of the flocculants. For example, at pH 6 the force required to remove 5 μm polystyrene spheres attached to glass with polyethyleneimine increased from 0.5 to 1 and then 2 nN when the polyethyleneimine molecular weight was increased from 2000 to 25,000 and then to 600,000, respectively.[34] An unexpected feature of some of the fundamental adhesion work was that particles were attached to surfaces by tethers as long as 5 μm. This distance exceeds the expected expansion of single polymer chains.[34,36] The implication is that in some cases polymers can exhibit multilayer adsorption.

We recently developed a new approach to measure directly the strength of flocs formed by the aggregation of colloidal particles with polymers. The technique we call micromechanics consists of supporting a floc between two suction micropipets. One pipet is slowly pulled while recording the deflection of the second pipet, which is shaped as a cantilever.[37] Figure 2.4 shows a floc supported between two pipets before and after the floc rupture experiment. In this case the floc was composed of

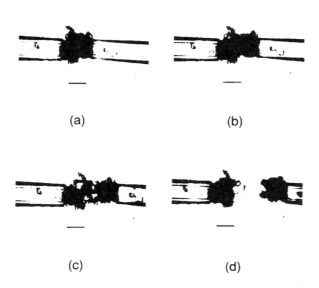

(a)   (b)

(c)   (d)

FIGURE 2.4   Sequence of photographs showing the rupture of a floc particle as the pipet on the right was pulled back.[37] The pipet on the left was the force transducer whose deflection provided a direct measure of the rupture force. The horizontal bars represent 10 μm.

precipitated calcium carbonate particles that were aggregated by a water-soluble polymer. The force required to rupture the floc was approximately 100 nN.

In papermaking, an important issue is whether flocs will reform after being broken by hydrodynamic forces. After pulling a floc into two pieces in a micromechanics experiment, the floc fragments can be pushed together giving the pieces a chance to reattach. For the system shown in Figure 2.4, a new bond formed when the fragments were pushed together; however, the tensile strength of the new bond was only 15% of the original floc. In many cases, floc fragments do not readhere when brought back together. The damage associated with floc rupture was demonstrated by Tanaka and co-workers[38] who have shown that if a polystyrene latex particle were deposited onto a cationic polyacrylamide-coated fiber and were subsequently removed by hydrodynamic forces, as much as 80% of the polyacrylamide was transferred to the latex. Furthermore, the molecular weight of the transferred polymer was reduced, indicating that the polyacrylamide chains were broken during particle detachment.

2.3.1 Polyelectrolyte Complex Formation

The attraction of oppositely charged polyelectrolytes to form complexes is an important part of paper technology, and the scientific aspects of complex formation have been summarized in a number of reviews.[39–43] The sequential addition of oppositely charged polymeric flocculants is an example. In another example, one of the strategies for removing the anionic polyelectrolyte components of the DCS in dirty pulps is to react the indigenous polymers with oppositely charged polymers to form polyelectrolyte complexes. The objective is to form small complex particles that then adsorb onto the wood pulp fibers so that the DCS is removed with the paper.

The driving force for polyelectrolyte complex formation is the electrostatic interaction of oppositely charged polymers. Because polyelectrolytes bear closely spaced charges, this process is cooperative and irreversible. Indeed, the tendency for low-molecular-weight linear polyelectrolytes to form quantitative polyelectrolyte complexes is the basis of a common analytical technique in the paper industry called *polyelectrolyte titration*, originally called the *colloid titration*.[44] In this procedure, a solution or a suspension is titrated with an oppositely charged standard polyelectrolyte solution. The endpoint occurs when the net surface charge on the suspended solids is reversed. Originally the end point was detected with a dye; however, a streaming current or streaming potential detector is usually used now.

Oppositely charged polyelectrolytes will form complexes over a broad range of stoichiometric ratios. However, the complexes tend to be water-soluble unless they are nearly stoichiometric because an excess of either positive or negative particle charge will confer water solubility. This behavior can be illustrated by the interaction of polyDADMAC, a linear cationic polyelectrolyte, with kraft lignin, which is a branched anionic phenolic polymer resulting from the decomposition of lignin in the kraft pulping process. The precipitated complex formation is illustrated schematically in Figure 2.5, and Figure 2.6 shows the mass of precipitated complex as a function of the concentration of the two polymers.[45] The maximum precipitate mass at pH 12

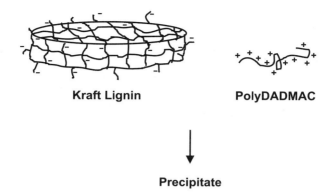

Kraft Lignin   **PolyDADMAC**

Precipitate

FIGURE 2.5   Formation of precipitated kraft lignin–polyDADMAC complex.

corresponded to the most concentrated polymer solutions mixed in the ratio of 0.53 g of polyDADMAC per gram of lignin. The amount of precipitated complex was decreased as the mixing ratio of the two polymers deviated from the optimum.

Figure 2.7 shows how the amount of polyDADMAC required to give maximum precipitation increased with the solution pH value. This reflects the fact that the charge content of lignin increased with pH due to the dissociation of carboxyl and phenolic groups. By contrast, polyDADMAC bears quaternary nitrogen groups, the charge content of which is independent of pH.

The upper curve in Figure 2.7 is the estimated mass ratio of the two polymers assuming charge balance. This ratio was computed from the measured charge contents

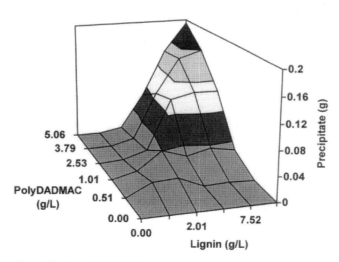

FIGURE 2.6   Mass of lignin–polyDADMAC precipitate at pH 12 as a function of concentrations of the two polymers.[45] The ridge corresponds to an optimum mixing ratio of 0.53 wt/wt polyDADMAC–lignin.

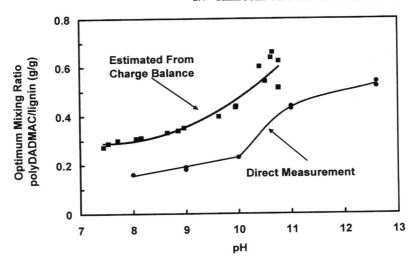

FIGURE 2.7   Optimum reactant mixing ratios of lignin and polyDADMAC reactants as a function of pH.[45]

of lignin as a function of pH. The experimental optimum mixing ratios were less than the estimates, indicating that many of the negatively charged groups on lignin were not accessible to polyDADMAC, presumably because of the network properties of some of the lignin molecules. Chemical analysis of the precipitated complex revealed that, at the optimum mixing ratio, about 80% of the added lignin ends up in the precipitate. By contrast, less than 35% of the added cationic polyDADMAC formed precipitate.[45] This means that the precipitates are not stoichiometric based on charge; 75% of the charges on the lignin are not bound to polyDADMAC. We propose that the 20% lignin that cannot be precipitated is present as small highly charged polymer fragments that form water-soluble complexes with most of the added polyDADMAC.

Clearly, kraft lignin is not ideally suited for fundamental studies of polyelectrolyte complex formation because lignin is a polydisperse polymer with a complicated structure. Most of the basic information comes from investigations of well-defined synthetic polymers—the work of Kötz and co-workers[46] has been particularly informative.

2.4   SELECTED RECENT DEVELOPMENTS

2.4.1   Polymer-enhanced Brownstock Washing

Described in this section is a new application for water-soluble polymers in pulp processing. Work done at McMaster University has shown that it is possible to improve the efficiency of pulp washing in the *kraft process* by the addition of cationic

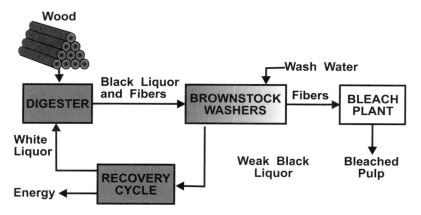

FIGURE 2.8   Simplified block diagram of the kraft pulp mill.

polyelectrolyte. Figure 2.8 shows a simplified flow diagram of a continuous kraft pulp operation for the production of chemical pulp. Wood chips are fed to the digester where they are impregnated with cooking chemicals at high temperature and high pH. The suspension leaving the digester contains wood pulp fibers and black liquor, a mixture of dissolved and suspended lignin decomposition products and spent cooking chemicals, all at high temperature (70°C) and high pH (approximately 11).

After pulping, the fibers suspended in black liquor are sent to the brownstock washers. The objective of *brownstock washing* is to separate the fibers from the black liquor. The washed fibers go to the bleach plant, and the weak black liquor that has been diluted with wash water is sent to recovery. The kraft recovery system is an ingenious sequence of operations that results in the production of energy by the combustion of concentrated black liquor, and the regeneration of the cooking chemicals while producing very little liquid or solid effluents. The evaporators are a particularly sensitive step in the kraft recovery cycle. In this operation, enough water is removed from the weak black liquor to produce a combustible mixture.

The efficiency of the brownstock washing operation has an enormous effect on the overall mill operations. Poor washing means that valuable cooking chemicals and lignin are carried forward to the bleach plant, causing increased consumption of bleaching chemicals, which, in turn, increases cost and environmental impact. On the other hand, if good washing is accomplished by adding an excess of wash water, the energy efficiency and the production capacity of the recovery system are reduced because more water must be evaporated per ton of pulp production.

In most kraft mills, brownstock watching consists of a sequence of between two and four vacuum drum washers operating in a countercurrent mode. Clean wash water enters the final washer, whereas weak black liquor leaves the first washer for the recovery cycle. A schematic illustration of a vacuum drum washer is shown in Figure 2.9. The heart of the washer is a large drum covered with a screen. The pressure on the interior side of the screen is maintained below atmospheric pressure by a vertical pipe (drop leg) that acts as a siphon. As the lower section of the drum rotates through

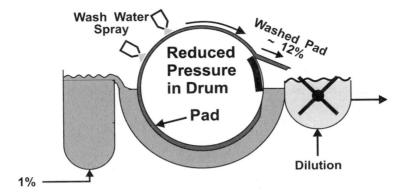

FIGURE 2.9 Vacuum drum pulp washer.

the vat containing the pulp suspension, a fiber pad forms on the face of the drum. Upon rotating out of the vat, wash water from the shower bars is gently sprayed upon the exterior surface of the fiber mat. The reduced pressure inside the drum causes the wash water to be drawn through the pad displacing the liquor. Thus the fundamental washing mechanism is a displacement operation. After the showers, the pad rotates to the 3 o'clock position where it is removed, diluted, and pumped to the next washer.

Although the optimization of washer operations is primarily an exercise in mechanical engineering, there is a role for the polymer industry. Defoamers are universally applied to improve brownstock washing.[47] The generally accepted mechanism is that air bubbles trapped in the pad interfere with washing by lowering the pad permeability[48] and defoamer addition lowers the content of dispersed air in the pulp.

The objective of our work was to determine if the efficiency of brownstock washers could be improved by the addition of chemicals other than defoamers. Lee[49] was the first to investigate the possibility of improving displacement washing efficiency by the addition of water-soluble polymers. He showed that high molecular weight polyethylene oxide or polyacrylamide could indeed improve the displacement of black liquor by washed liquor in a pulp pad. Lee proposed that the mechanism involved the retardation of viscous fingering during displacement. This work has had very little commercial impact because the polymers that improve displacement washing efficiency also reduced the liquor flow rate through the pad. In other words, better washing is achieved at the expense of a significant reduction in production rates. This effect arises because of the high extensional viscosity of the very high molecular weight water-soluble polymers.

In the case of ideal displacement washing, the minimum volume of wash water for complete washing is equal to the volume of liquor initially between the fibers. Real brownstock washers, however, are much less efficient. An important cause for poor washing efficiency is that the fiber pad formed on the face of the washer drum is not uniform. Poor pad uniformity means that some areas of the pad will be much more permeable than others. Thus the fiber pad on washer drums contains channels through which more wash water can pass than through the less permeable parts of the pulp pad.[50,51] In the extreme case, clean wash water passes through these channels, doing the washing, while other areas of the pad are not exposed to wash water at all. The following paragraphs describe our new technology, *polymer-enhanced washing*, a process designed to plug the most open channels.

Commercial brownstock washers are not well suited for experimentation; they are large and they experience a lot of process variability. A small-scale laboratory washing apparatus is a more practical approach to washing research, although a meaningful experimental protocol must include a way to generate nonuniform pads in a reproducible manner. We chose to replace the fiber mat with a more reproducible porous medium. Pulp pads were modeled by a packed bed of uniform fine glass beads. High-permeability regions in pulp pads were modeled by a single vertical channel of coarse beads that have a permeability ~20 times greater than that of the fine beads. Figure 2.10 shows a flat cell version of our laboratory washing apparatus. In a typical washing experiment, the cell is packed with glass beads that are then saturated with black liquor. The displacement washing experiment is then initiated by adding wash

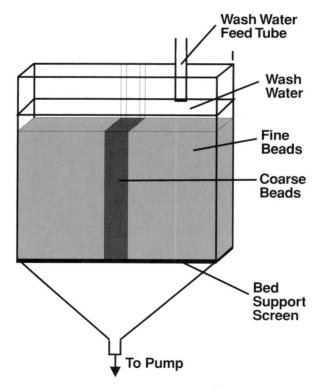

FIGURE 2.10 Laboratory displacement washing apparatus.[52] A vertical channel of coarse (640-mm-diameter) beads is surrounded by fine (120-μm) beads. The bed is initially saturated with black liquor and wash water is introduced from the top.

water to the top of the packed bed while draining black liquor eluate from the bottom. Transparent sides in the apparatus permit visualization of the flow front during a displacement washing experiment. Figure 2.11 shows a series of pictures of the center portion of the flat cell during a displacement washing experiment using pure wash water as the washing fluid.[52] Within the first 8 s the dark black liquor solution was displaced from the center channel while much more time was required for the displacement front to move down the fine bead zones on either side of the center channel.

Lignin in black liquor is present as soluble polymer molecules and microgel particles that are both negatively charged polyelectrolytes. In polymer-enhanced washing, the displacing wash liquor is a dilute aqueous solution of cationic polyelectrolyte. Figure 2.12 shows the washing experiment in which polyDADMAC, a cationic polyelectrolyte, was present in the wash water. In contrast to Figure 2.11, dark regions form and remain in the coarse-bead center channel at the end of the washing experiment. These dark regions are precipitated polyDADMAC–lignin complex (see Fig. 2.5). The precipitate has the effect of lowering permeability of the channel by about a factor of 2, causing less wash water to be wasted by flowing through the center channel. The remarkable feature of this experiment is that the lignin–polymer complex selectively formed in the coarse-bead channel. If complex precipitates had formed in the entire bed, the pulp would be contaminated with precipitate and displacement would be inhibited.

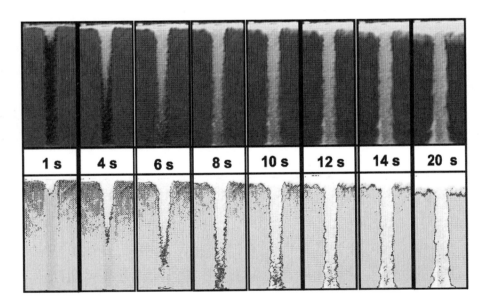

FIGURE 2.11 Time series of video frames showing the displacement of black liquor with water from the apparatus shown in Figure 2.10.[52] The top contains the original photographs and the bottom row are the same pictures enhanced by image analysis.

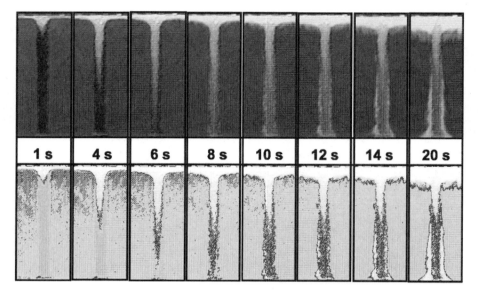

FIGURE 2.12 Time series of video frames showing the displacement of black liquor with cationic polyDADMAC.[52] Dark regions in the coarse-bead region are lignin–polymer complex.

An automated cylindrical washing cell, shown in Figure 2.13, was developed to collect quantitative washing data. Pressure drops and flow rates were continuously monitored, and the lignin concentration in the eluate stream was measured with a conductivity sensor. Because the ionic strength of black liquor is orders of magnitude greater than the ionic strength of the wash water, solution conductivity was shown to be a linear function of the lignin concentration.[53] Detailed information about the movement of the displacement front through the cell was obtained from microconductivity sensors placed throughout the bed.

The results of four displacement washing experiments are shown in Figure 2.14. The results are plotted as breakthrough curves; the y axis is the eluate conductivity divided by the conductivity of the initial black liquor. The x axis is the dimensionless time; the value of 1 corresponds to the time required to collect an eluate volume equal to the initial volume of black liquor in the bed. The curves labeled "homogeneous bed" were obtained using a uniform bed of fine beads, and the results for both washing with water and dilute polymer (polyDADMAC) solution correspond to near ideal displacement washing. In the ideal case, the breakthrough curves are a step function with conductivity changing at the dimensionless time 1.

More than three times as much water was required to displace the black liquor from the channel bed than from the homogeneous bed. The channel bed showed two breakthroughs with a characteristic plateau separating them. The first breakthrough at a dimensionless time of about 0.25, corresponded to the complete displacement of lignin from the coarse-bead center channel. The second breakthrough, occurring at

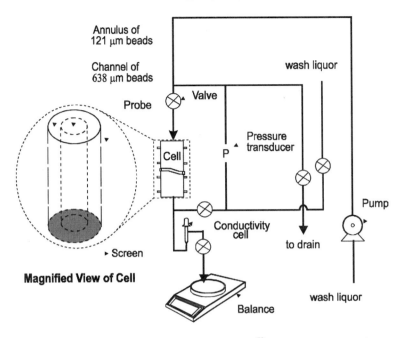

FIGURE 2.13 Automated displacement washing apparatus.[53] Microconductivity probes map the displacement front in the bed.

dimensionless times 3.25 to 3.75, corresponded to the displacement front reaching the end of the fine-bead portion of the bed.

The presence of polyDADMAC in the wash water had an enormous effect on the displacement experiments. The conductivity of the plateau portion of the curve was about twice that of the corresponding curve without polymer. This higher conductivity indicates that there was less wash water diluting the eluate stream. Also, the second breakthrough occurred at a much earlier dimensionless time; about half as much wash solution was required to completely displace the black collector liquor when polyDADMAC was present in the wash solution.

The displacement washing experiments, shown in Figure 2.14, were conducted under conditions of constant eluate flow rate. The corresponding pressure profiles, measured across the length of the bed, are shown in Figure 2.15. The channel bed pressures were lower than the homogeneous bed pressures because the overall permeability of the channel bed was higher. However, higher pressures were observed in the channel bed when polymer was present in the wash water. This observation serves as additional evidence for the partial plugging of the coarse center channel with lignin–polyDADMAC precipitate.

The essential requirement for polymer-enhanced washing is that precipitate selectively forms deposits in the most open channels of the fiber bed. We believe that

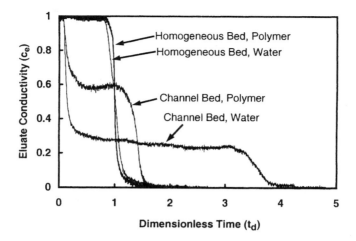

Figure 2.14   Breakthrough curves from homogeneous and channel beds displaced with water or 29 g/L polyDADMAC.[53]

the selectivity is achieved because only in the higher flow regions of the bed is there enough mixing of cationic and anionic polymer to form precipitates.

The residence time of a pulp pad on the face of a washer drum is less than 10 s. For the polymer-enhanced washing concept to function on a real washer, precipitate formation and deposition in the fiber bed must be rapid. The kinetics of precipitate formation from very dilute solution under quiescent conditions was measured by dynamic light scattering. Figure 2.16 shows the average diffusion coefficients of

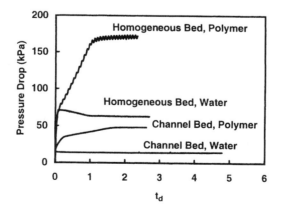

FIGURE 2.15   Pressure drops across the cylindrical beds during displacement experiments conducted at constant flow rates.[53] Pressure oscillations were caused by the peristaltic pump used to push the wash fluid through the cell.

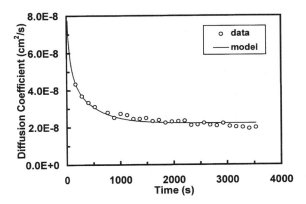

FIGURE 2.16   Experimental and theoretical diffusion coefficients of lignin–polyDADMAC precipitate particles as a function of time from polymer mixing.[45]

particles measured by dynamic light scattering as a function of time. The solid line in Figure 2.16 shows the results of a diffusion-limited aggregation model for uniform hard spheres published by Vermold and Hartl.[54] The model involves two fitting parameters, the radius of the initial hard sphere suspension and a time constant that is the half-life of the suspension based on Smoluchowski flocculation kinetics. The data fit, shown in Figure 2.16, was based on a 25-nm primary particle radius and a time constant of 34.5 s, which reflects the very dilute nature of the suspensions. Since the model fitted the data, we concluded that the formation of polyDADMAC–lignin precipitates is a mass-transport-controlled process.

Polymer-enhanced washing has been evaluated in commercial mill trials. Our initial work involved the addition of Channel Block, commercial cationic polymer especially formulated for this application, to the showers of a brownstock washer. This work was hampered by the difficulties in obtaining meaningful data from short-term brownstock washer trials. Because of multiple feedback loops and the inherent noise in the continuous kraft pulping process, it was difficult to obtain reproducible results. To circumvent these difficulties, Channel Block was applied with a special device called the minishower bar that only treated part of the washer (see Fig. 2.17). The advantage of this approach is that the effectiveness of polymer-enhanced washing can be evaluated by comparing the quality of washing in the treated areas against the untreated areas.[55] Fluctuations in the feed pulp suspension properties are accounted for by continually measuring washing efficiency in the untreated (control) zone of the washer.

Figure 2.18 shows lignin removal on the second washer as a function of the Channel Block concentration in the solution sprayed from the minishower bar. Washing is expressed as a relative displacement ratio that compares lignin removal in the treated zone with lignin removal in the untreated zone. The dashed line labeled RDR_{base} (in Fig. 2.18) shows the results when no polymer was added to the minishower bar. The dashed line labeled RDR_{max} corresponds to the complete removal

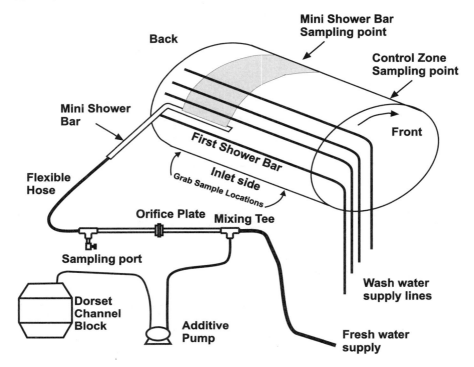

FIGURE 2.17   Illustration of apparatus used to spray cationic polymer onto part of a vacuum drum washer using the minishower bar.

of lignin from the path. Clearly, the addition of Channel Block improved washing efficiency.

2.4.2   Nonionic Flocculants

Papermaking with mechanical pulps presents difficult problems for the application of water-soluble polymers because of the very high concentrations of DCS. Retention aids are the main use of water-soluble polymers in mechanical-pulps-based papers because strength-enhancing polymers are usually not used in these grades. Newsprint manufacture, in particular, is a challenging application for retention aids because the pulps are dirty, the machines are fast with very aggressive dewatering, and because newsprint is a relatively low-cost product. All of these factors limit the possibilities for chemical addition. Until the 1980s there was very little interest in the use of retention aids for newsprint manufacture. However, the replacement of groundwood pulps with thermomechanical pulps and deinked pulps, both of which contain less fines, as well as increased environmental regulations have given the newsprint papermakers much more incentive to increase retention. Today most newsprint machines use water-soluble retention aids.

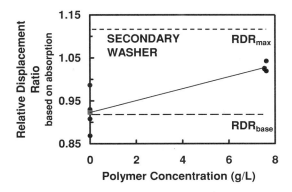

FIGURE 2.18 Lignin removal from the treated zone on the washer divided by the corresponding lignin removal in the control zone. RDR_{max} corresponds to complete lignin removal whereas RDR_{base} was obtained by adding only water to the minishower bar.[55]

At present, three main types of retention aids are used in newsprint manufacture: (1) a low-molecular-weight, highly charged cationic coagulant, whose role is to complex with and remove the DCS component, followed by a very high molecular weight cationic flocculant that deposits the small particles onto the fibers; (2) an activated bentonite; and (3) an anionic water-soluble phenolic resin followed by the addition of very high molecular weight polyethylene oxide (PEO). The following sections review some of our recent work in the PEO-based systems.

Many years ago, Pelton and co-workers[56] showed that very high molecular weight PEO was an effective retention aid in some newsprint furnishes. Based on the teaching of a Sandoz patent,[57] it was shown that the utility of PEO could be extended to include nearly all mechanical pulp furnishes if the stock was pretreated with a phenolic polymer. The use of a phenolic polymer, which we call a cofactor, followed by very high molecular weight PEO is now an established part of paper chemistry technology.

To be effective, a flocculant must be able to adsorb onto the surfaces upon which it is to act. Aqueous PEO is not a strongly adsorbing molecule; it has no electrically charged groups and it does not adsorb onto pure cellulose from water, whereas it will interact with a lignin-coated surface.[58] The chemical structure of PEO is very simple; only the polyether oxygen is capable of hydrogen bonding with proton donors.[59,60] The pair of methylene groups acting as spacers between the polyether oxygen atoms are hydrophobic and are capable of interacting with surfactants as well as causing PEO to adsorb at the air–water interface.

We believe that PEO-based retention aids are a dual-polymer retention aid system. In a few particularly dirty newsprint pulps the first polymer, probably a water-soluble lignin derivative, is naturally present. However, synthetic phenolic cofactors are usually employed before PEO addition to ensure process stability. The mechanism of the PEO–cofactor system has been studied by a number of groups, and there is general agreement about some aspects; however, others remain controversial. The fundamental interaction between PEO and cofactor was proposed to be hydrogen

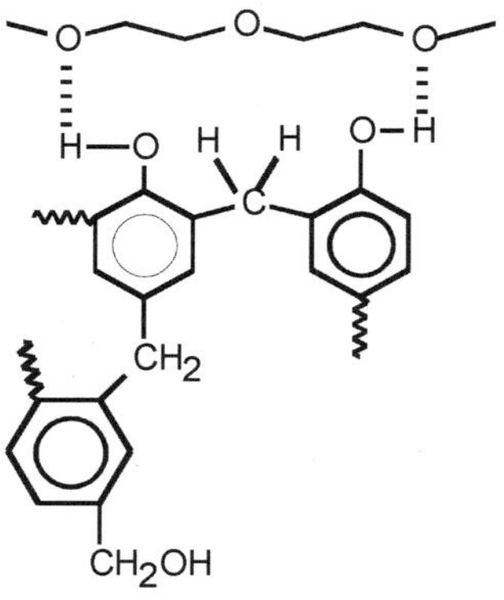

FIGURE 2.19   Structure of the phenol resin–PEO complex proposed by Stack et al.[61]

bonding by Pelton and co-workers.[9] The structure of the hydrogen-bonded complex, shown in Figure 2.19, was first given by Stack et al.[61] who suggested that only every other polyether oxygen could participate in hydrogen bonding for steric reasons. This part of the mechanism seems to be generally agreed upon, although all the evidence supporting it is indirect. The mixing of aqueous high-molecular-weight PEO with some cofactors results in precipitate formation—clearly some form of interaction is taking place. When the pH is raised to a sufficiently high value such that the phenolic hydroxyl groups are partially disassociated, no precipitation occurs. Similarly, the PEO–cofactor retention aid system is not effective at high pH values, presumably because the PEO and cofactor are not interacting.

There are, however, two curious aspects that may suggest more is involved than simple hydrogen bonding of PEO with cofactor. First, although there are several effective cofactor structures including phenolic resins, poly(vinyl phenol) and its copolymers, and tannic acid, all of these materials contain phenolic hydroxyl groups. On the other hand, there are a number of polymers known to form H-bonded complexes with PEO including poly(acrylic acid) and poly(methacrylic acid)[61] that do not act as cofactors.[58] The ability of a cofactor to hydrogen bond seems to be a necessary but not sufficient criterion for an effective cofactor. The reason why effective cofactors appear to require aromatic groups is not known.

Second, we have conducted molecular modeling simulations of the interaction of short polyethylene oxide chains with poly(p-vinyl phenol).[58] Calculations were initiated by forcing the polyether oxygens to be in perfect registration with the phenolic hydroxyl groups. When the system was allowed to relax, only every fourth or fifth polyether oxygen was hydrogen bonded to poly(vinyl phenol) in the final complex. A sample structure from the calculations is shown in Figure 2.20. Great caution must be exercised in the interpretation of such calculations; nevertheless, this work suggests that steric effects may prevent the registration of the polyether chain with poly(vinyl phenol). Future calculations undoubtedly will clarify these initial results.

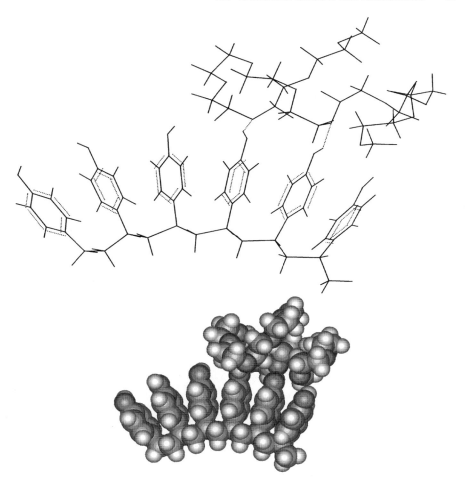

FIGURE 2.20 Structure of the hydrogen bond complex between PEO and isotactic poly(*p*-vinyl phenol) oligomer.[58] Adjacent phenolic protons were initially bonded to every fourth (1,5) ether oxygen, however, after minimization only two hydrogen bonds (dotted lines in shaded area of stick structure) remained.

In summary, hydrogen bonding does seem to be the major interaction between PEO and cofactor. The requirement for aromatic phenolic groups in the cofactor may reflect secondary hydrophobic interactions between PEO and cofactor.

The flocculation mechanism of the PEO cofactor system has been studied by a number of groups. Lindström and Glad-Nordmark[62] postulated a network flocculation mechanism in which PEO interacts with cofactor to form a macroscopic network that mechanically traps colloidal material to form larger structures that are then captured in the pad during sheet formation. Alternative mechanisms have been proposed by our group[63] and by van de Ven's group.[64] Our mechanism is illustrated in Figure 2.21, which shows the flocculation of polystyrene latex particles in the presence of wood

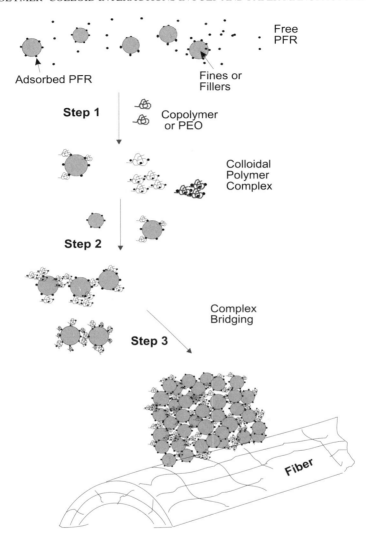

FIGURE 2.21   Proposed flocculation mechanism for latex with phenolic resin–PEO system in the presence of wood fibers.[65]

pulp fibers.[65] It is important to appreciate the different size scales of the components in this mixture; Figure 2.21 has been drawn to scale. The phenolic resin molecules, labeled PFR, are about one tenth the size of individual PEO coils. The PEO is in turn about a third of the size of the latex particles, which are themselves about one fourth of the thickness of a pulp fiber wall. The phenolic resin is added first, and some of it will adsorb onto the surfaces of latex particles and onto the wood fibers.[66] The driving force for adsorption of negatively charged phenolic resin onto negatively charged polystyrene latex is probably hydrophobic interactions. However, it is less clear why the resin adsorbs onto the hydrophilic wood pulp fibers.

In step 1 (see Fig. 2.21) PEO addition leads to complexes formed between PEO and phenolic resin both in solution and on the latex particle surface. The dispersed complexes are not colloidally stable, and aggregation occurs capturing both dispersed complex and latex particles in colloidal flocs (step 2). With time, rather large aggregates are formed and deposit onto fibers surfaces (step 3). An example is shown in Figure 2.22.[65]

Recent years have seen considerable efforts toward the development of better cofactors. The original types of phenolic resins used with PEO were marginally water soluble so the PEO–cofactor complex tended to collapse followed by precipitation. The precipitated complex has no flocculation efficacy; therefore precipitation competes with the desired flocculation mechanism shown in Figure 2.21. Modern cofactors tend to be more hydrophilic, usually because of a higher concentration of charged groups. Polyethylene oxide complexes formed with the more hydrophilic cofactors do not form precipitates in water.

In the earliest PEO work, it was shown that only the very highest molecular weight polymers were effective flocculants[56]; this finding has been verified by many others. Nevertheless, the reason for the molecular weight sensitivity remains a mystery. It is generally observed that higher molecular weights are better for bridging flocculants. For example, the best flocculants based on cationic copolymers of polyacrylamide have molecular weights above one million. However, the molecular weight sensitivity of PEO seems to be extreme. Retention testing shows that there are big differences between 4 million and 8 million molecular weight PEOs. Recognizing that the molecular weight of the repeat unit is only 42, a 4-million-molecular-weight PEO chain is extremely long and should be a good flocculant. The conventional argument for high-molecular-weight bridging flocculants is that thickness of the adsorbed layer must expand beyond the effective barriers of electrostatic repulsion. However, a 500,000 or one million molecular weight PEO should be able to satisfy this requirement, and yet these polymers are poor flocculants in the PEO phenolic resin system.

Pelton and co-workers[58] proposed that when very high molecular weight PEO molecules formed a complex with cofactor, the rate of syneresis and complex precipitation was slow relative to the competing rates of flocculation. Whereas with short PEO chains, PEO–cofactor syneresis was so fast that there was not enough time for the competing flocculation processes. The problem with this explanation is that it does not explain why the newer nonprecipitating cofactors also appear to require very high molecular weight PEO.

Based on gel permeation chromatography (GPC), static light scattering, and dynamic light scattering experiments, Polverari and van de Ven[67] have supported earlier propositions[68] that PEO molecules associate in water to form clusters containing a few hundred polymer chains. The possible role of the clusters in flocculation has not been clarified. Indeed, the existence of cluster formation remains controversial.[69,70]

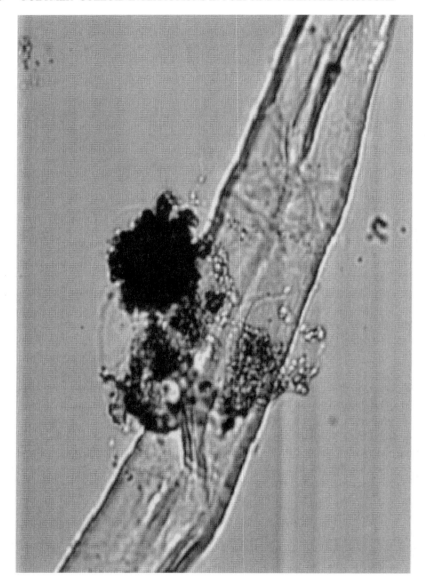

FIGURE 2.22 Optical micrograph showing a polystyrene latex floc attached to a bleached kraft pulp fiber. The floc contains hundreds of submicron latex particles.[63]

Very high molecular weight polyethylene oxide is expensive relative to other high-molecular-weight polymers and is susceptible to degradation by oxidation or shear. To circumvent these problems Xiao et al.[71] prepared a series of comb copolymers consisting of a polyacrylamide backbone supporting a few, short

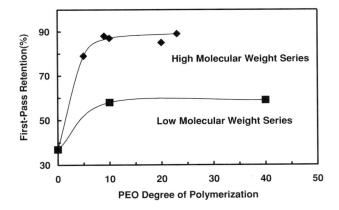

FIGURE 2.23 Effects of pendant PEO chain lengths on fines retention.[65] The molecular weights of the two series were approximately 1 and 4 million. The dynamic drainage jar measurements were conducted at 50°C with propeller speed 250 rpm.

polyethylene oxide pendant chains. The polyacrylamide support chain is less sensitive to degradation than PEO. Figure 2.23 shows DDJ newsprint flocculation results as a function of the length of the pendant PEO chains on the comb copolymers. Results for two series of polymers are shown. The higher molecular weight copolymers gave better flocculation, and, in both cases, good retention was observed when the PEO pendant chains were as short as 10 repeat units. These results suggest that there may exist better structures than PEO for retention aid systems for dirty pulps.

2.5 CONCLUDING REMARKS

This chapter has given an overview of polymer applications in pulp and paper with a strong bias to the systems studied at McMaster University. No new science needs to be invoked to explain most of the observations when polymers are employed in paper systems. On the other hand, quantitative modeling of even relatively simple experiments such as the adsorption of a single type of polymer onto clean fibers is difficult because fibers have a complicated structure.

Most new polymer-based papermaking technologies come from the laboratories of multinational polymer companies. New concepts are first described in the patent literature and in most cases, new polymer technologies are used in the marketplace before a detailed understanding of the relevant mechanisms has been developed. This situation presents many opportunities for academic researchers to advance paper science.

REFERENCES

1. L.H. Allen, *Colloid Polym. Sci.* **257**, 533 (1979).

2. R.H. Pelton, Jordan, and L.H. Allen, *Tappi* **68**(2), 91 (1985).

3. O.L. Forgacs, The Characterization of Mechanical Pulps, CPPA Trans. Tech. Sect., T89-116 (1963).

4. L. Wågberg and M. Björklund, *Nordic Pulp Paper Res. J.* **4**(8), 399 (1993).

5. J. Nylund, H. Byman-Fagerholm, and J.B. Rosenholm, *Nordic Pulp Paper Res. J.* **2**, 280 (1993).

6. J. Nylund, O. Lagus, and C. Eckerman, *Colloids Surfaces A*, **85**, 81 (1994).

7. M. Kleen and T. Lindström, *Nordic Pulp Paper Res. J.* **2**, 111 (1994).

8. A. Sundberg, R. Ekman, B. Holmborn, K. Sundberg, and J. Thornton, *Nordic Pulp Paper Res. J.* **1**, 226 (1993).

9. R.H. Pelton, L.H. Allen, and H.M. Nugent, *Svensk Papperstidning* **83**, 251 (1980).

10. K. Sundberg, J. Thornton, R. Ekman, and B. Holmborn, *Nordic Pulp Paper Res. J.* **2**, 125 (1994).

11. L. Wågberg and G. Annergren, in Proceedings of "The Fundamentals of Papermaking Materials," 11th Fundamental Research Symposium in the Oxford and Cambridge Series, 1–82, 1997.

12. A.J. Kerr and D.A.I. Goring, *Cellulose Chem. Tech.* **9**, 563 (1975).

13. J.E. Stone and A.M. Scallan, *Tappi* **50**(10), 496 (1967).

14. A.M. Scallan, *Wood Science* **6**, 266 (1978).

15. L. Ödberg, H. Tanaka, and A. Swerin, *Nordic Pulp Paper Res. J.* **8**(1), 6 (1993).

16. R. Pelton, *Nordic Pulp Paper Res. J.* **8**(1), 113 (1993).

17. C.J. Biermann, *Handbook of Pulping and Papermaking*, Academic, New York, (1996), p. 210.

18. T.G.M. van de Ven, *J. Pulp Paper Sci.* **10**(3), 57 (1984).

19. D.H. Page, *Tappi* **52**(4), 674 (1969).

20. J.W. Swanson, *Tappi* **44**(1), 142 (1961).

21. R.C. Howard and C.J. Jowsey, *J. Pulp Paper Sci.* **15**(6), J225 (1989).

22. M. Georgeson, in C.O. Au and I. Thorn, eds., *Applications of Wet-End Paper Chemistry*, Blackie Academic and Professional, London, 1995, p. 76.

23. H.G.M. Van de Steeg, A. de Keizer, and B.H. Bijsterbosch, *Nordic Pulp Paper Res. J.* **2**, 173 (1989).

24. S.N. Jenkins, in C.O. Au and I. Thorn, eds., *Applications of Wet-End Paper Chemistry*, Blackie Academic and Professional, London, 1995, pp. 91–94.

25. L.L. Chan, *Wet-Strength Resins and Their Application*, Tappi, Atlanta, 1994.

26. D. Eklund and T. Lindström, *Paper Chemistry an Introduction*, DT Paper Science Publications, Grankulla, Findland, 1991.

27. G.J. Fleer, M.A. Cohen Stuart, J.M.H.M. Scheutjens, T. Cosgrove, and B. Vincent, *Polymers at Interfaces*, Chapman & Hall, London, 1993.

28. L. Winter, L. Wågberg, L. Ödberg, and T. Lindström, *J. Colloid Interface Sci.* **111**(2), 537 (1986).

29. L. Wågberg, L. Ödberg, T. Lindström, and R. Aksberg, *J. Colloid Interface Sci.* **123**, 287 (1988).

30. R. Pelton, *J. Colloid Interface Sci.* **111**(2), 475 (1986).

31. P.A. Tam Doo, R.J. Kerekes, and R.H. Pelton, *J. Pulp Paper Sci.* **10**, J80 (1984).

32. K.W. Britt and J.E. Unbehend, *Tappi* **59**(2), 67 (1976).

33. M.A. Hubbe, *Colloids Surfaces* **16**, 227 (1985).

34. R.H. Pelton and L.H. Allen, *J. Colloid Interface Sci.* **99**(2), 387 (1984).

35. T.G.M. van de Ven, *Colloidal Hydrodynamics*, Academic, New York, 1989, p. 490.

36. T.G.M. van de Ven, T. Dabros, and J. Czarnecki, *J. Colloid Interface Sci.* **93**(2), 580 (1983).

37. A.K.C. Yeung and R.H. Pelton, *J. Colloid Interface Sci.* **184**, 579 (1996).

38. H. Tanaka, A. Swerin, and L. Ödberg, *J. Colloid Interface Sci.* **153**(1), 13 (1992).

39. A.S. Michaels, *Ind. Eng. Chem.* **57**(10), 32 (1965).

40. E. Tsuchida and K. Abe, *Interactions between Macromolecules in Solution and in Intermacromolecular Complexes*, Advances in Polymer Science, Vol. 45, Springer Berlin, 1982.

41. B.I. Philipp, H. Dautzenberg, K. Linow, J. Kötz, and W. Dawydoff, *Prog. Polym. Sci.* **14**, 91 (1989).

42. Y. Li and P.L. Dubin, *ACS Symp. Ser.* **578**, 320 (1994).

43. E.A. Bekturov and L.A. Bimendina, *Rev. Macromol. Chem. Phys.* **C37**(3), 501 (1997).

44. H. Terayama, *J. Polym. Sci.* **VIII**(2), 243 (1952).

45. R.E. Lappan, R. Pelton, I. McLennan, J. Patry, and A.N. Hrymak, *Ind. Eng. Chem. Res.* **36**(4), 1171 (1997).

46. J. Kötz, *Nordic Pulp Paper. Res. J.* **8**(1), 11 (1993).

47. R. Pelton, *Pulp Paper Canada* **90**(2), T61 (1989).

48. A.K.T. Chan, R.H. Pelton, S. Zhu, and M.H.I. Baird, *Can. J. Chem. Eng.* **74**(2), 229 (1996).

49. P.F. Lee, *Tappi* **67**(11), 100 (1984).

50. R.H. Crotogino, N.A. Poirier, and D.T. Trinh, *Tappi* **70**, 95 (1987).

51. H. Dahllöf and U. Gren, "Displacement Washing of Pulp Fibre Beds—Comments on the Role of Permeability," Pulp Washing '96, Preprints CPPA/TAPPI Meeting, Vancouver, Oct. 7–10, 9 (1996).

52. R. Pelton and B. Grosse, *J. Pulp Paper Sci.* **20**(3), J91 (1994).

53. D. De, A.N. Hrymak, and R. Pelton, *AIChE J.* **43**(10), 2415 (1997).

54. H. Vermold and W. Hartl, *J. Chem. Phys.* **79**, 4006 (1983).

55. R.E. Lappan, A.N. Hrymak, and R. Pelton, *Tappi* **79**(11), 170 (1996).

56. R.H. Pelton, L.H. Allen, and H.M. Nugent, *Pulp Paper Canada* **81**(1), T9 (1980).

57. J.P. Carrard and H. Pummer, U.S. Pat. 4,070,236 (1978).

58. R. Pelton, H. Xiao, M.A. Brook, and A. Hamielec, *Langmuir* **12**(24), 5756 (1996).

59. I. Illopoulos and R. Audebert, *Macromolecules* **24**, 2566 (1991).

60. K. Abe, M. Koide, and E. Tsuchida, *Macromolecules* **10**, 1259 (1977).

61. K.R. Stack, L.A. Dunn, and N.K. Roberts, *Colloids Surfaces* **61**, 205 (1991).

62. T. Lindström and Glad-Nordmark, *Colloids Surfaces* **8**, 337 (1984).

63. H. Xiao, R. Pelton, and A. Hamielec, *J. Pulp Paper Sci.* **22**(12), J475 (1996).

64. T.G.M. van de Ven and B. Alince, *J. Pulp Paper Sci.* **22**(7), J257 (1996).

65. H. Xiao, R. Pelton, and A. Hamielec, *Tappi* **79**(4), 129 (1996).

66. K. Johansson, A. Larsson, G. Ström, and P. Stenius, *Colloids Surf.* **25**, 341 (1987).

67. M. Polverari and T.G.M. van de Ven, *J. Chem. Phys.* **100**, 13687 (1996).

68. W. Brown, *Macromolecules* **17**, 66 (1984).

69. K. Devanand and J.C. Selser, *Nature* **343**, 739 (1990).

70. B. Porsch and L.O. Sundelöf, *Macromolecules* **28**, 7165 (1995).

71. H. Xiao, R. Pelton, and A. Hamielec, *Polymer* **37**, 1201 (1996).

3   Dual-Addition Schemes

GUDRUN PETZOLD

Institute of Polymer Research, Hohe Strasse 6, 01069 Dresden, Germany

3.1   INTRODUCTION

In the production of paper, a mixture of water, cellulose, filling material (such as clay), and dissolved materials (such as anionic oligomers and polymers, the so-called anionic trash), is drained on a wire mesh running at high speed. Flocculants are applied to accelerate the drainage process and to increase the retention of the finely dispersed material in the paper. Flocculation is one of the most important factors that influences both the machine runability and paper quality. Many chemical and physical factors influence flocculation in a complex way.

3.1.1   Types of Flocculants

Usually, high molecular weight cationic polymers are employed for flocculating the anionically charged suspension. However, the improved effect of *dual systems* is now also appreciated. The application of two-component flocculants is not new. Two or even more components, added in sequence as flocculant systems, produce synergistic effects on the flocculation. Many different systems have been applied in the paper industry and also in other fields such as peat dewatering,[1] flocculation of harbor sediments,[2] wastewater,[3] or sugar beet washings.[4]

3.1.2   History

It was first described in the literature of the 1970s that a so-called dual system, consisting of *two oppositely charged polymers*, can produce distinct improvements in retention and dewatering.[5-7] From the 1980s until now much work was done to

Colloid–Polymer Interactions: From Fundamentals to Practice, Edited by Raymond S. Farinato and
Paul L. Dubin
ISBN 0-471-24316-7   © 1999 John Wiley & Sons, Inc.

improve our understanding of such systems. This has been especially true in research areas such as paper forming and floc shear stability. A new type of dual system, the so-called *microparticle–containing system*, has been recently developed. In these systems, the addition of a cationic polymer is followed by the addition of an anionic submicron particle suspension. Examples of this type of retention aid system are cationic starch used in conjunction with anionic colloidal silica or anionic colloidal aluminum hydroxide, and cationic copolymers of acrylamide used together with sodium montmorrilonite. Several systems of this type are commercially available. They are said to be very efficient flocculants that give smaller flocs at an equal degree of flocculation compared to single-component systems. Very good reviews that include a survey of some aggregation mechanisms[8] and comparisons between different dual systems are available.[9–16]

This chapter describes *polymer–polymer systems* mainly in the area of paper retention aids. Finally, current approaches used in the field of flocculation will also be mentioned.

3.2 POLYMER–POLYMER SYSTEMS

The following discussion will focus on methods of investigation and procedures for optimization. The various factors include the nature of the polycation and polyanion, anionic trash content, salt content, and shear forces.

An example of an early and very detailed investigation of such a system is the work of Moore,[17] in which different types of cationic polymers were combined with hydrolyzed polyacrylamides at various alum concentrations. In contrast to other workers dealing with two-component systems, Moore studied the charge relationship of the various charged species. He discovered that a combination of cationic and anionic polymers can give very high levels of retention with high shear resistance only in the case of a proper balance of charges and concentrations. He noted, however, that achieving this in practice would be most difficult.

Müller and Beck[18] have investigated cationic polyethylene imine or polyamidoamine in combination with an anionic polyacrylamide. They explained that under conditions of optimum performance, two mechanisms are operating: charge patch formation and bridging. The relatively short-chain polycation produces a very fine flocculation of the particles via a charge patch destabilization mechanism. If a long-chain polyacrylamide is then added to the stock, the negatively charged chains "get a good grip" on the positive patches of the primary floc and bring about further linkages by forming bridges. Other aspects of the floc formation mechanism were studied by Petäjä,[19] including the influence of the type and amount of polycation, the time delay between cationic and anionic addition, and the degree of turbulence. It was shown that the agitation level and control of floc formation process after the cation addition is very important for good sheet formation.

In Table 3.1 a wide variety of examples of such polymer–polymer systems from the literature are listed. Despite the common use of dual-component retention aid

TABLE 3.1 Different Types of Polymer–Polymer Systems[a]

Type of Polycation	Type of Polyanion	Remarks (Reference)
PEI or PAAm; LMW 0.2–0.3%	Anionic PAA; HMW 0.02–0.04% (for waste paper)	Optimum polymer ratio necessary (18)
PEI or other polycation LMW up to HMW	Polyanion	Dual systems are not the solution to all retention problems (19)
Dimethylaminoepichlorohydrin resin; LMW, 0.07%	Anionic PAA; HMW (medium charge) 0–0.12%	20
PDADMAC; MMW high charge density	Anionic PAA; HMW (medium charge)	20
LMW polymers, e.g., polyamines (highly cationic) or cationic PAA	Anionic PAA	Dual systems need better control in terms of optimum polymer ratio(21)
PEI or PAAm-type 0.17–0.19%	PAA or "modified" PAA 0.03%	22
Cationic PAM 5–7 × 10^5 g/mol	Anionic PAA (medium or HMW)	23
Cationic starch 2%	Anionic PAA 0–0.08%	24

[a]HMW, high molecular weight; MMW, medium molecular weight; LMW, low molecular weight; PEI, polyethylenimine; PAAm, polyamidoamine; PAA, polyacrylamide; PDADMAC, poly(diallyldimethylammonium chloride).

systems, there is still a lack of fundamental understanding of their mode of action and of the parameters controlling their efficacy.

The most commonly used systems are those in which the polycation is added prior to a high-molecular-weight polyanion. It was confirmed by different authors that this order of addition is essential for efficacy. Drainage and retention are significantly increased if the polycation is added before the polyanion, a treatment usually superior to the addition of a single polycation. However, the dose rates of such systems are said to be often higher than those for single polymers.[14] It may be that the optimum polymer balance was not reached in this case.

Table 3.1 demonstrates the large differences in the compositions of dual systems. The quantity of polymers is given in weight percent without consideration of the charge content of the systems to be flocculated.

The aim of this work was to attain a deeper insight into the mechanism of such polymer–polymer systems by understanding the polymer–polymer interaction as well as the interaction between the polymers and the solid surfaces. Model systems relevant to those in flocculation practice in the papermaking industry were investigated.

This work will deal with dual systems having the same polycation but different polyanions. After investigating the reaction between the polycation and the cellulosic suspension, we studied the influence of adding polyanion. The most important objectives were to determine the influence of the properties of the two oppositely charged polyelectrolytes (e.g., molecular weight, charge ratio, and charge density), their concentrations, and the character of the suspension (anionic trash content) on flocculation. Several different methods were used to investigate and characterize the flocculation process.

3.3 EXPERIMENTAL

3.3.1 System

The paper stock contained a mixture of sulfite-treated spruce cellulose and clay. The cellulose, beaten in an Escher-Wyss refiner to a beating degree of 53°SR, was centrifuged and stored in a refrigerator until used. To prepare suspension I (2 g/L), the cellulose was stirred together with clay in distilled water for 30 min. To simulate a high anionic trash[25] content (suspension II), a certain volume of wood extract (20 g of chemothermomechanical pulp (CTMP), washed in water at 50°C for 30 min) was added to the suspension. For experiments with the Fibre Optical Flocculation Sensor, a special mixture (suspension III) was necessary.[26] Details and characterization of these systems are given in Table 3.2.

3.3.2 Polymers

The cationic polymer poly(diallyldimethylammonium chloride) (PDADMAC) was obtained from Katpol Chemie Bitterfeld (Germany) as aqueous solutions with molecular weights of about 9000, 35,000, and 130,000 g/mol. In addition, three

TABLE 3.2 Composition and Characterization of the Cellulosic Suspensions

	Suspension I (with low anionic trash content)	Suspension II (with high anionic trash content)	Suspension III (for FOFS experiments)
Sulfited spruce cellulose	70%	70%	Special mixture of cellulose, clay and CTMP, beaten together according to Ref. 26
Clay FKS 84	30%	30%	
Wood extract	—	200 mL	
Concentration of suspension	2 g/L	2 g/L	3 g/L
pH	7–8	7–8	6.3–7.3
Zeta potential (mV)	−4.5	−18	−36
Cationic demand[a] (μmol/g)	8	25	30

[a]Measured by polyelectrolyte titration.

TABLE 3.3   Polyanions of the Very High Molecular Weight Anionic Polyacrylamide Type Used

Product	Produced by	Optimum pH[a]
Laboratory product A	BASF, Germany	—
Commercial product B (Praestol)	Stockhausen, Germany	3–13 at low pH 6–13 at high CD[b]
Commercial product C (Chupamid)	Chupa, Switzerland	1–9

[a]According to the manufacturer.
[b]CD, charge density.

high-molecular-weight (HMW) polyacrylamide-type polyanions (product A, B, and C), produced by different plants, were used (see Table 3.3). For investigating the influence of anionic charge density, three different Praestol polymers (products of plant B) with charge densities of about 2, 3.2, and 5.7 meq/g were used.

3.3.3   Methods

Fibre Optical Flocculation Sensor   Flocculation was characterized using classical paper industry methods, as well as a computer-controlled Fibre Optical Flocculation Sensor (FOFS), developed at BASF (Germany).[27] The FOFS enables a rapid in situ measurement of the degree of flocculation as a function of the polymer dose. Floc formation takes place under well-defined shear conditions in laminar tube flow. The F value (an index from the FOFS measurement that characterizes the size of the flocs) typically increases with increasing polymer concentration. For studies of the two-component system, a certain quantity of polycation was added to the suspension, followed by computer-controlled dosing of the polyanion. Shear forces and hence the kinetics of the process were influenced by the flow velocity and the length of the tube.

Drainage Test   In addition, the drainage behavior was tested with a Schopper–Riegler tester. This test procedure is a laboratory method that simulates the papermaking process. Normally this method is used for characterizing the drainage behavior of pure cellulosic suspensions[28] (the so-called beating degree). Exactly 1 L is poured over the wire of the Schopper–Riegler tester and left to drain in two graduated flasks. The filtrate volume, V_{SR}, collected at the side pipe of the apparatus typically rises with the drainage behavior. We have used this equipment to characterize different polymer combinations as flocculants. After adding the polymer, the suspension was stirred for 30 s, poured over the tester, and left to drain. For dual systems, the polycation was added first, followed by the polyanion.

Additional Methods The retention of the system was calculated by measuring the residual turbidity of the collected water. In addition, we measured the polymer concentration (polyanion or polycation) after drainage by means of a polyelectrolyte titration procedure that used a streaming current detector (PCD 02, Mütek, Germany). The zeta potential of the flocculated suspension was measured with a System Zetapotential SZ 2 (Magendans, Netherlands).

3.4 RESULTS AND DISCUSSION

The process of flocculation was studied systematically by measuring the cellulosic suspension dewatering behavior with a Schopper–Riegler tester to study first the interaction between polycation and suspension and then the influence of additional polyanion. Three suspensions, differing in their anionic character, were investigated (Table 3.2).

3.4.1 Flocculation with Polycation

Figure 3.1 shows that the filtrate volume V_{SR} increased with increasing polymer concentration up to 0.5 mg/g PDADMAC (MW = 35,000 g/mol) in suspension I. This amount is necessary for good flocculation. The charge of the residual solution, measured by polyelectrolyte titration, was slightly anionic at this point. This result was confirmed by measurements of the zeta potential of the suspension, which increased from − 4.5 mV (without PDADMAC) to zero, when 1 mg/g of polycation was added. The nonzero zeta potential at the point of optimum flocculation indicates a charge patch flocculation mechanism: cationic sites, or "islands," with a high charge density are formed on the solid surfaces (fiber and filler) by the adsorption of the polycation. Flocculation will then take place through electrostatic attraction between the oppositely charged parts on the particles. The zeta potential need not be zero since the interaction energy is strongly dependent on the local charge patches.[8]

Influence of Anionic Trash For suspension II with higher anionic trash content, the cationic demand shifted to about 2.4 mg/g PDADMAC, but the drainage behavior was not influenced (Fig. 3.2). The maximum V_{SR} was the same as for the system with low anionic trash content. The quantities of polycation necessary for optimum flocculation of the systems are of great importance for the optimization of a dual system and are given in Table 3.4.

Influence of Molecular Weight The influence of polycation molecular weight on flocculation was investigated with FOFS, using suspension III (Fig. 3.3). The sizes of flocs were followed with increasing polymer addition under well-defined shear conditions. It was determined from the dewatering behavior (Fig. 3.4) that the low-molecular-weight polycation (MW of 9,000 g/mol) was inferior to those with molecular weights in the range from 35,000 to 130,000 g/mol. This result is consistent

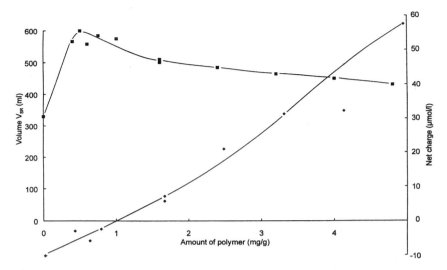

FIGURE 3.1 Flocculation of cellulosic suspension I (2 g/L) by PDADMAC, $M_w = 35{,}000$ g/moL; dependence of the filtrate volume V_{SR} (■) and net charge of the residual solution (♦) on the amount of PDADMAC.

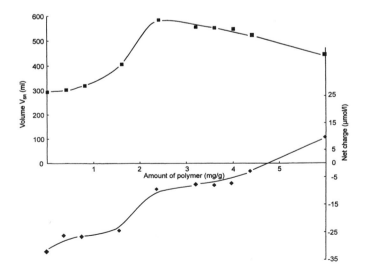

FIGURE 3.2 Flocculation of cellulosic suspension II (2 g/L) by PDADMAC, $M_w = 35{,}000$ g/mol; dependence of the filtrate volume V_{SR} (■) and net charge of the residual solution (♦) on the amount of PDADMAC.

TABLE 3.4   Summary of Important Parameters for Optimization of the Dual Systems

	Suspension I	Suspension II
Cationic demand of the suspension[a]	8 μmol/g	25 μmol/g
Zero point of charge[b]	1.0 mg/g PDADMAC	4.8 mg/g PDADMAC
PDADMAC, necessary for optimum flocculation[c]	0.5–0.8 mg/g	2.4 mg/g

[a]Measured by polyelectrolyte titration of the suspension (Mütek).
[b]Determined from zeta potential measurement (Magendans).
[c]Determined with Schopper–Riegler tester.

with expectations based on adsorption behavior.[29] Due to the rapid penetration of cationic polymer into the porous cell wall of the cellulosic fibres, the molecular weight of the cationic polyelectrolyte must not be too low.

3.4.2   Influence of Added Polyanion

According to the literature, the process of flocculation will be improved by the addition of polyanion. This result was confirmed by our own investigation with the FOFS. In Figure 3.5 we show that the F value was influenced by the quantity of polycation as well as by the amount of polyanion.

To investigate this process in detail quickly becomes very complicated. Different sources provide very different optimum composition data. We previously[30,31] saw that

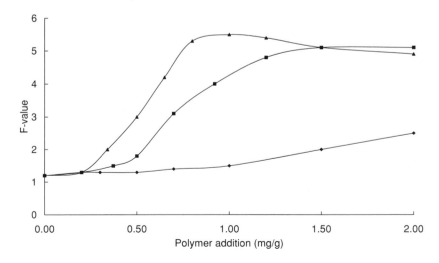

FIGURE 3.3   Flocculation of suspension III (3 g/L) with PDADMAC of different molecular weights; F value vs. the concentration of polycation; M_w = 9,000 (♦), 35,000 (■), and 130,000 (▲).

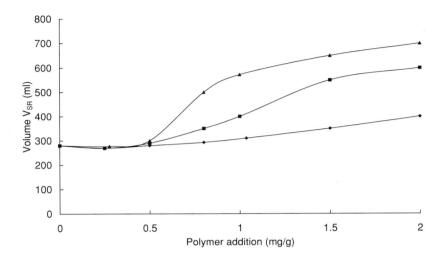

FIGURE 3.4 Flocculation of suspension III (3 g/L) with PDADMAC of different molecular weights; filtrate volume V_{SR} vs. the concentration of polycation; symbols as in Fig. 3.3.

the ratio of charges n^-/n^+ (n^- = polyanion, n^+ = polycation) at constant amount of polycation was of great importance and was a useful parameter for describing the system. The charges of the polycationic and polyanionic solutions were measured by charge-compensating polyelectrolyte titration[30,31] using the particle charge detector (PCD 02, Mütek, Germany). If ionic polymers of opposite charges are tested in this way, they react by complexing. The streaming potential induced was compensated by titrating either with PDADMAC or potassium poly(vinyl sulfonate) depending on the type of charge induced. The quantity of polymer consumed to compensate the buildup charge corresponded to the amount of polymer present in solution. The type of polymer needed for the liberation indicated the type of surface charge (anionic or cationic).

The dependence of the flocculation of paper stock on the ratio n^-/n^+ at a constant concentration of polycation was investigated. The molecular weight of the polycation used was 35,000 g/mol in this case.

Influence of the Polyanion Molecular Weight Whereas the influence of the molecular weight of the polycation (in the range investigated) was shown to be weak, the influence of the molecular weight of the polyanion on flocculation was of great importance. It was shown[32] that the combination of two oppositely charged polyelectrolytes with low molecular weights of the same order of magnitude caused the formation of a polyelectrolyte complex with poor dewatering characteristics, while a strong flocculation was achieved with a dual system consisting of a

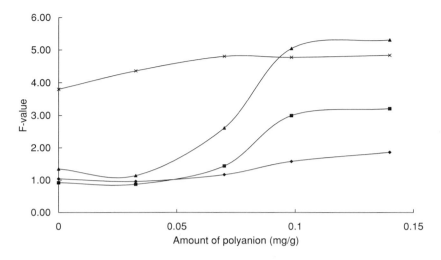

FIGURE 3.5   Flocculation of suspension III (3 g/L) with dual systems of different composition; influence of the quantity of polyanion on F value at different dosage levels of PDADMAC (35,000 g/mol), (♦) 0.12 mg/g, (■) 0.2 mg/g, (▲) 0.4 mg/g, (×) 1.2 mg/g.

polycationic PDADMAC (35,000 g/mol) and a polyanion of very high molecular weight (several millions). Figure 3.6 shows the dependence of the filtrate volume V_{SR} on the ratio n^-/n^+ at a constant concentration of PDADMAC. For flocculation of suspension I, 0.8 mg/g PDADMAC were added. After stirring for 30 s, certain quantities of the polyanion (according to the ratio n^-/n^+) were added and the Schopper–Riegler test was carried out. The filtrate volume increased from 600 mL (only polycation) to about 800 mL with an optimum at an n^-/n^+ ratio of about 0.5. This behavior was the same for different polyanions (A, B, C) of the anionic polyacrylamide type (medium charge density), produced by different plants. We believe the reason is that the primary flocs formed are linked together by bridging. It is essential for bridging that long loops and tails are formed by the adsorbed polymer, which in turn depends on the type of polymer, contact time, and surface properties. It is evident that a polymer with a high molecular mass is most suited for bridging.[8] In comparison, the application of a low-molecular-weight (LMW) poly(maleic acid-*co*-α-methylstyrene), MW ≈ 24,000 g/mol, was not successful as demonstrated in Petzold et al.[32] and seen in Figure 3.6.

Optimization   Figure 3.7 shows that even a small quantity of polycation, combined with the HMW polyanion (medium charge density) could increase the filtrate volume to more than 700 mL for suspension I. The optimum amount according to Table 3.4 is 0.8 mg/g of polycation. A further increase in V_{SR} was obtained with 1.6 mg/g PDADMAC, but a higher content of PDADMAC did not improve the drainage behavior. In the case of suspension II (Fig. 3.8) at least 2.4 mg/g PDADMAC (see also

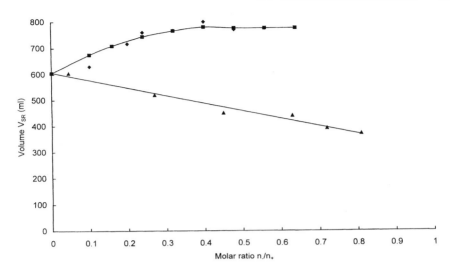

FIGURE 3.6 Flocculation of suspension I (2 g/L) with dual systems consisting of 0.8 mg/g PDADMAC (35,000 g/mol) and different polyanions; (♦) A (HMW), (■) B (HMW), (▲) LMW copolymer.

Table 3.4) was necessary for the creation of "cationic islands." The same good flocculation behavior as in Figure 3.7 was obtained. Further increase of the filtrate volume with polycation concentration was not observed. Figure 3.9 shows the influence of the total amount (polycation and polyanion) on flocculation. An increase

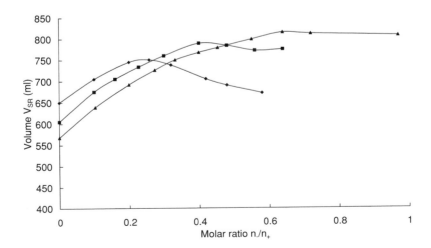

FIGURE 3.7 Flocculation of suspension I (2 g/L) with dual systems; filtrate volume V_{SR} vs. ratio n^-/n^+ at different dosage levels of PDADMAC (35,000 g/mol); (♦) 0.4 mg/g, (■) 0.8 mg/g, (▲) 1.6 mg/g.

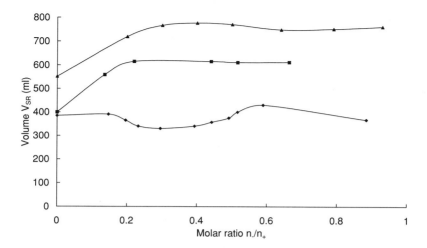

FIGURE 3.8   Flocculation of suspension II (2 g/L) with dual systems; filtrate volume V_{SR} vs. ratio n^-/n^+ at different dosage levels of PDADMAC (35,000 g/mol); (♦) 0.8 mg/g, (■) 1.6 mg/g, (▲) 2.4 mg/g.

in polymer concentration above the optimum value did not result in an improvement of the flocculation behavior.

Influence of the Charge Density   To investigate the influence of the charge density of the high-molecular-weight polyanion, three products (plant B) with molecular

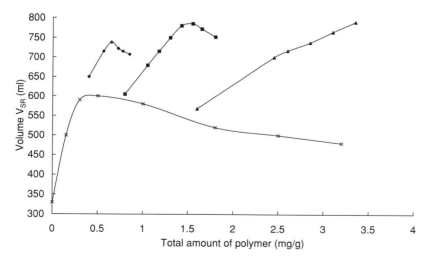

FIGURE 3.9   Flocculation of suspension I (2 g/L); filtrate volume V_{SR} vs. the total amount of polymer (PDADMAC and polyanion) at different dosage levels of PDADMAC (35,000 g/mol); (♦) 0.4 mg/g, (■) 0.8 mg/g, (▲) 1.6 mg/g, (×) only PDADMAC.

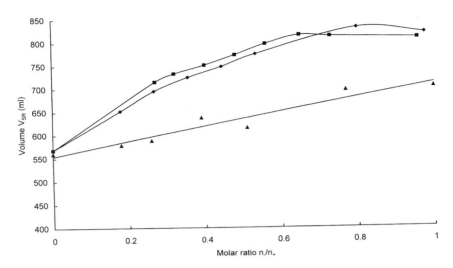

FIGURE 3.10 Flocculation of suspension I (2 g/L) with dual system; 1.6 mg/g PDADMAC (35,000 g/mol); filtrate volume V_{SR} vs. the ratio n^-/n^+ at different charge densities of the polyanion; (♦) low, (■) medium, (▲) high.

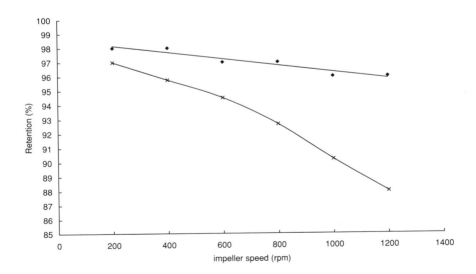

FIGURE 3.11 Effect of shear forces on the retention in cellulosic suspension I (2 g/L) flocculated with (×) PDADMAC (35,000 g/mol) or an (♦) optimized dual system.

weights of about 7 million but with different charge densities (low, 2; medium, 3.2; and high, 5.7 meq/g) were used in combination with PDADMAC. There was no difference between polyanions of low and medium charge density, but highly charged anionic polymers were not as effective (Fig. 3.10). Therefore, it seems that the charge density of the polyanion is of crucial importance for the dewatering behavior. If only the amount of polyanion were important for bridging, then we should achieve the same maximum filtrate volume but at different molar ratios. However, dewatering was better with anionic polymers having low and medium charge density because in this case they were not adsorbed on the positively charged surfaces in a flat conformation and were more available for bridging.

Addition Time   The maximum volume at a certain composition decreased (600 mL instead of 650 mL) if the time between the addition of the polycation and polyanion was 3 min instead of 30 s. According to Ödberg and others,[29,32] the reduced interaction between the adsorbed polycation and polyanion occurs because the cationic polymer transports to the internal surfaces of the fiber very quickly and is no longer accessible to the high-molecular-weight polyanion.

Shear Stability   The shear stability of the flocs was tested using a Britt dynamic drainage jar. In contrast to the pure polycation, the retention of an "optimized" dual system was more resistant to high shear forces, which were controlled by the speed of the impeller (Fig. 3.11).

3.5   RELATED APPROACHES

Two-component systems are in common use. Several systems are commercially available and would be highly effective if applied in paper mills.[34] Here some examples for new approaches in flocculation systems will be given.

3.5.1   Dual Systems Employing Microparticles

The efficiency of the microparticle system is enhanced by using a three-component system. A model that quantitatively predicts the flocculation behavior in such systems has been presented. A low-molecular-mass polymer was added prior to the microparticle retention aid system. The low-mass polymer acted as a site-blocking agent and maximized the extension of the retention aid polymer.[35,36]

   Other areas of research include the improvement of anionic colloidal systems and the application of new systems. One example is a combination of anionic and cationic aluminum compounds[37] as an improvement on the classical silica systems.[38] The main scope of this investigation was to quantify the impact of certain properties of this class of silica on the retention of fillers and fines. Model anionic colloidal silica (ACS) was used. The size, charge, and shape were shown to be important for retention.

New aqueous suspensions of colloidal particles have been described.[39] They include both silica-based anionic particles and hydrated particles of clays of the smectic type, which are expandable in water. The suspensions have good retention and drainage properties in papermaking when used with cationic and amphoteric polymers.

Some advantages of using a cationic polymeric microparticle system instead of an anionic microparticle system as a retention aid are mentioned in Ono and Deng.[40] These particles are efficient flocculants for positively charged particles such as precipitated calcium carbonate if they are used in combination with an anionic polyacrylamide. Several new all-polymeric systems are now also available.[41–43]

3.5.2 Polyelectrolyte Complexes as Flocculants

The application of premixed polyelectrolyte complexes as flocculants is a field of academic work that has not found a practical application yet. But we previously saw[44] that such complexes made from oppositely charged polymer solutions have interesting properties and are very effective for the flocculation of finely dispersed inorganic particles. A reflocculation was not noticed in a wide range of concentration.

Somasundaran and co-workers[45] found that combinations of polystyrene sulfonate and cationic polyacrylamide enhanced the flocculation of (positively charged) alumina suspensions. The weight ratio of the two polymers was kept at 1:1 for all experiments. Premixing of the two polymers was not as good as the results obtained when the polymers were added step by step. The formation of polymer complexes has been proposed to be central to the mechanism for flocculation in the following cases. The drainage behavior of pulp suspensions with a two-component system was analyzed with attention to polyion complex formation.[46] With cationic and anionic polyacrylamides (MW 3,000,000 and 1,000,000 g/mol) added individually a polyion complex with an irregular three-dimensional structure was precipitated on the fiber surface. This reduced the homogeneity of the paper sheets. In contrast to this a premixed system of an anionic and cationic polyacrylamide[47] resulted in an improvement in pulp retention and paper quality (paper strength).

Polyelectrolyte complexes have been applied to the flocculation of different suspensions in salt solutions.[48] Polymer–surfactant complexes can also be used as flocculants, for example, in montmorrilonite dispersions.[49] An anionic surfactant is combined with a cationic polymer, which is preadsorbed on the clay particles. Such interactions may lead to a flocculation mechanism that combines polymer adsorption, charge neutralization, and hydrophobic interactions.

3.5.3 Filler Modification

Different groups are exploring filler modification. The retention of $CaCO_3$ filler in paper has been increased by a treatment with a specially modified polyethyleneimine before flocculation.[50] Seppänen et al.[51] developed a new technique for preparing small, strong, and porous microflocs of clay. This preflocculated clay was successfully used to improve the optical properties of paper sheets.

3.6 SUMMARY AND CONCLUSION

In summary, an understanding of colloid–polymer interaction has led to many solutions of different problems in the paper industry and other technologies. Application of flocculants with two or more components is of growing importance for processes involving solid–liquid separation.

In contrast to microparticle-containing systems, "classical" dual retention aid systems (polymer–polymer) are said to give high retention combined with a poor dewatering.[24] We were able to show that an optimized polymer–polymer system of the type mentioned above has many advantages such as good dewatering, superior retention, and shear-resistant flocs. These effects are obtained by a combination of charge patch formation and bridging. The behavior especially depends on the concentration of polycation (according to the anionic character of the suspension) and the molar ratio of anionic and cationic charges.

The use of dual- and multicomponent retention systems is increasing. The present disadvantage of some of these systems is higher cost relative to single-polymer systems. But current trends in the paper industry to use more contaminated pulp and higher filler loads, to increase water system closure, to discharge cleaner effluent, and to maintain high levels of production efficiency will further encourage the use of more complex retention technology.[14] Therefore, further research into the details of the very complex processes will be needed to maximize the effectiveness of process chemicals.

REFERENCES

1. L. Ringqvist and P. Igsell, *Energy & Fuels* **8**, 953 (1994).

2. W.-M. Kulicke, S. Lenk, D. Detzner, and T. Weiß, *Chem.-Ing.Tech.* **65**(5), 541 (1993).

3. N. Böhm and W.-M. Kulicke, *Colloid Polym. Sci.* **275**, 73 (1997).

4. X. Zhang, *Water Treatment* **7**, 33 (1992).

5. J.N. Arno, W.E. Frankle, and J.L. Sheridan, *Tappi* **57**(12), 97 (1974).

6. K.W. Britt and J.E. Unbehend, *Tappi* **57**, 81 (1974).

7. E.E. Moore, *Tappi* **56**, 71 (1973).

8. T. Lindström, in G.F. Baker and V.W. Punton, eds., *Fundamentals of Papermaking*, Trans. Ninth Fundam. Res. Symp. at Cambridge 1989, Mech. Eng. Publ., London, 1989, pp. 311–411.

9. K. Andersson, A. Sandström, K. Ström, and P. Barla, *Nordic Pulp Paper Res. J.* **2**, 26 (1986).

10. S. Wall, P. Samuelsson, G. Degerman, P. Skoglund, and A. Samuelsson, *J. Coll. Interface Sci.* **151**(1), 178 (1992).

11. A. Swerin, L. Ödberg, and U. Sjödin, *Paper Tech.* **33**, 28 (1992).

12. J.G. Langley and E. Litchfield, Proc. Tappi Papermakers Conference, New Orleans, 1986, pp. 89–92.

13. A. Swerin, U. Sjödin, and L. Ödberg, *Nordic Pulp Paper Res. J.* **4**, 389 (1993).

14. R.I.S. Gill, paper presented at Chemistry of Papermaking Conf., 30–31 Jan. 1991 at Solihull, UK.

15. A. Swerin and L. Ödberg, *Nord. Pulp Paper. Res. J.* **11**(1), 22 (1996).

16. D. Eklund and T. Lindström, *Paper Chemistry*, DT Paper Science Publications, Grankulla, Finland, 1991, pp. 145–191.

17. E.E. Moore, *Tappi* **59**(6), 120 (1976).

18. F. Müller and U. Beck, *Das Papier* **32**(10A), V25 (1978).

19. T. Petäjä, *Kemia-Kemi* **5**, 261 (1980).

20. L. Wagberg and T. Lindström, *Nordic Pulp Paper Res. J.* **2**, 49 (1987).

21. R.I.S. Gill, *Paper Tech.* **32**(8), 34 (1991).

22. J. Reuter, The Chemistry of Papermaking: Symposium 28–29 Jan. 1992, Pira, Leatherhead, paper 12.

23. B. Krogerus, *Nordic Pulp Paper Res. J.* **8**(1), 135 (1993).

24. F. Hedborg and T. Lindström, *Nordic Pulp Paper Res. J.* **4**, 254 (1996).

25. F. Linhart, W.J. Auhorn, H.J. Degen, and R. Lorz, *Tappi* **70**(10), 79 (1987).

26. G. Petzold, G. Kramer and K. Lunkwitz, *Das Papier* **49**(6), 325 (1995).

27. L. Eisenlauer and D. Horn, *Coll. Surf. A* **25**, 111 (1987).

28. E. Gruber, *Das österreichische Papier* **1**, 22 (1981).

29. L. Winter, L. Wagberg, L. Ödberg, and T. Lindström, *J. Coll. Interface Sci.* **111**(2), 537 (1986).

30. G. Kramer, H.-M. Buchhammer, and K. Lunkwitz, *J. Appl. Polym. Sci.* **65**, 41 (1997).

31. G. Petzold and K. Lunkwitz, *Coll. Surf. A* **98**, 225 (1995).

32. G. Petzold, H.-M. Buchhammer, and K. Lunkwitz, *Coll. Surf. A* **119**, 87 (1996).

33. L. Ödberg and L. Wagberg, *Das Papier* **43**, V37 (1989).

34. B.E. Doiron, paper presented at 1994 Papermakers Conf. held at San Francisco, USA, Tappi Press, Atlanta, 1994, 2 vols., Vol. 2, pp. 603–607.

35. L. Wagberg, M. Bjoerklund, I. Asell, and A. Swerin, *Tappi J.* **79**(6), 157 (1996).

36. A. Swerin, L. Ödberg, and L. Wagberg, *Coll. Surf. A* **113**, 25 (1996).

37. R. Moffett, U.S. Pat. 5,595,630.

38. K. Andersson and E. Lindgren, *Nordic Pulp Paper Res. J.* **11**(1), 15 (1996).

39. Anon., *Res. Disclosure* **375**, 467 (1995).

40. H. Ono and Y. Deng, *J. Coll. Interface Sci.* **188**, 183 (1997).

41. D.S. Honig, E.W. Harris, L.M. Pawlowska, M.P. O'Toole, and L.A. Jackson, *Tappi J.* **76**(9), 135 (1993).

42. D.S. Honig and E. Harris, U.S. Pat. 5,167,766 (1992).

43. R.G. Ryles, D.S. Honig, E.W. Harris, and R.E. Neff, U.S. Pat. 5,171,808 (1992).

44. G. Petzold, A. Nebel, H.-M. Buchhammer, and K. Lunkwitz, *Coll. Pol. Sci.* **276**(2), 125 (1998).

45. P. Somasundaran and Y. Xiang, *Coll. Surf. A* **81**, 17 (1993).

46. F. Onabe, A. Yamazaki, M. Usuda, and T. Kadoya, *Mokuzai-Gakkai-shi* **29**, 60 (1983).

47. T. Nagata, *Jpn. Tappi J.* **45**(2), 29 (1991).

48. DD Pat. 301 052 A7 (1992), J. Kötz, U. Gohlke, B. Philipp.

49. S. Magdassi and B.-Z. Rodel, *Coll. Surf. A* **119**, 51 (1996).

50. D. Moench, P. Lorencak, and F. Linhard, *Wochenbl. Papierfabr.* **122**(17), 659 (1994).

51. R. Seppänen, J. Elftonson, and G. Strom, *Helsinki Univ. Technol. Lab. For. Prod. Chem. Rep., Ser. C* **5**, 5 (1994).

4 Role of Polymers in Particle Adhesion and Thin Particle Layers

MARCEL R. BÖHMER, WILLEM HOOGSTEEN, and
GERALD F. BELDER
Philips Research Laboratories Eindhoven, (WA04) Prof. Holstlaan 4, 5656 AA
Eindhoven, The Netherlands

4.1 INTRODUCTION

The application of colloidal particles in industry is often intimately connected to the use of polymers. Polymers are widely used to stabilize or flocculate suspensions. In the case of stabilization, polymer chains create repulsive interactions between the suspended particles by a steric or electrosteric stabilization mechanism.[1] For application in water purification, addition of small amounts of polymer results in attractive interactions, causing big flocs to be formed by bridging flocculation. In the food and cosmetics industries a suspension or emulsion is usually the end product. However, in the electronics or in the paint industry a suspension is only an intermediate product, the final product being often a (sintered) layer of particles on a surface. To arrive at such particle layers, the colloidal interactions in suspension have to be controlled. In the case of thin particle layers, polymers can also be used to control the interaction between the particles and the macroscopic surface. Just as polymers can be used as efficient flocculants for suspensions, they can also be used to accomplish efficient adhesion of particles onto macroscopic surfaces, a process that can be considered as flocculation between two different particles: a small one and an infinitely large one.

Apart from establishing efficient adhesion, polymers may also be used to prevent particle adhesion on a surface. In industry the interest is usually in the extremes, that is, a surface with no particles at all or one that is completely covered. These two limiting situations are combined in patterned particle layers. In this case high

Colloid–Polymer Interactions: From Fundamentals to Practice, Edited by Raymond S. Farinato and Paul L. Dubin
ISBN 0-471-24316-7 © 1999 John Wiley & Sons, Inc.

selectivity is a prerequisite. An example is the adsorption of a polymer-stabilized nanometer-sized palladium sol that serves as a catalyst for electroless deposition of metals on nonconducting surfaces such as glass.[2] In order to make conducting tracks on the surface, the Pd particles have to adhere strongly to some parts of the surface and not at all to other parts.

Another application of selective adhesion is the formation of a high resistivity layer of Cr_2O_3 onto a glass substrate containing holes. Cr_2O_3 should cover the front side and the inner sides of the holes but not the back side. It is hard to reach the inner sides of the holes with conventional techniques such as evaporation or sputtering. Much more convenient is the application of a particle layer by suspension spraying.[3] Alternatively, the glass surface is pretreated with an adsorbing polymer, which is then removed from one side by cleaning, for example, with ultraviolet (UV)-ozone, the final layer then selectively applied by particle adsorption on the polymer-containing surface. Particle adsorption from stable suspensions is often limited to a (sub)monolayer level.[4–6] However, by repeating the pretreatment with polymer, followed by a subsequent particle adsorption step, thicker layers can be applied. This technique, pioneered by Iler,[7] currently attracts much attention.[8,9]

Selective particle adsorption also plays a prominent role in the processing of "phosphors," the light-emitting powders used on the screens of televisions and computer monitors. Phosphors for screen applications consist of ZnS, Y_2O_3, or Y_2O_2S particles (around 5 µm in diameter) and are doped to tune the color. In this way red, green, and blue phosphors can be prepared. Submonolayer particle coatings of iron oxide (red) or cobalt aluminate (blue) are often applied on these particles in order to increase color contrast and to improve handling. Polymers are crucial to the adhesion of the phosphor particles onto the screen. The phosphor layers are made by coating a phosphor suspension onto the screen. To this suspension a UV-sensitive water-soluble polymer system is added. The phosphor layer is then exposed to UV light through a shadow mask. An insoluble network is formed at the exposed areas, holding the particles together as a line or dot (typical dimension around 200 µm). The screen is subsequently developed by washing away the particles, together with the water-soluble polymer, from the nonexposed areas. It is important that no phosphor particles are left at these areas since this is where the phosphor particles of the next color are to be applied. The presence of phosphor particles of the wrong color leads to color contamination. Polymer precoats on the glass have to be optimized to reduce this unwanted adherence, which is termed haze. On the other hand, the phosphor lines or dots should adhere strongly to the screen. Photochemical reactions are used to achieve this.[10]

In this chapter we first focus on the adhesion of model silica (SiO_2) particles on surfaces and consider the effect of salt concentration. Some aspects of the adsorption of polymers relevant to particle adhesion are treated. We then consider the influence of surface pretreatment and the presence of free polymer on the adhesion of particles. Some attention is paid to the buildup of particle multilayers. The measurements on these model systems are obtained by reflectometry in stagnation point flow.[11,12] We next present results for more applied systems such as a high resistivity layer of colloidal Cr_2O_3 particles on glass. The effects of suspension stability, adsorption time, salt concentration, and particle multilayer formation will be discussed, partly based on insight gained from the SiO_2 system. Subsequently results for much larger phosphor particles will be presented, where we will again focus on surface pretreatment and the presence of free polymer. A comparison will be made with the results for the model SiO_2 colloids. Finally, some sedimentation results will be presented for the packing density of thin particle layers, and the effects of different surface pretreatments will be described.

4.2 PARTICLES ON SURFACES

Particles flocculate in suspension if the repulsive barrier to particle approach is negligibly small or nonexistent. At small interparticle distances the interparticle attraction will be most pronounced if the particles are oppositely charged or if an adsorbing polymer is able to bridge the interparticle gap. Similar interactions occur when particles adsorb onto a macroscopic surface. If a glass surface has a positive charge, established, for example, by adsorption of an amino-silane or by adsorption of a positively charged polyelectrolyte, negatively charged particles will adsorb onto it strongly. In this case a "hit-and-stick" mechanism applies, meaning that every particle that touches the surface gets irreversibly bound. This corresponds to fast coagulation in a suspension, when every collision between two particles leads to the formation of a doublet.[13] Slow flocculation may occur in suspensions at salt concentrations below the critical coagulation concentration. A similar situation can in principle also occur for particle adsorption onto a macroscopic surface.[5,14] An overview of particle deposition was recently given by Elimelech et al.[15]

The random sequential adsorption (RSA) model is widely used to describe irreversible adsorption of particles. An extensive review of this model has recently been published.[5] The RSA model assumes the particles approach the surface one by one. If they find an empty spot upon their first contact with the surface, they adsorb irreversibly. However, if overlap with an already deposited particle occurs, then they are removed from the system. The RSA model leads to a jamming limit of 54.6% for spherical particles. The configurations of the particles on the surface have been calculated and characterized by correlation functions.[16,17] In RSA the particles only interact with each other through their excluded volume; in other words only hard sphere interactions are assumed. For the aqueous systems we consider in this chapter the particles also undergo electrostatic interactions. Adamczyk et al.[5] used an effective hard sphere model in which these interactions are modeled by assuming that the particles have a larger effective radius. The parameter H^* is used to take the electrostatic repulsion into account:

$$H^* = \xi / \kappa a \qquad (4.1)$$

where ξ is a proportionality constant depending on the logarithm of the surface potential of the particles, $1/\kappa$ is the Debye length, and a is the particle radius. The jamming limit in the presence of double-layer interactions (θ_{mx}) is then related to the jamming limit in the absence of interactions (θ_{∞}) by

$$\theta_{mx} = \theta_{\infty} / (1 + H^*)^2 \qquad (4.2)$$

Adamczyk et al.[5] have given experimental results on the jamming limit as a function of particle size and salt concentration, which confirm Eq. (4.2). Experimental data are also available from the work by Johnson and Lenhof.[6] Böhmer et al.[18] measured the surface coverage of nanometer-sized silica particles by optical reflectometry in stagnation point flow geometry. In reflectometry the intensity ratio of parallel and perpendicularly polarized light reflected off an interface is measured and converted to the adsorbed amount. For more details about reflectometry we refer the reader to Chapter 13. For small particles, different optical models, such as a homogeneous slab model or a particulate model,[18,19] can be used to obtain the surface coverage from the relative change in intensity ratio ($\Delta S/S$). Surface coverages for two different particle sizes (from Böhmer et al.[18]) are compared in Figure 4.1 with calculations using Eq. (4.2) with $\xi = 2.3$. The theoretically expected increase of the surface coverage with salt concentration and with particle size is indeed found. A compilation of many of these data was recently given by Semmler et al.[20]

The kinetics of particle deposition processes has also received attention. Hydrodynamic interactions between particle and surface play a very important role, especially for somewhat larger colloids. A setup that is often used in particle

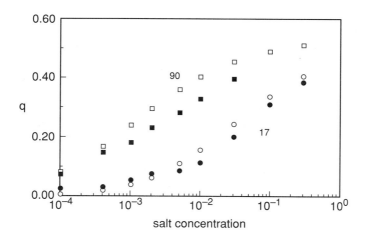

FIGURE 4.1 Surface coverage obtained by reflectometry as a function of the salt concentration for 17 nm (Ludox SM, Dupont) and 90 nm (Monospher 100, Merck) silica particles adsorbed onto a macroscopic silica surface (open symbols). The silica surface is a 100-nm layer on a silicon wafer pretreated with a cationic polymer to ensure effective particle adsorption. Filled symbols: calculations using Eq. (4.2) with $\xi = 2.3$ (data taken from Ref. 17).

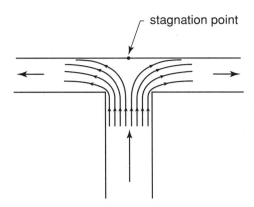

FIGURE 4.2 Streamlines in a stagnation point flow cell.

deposition studies is the stagnation point flow cell,[5,18,21–25] schematically depicted in Figure 4.2. In such a cell a laminar stream of fluid (usually a dilute suspension) flows through a cylindrical channel, impinges on the surface, and then flows out in the radial directions. However, the fluid flow vanishes at one point on the surface, located exactly above the center of the flow channel. This is the so-called stagnation point. For most practical purposes a circular region centering around the stagnation point can be effectively regarded as the stagnation point. In this region the transport of particles to the surface is completely controlled by diffusion.

Stagnation point flow cells have been combined with optical microscopy, which restricts the study to fairly big particles but has the advantage that the distribution of particles on the surface can be directly studied. For smaller particles stagnation point flow has been combined with evanescent wave spectroscopy[25] and optical reflectometry[18] to detect the particles on the surface. However, information on the distribution of particles is not accessible with these detection methods. Under different flow conditions, such as in laminar flow in a parallel plate flow cell, as used by Meinders and Busscher,[26] the distribution of particles may become asymmetric because an adsorbed particle changes the fluid streamlines and therefore hinders other particles from coming close to the surface in the direction of the flow. These "shadow" effects may extend up to 10 times the particle radius.

Obtaining information on the particle distribution on the surface from dried samples is difficult. It has often been observed that capillary forces during drying drastically change the distribution of the particles.[4,27,28] Effects of drying can be easily visualized using an optical microscope and fairly big, spherical particles. During the drying process particles apparently attract each other over large distances, due to capillary forces. It is possible to fix the particles on the surface, for example, by immersing the substrate in a solution of nitrocellulose in acetone, which gels during drying and therefore fixes the particles in their original position.[4]

4.3   ADSORPTION OF POLYMERS

Polymer adsorption on both particles and macroscopic surfaces can be used to influence the interparticle and particle–surface forces. A comprehensive treatment on polymer adsorption can be found in Fleer et al.[29] A particular feature of polymer adsorption is the absence of desorption upon dilution. This makes them suitable for the pretreatment of surfaces from which the excess can subsequently be removed by rinsing. Another aspect that makes polymers attractive in promoting particle adhesion is that they often exhibit high-affinity adsorption isotherms, especially if they are monodisperse or highly charged. Adsorption isotherms of poly(vinyl imidazole) (PVI) of different degrees of methylation, taken from Böhmer et al.,[30] are given in Figure 4.3. It is clear from the figure that the 50% methylated PVI gives rise to a high-affinity isotherm, but that a rounded isotherm is found at low degrees of methylation.

If the polymer is used in the suspension to adsorb onto the particles, many different effects may occur, as schematically depicted in Figure 4.4. The polymer may flocculate the suspension by, for example, bridging flocculation or (partial) charge neutralization (Fig. 4.4a). This causes the adsorption of aggregates of particles on the surface. If no flocculation occurs, adsorption of polymers on the particle may change the affinity of the particle for the surface. For example, if both particle and surface are negatively charged, particle adsorption is very weak or absent. But if a positively charged polymer is adsorbed onto the particles, their affinity for the surface will increase drastically, as seen in Figure 4.4b. If this is a situation, however, where the free polymer also has an affinity for the macroscopic surface, free polymer will compete with the particles for the surface, as shown pictorially in Figure 4.4c. This obviously affects the particle adsorption process.[23,31,32]

The effect of free polymer in dilute suspensions is hard to study for polymers that also adsorb on the particles. Therefore we studied a system of small silica particles

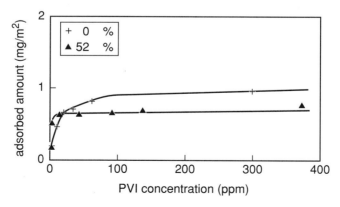

FIGURE 4.3   Adsorption isotherms of partially quaternized poly(vinyl imidazoles) on silica (Ludox AS) at pH 10 for 10^{-2} M aqueous solutions. Reprinted from Ref. 30 with permission from Elsevier Science.

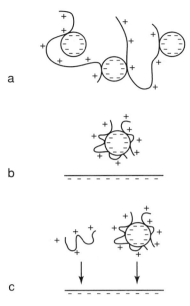

FIGURE 4.4 Possible polymer effects on particle adsorption on macroscopic surfaces: (*a*) flocculation of the suspension, (*b*) adsorption on negative particles by a cationic polyelectrolyte increasing the affinity for the surface, and (*c*) effect of free polymer in the solution.

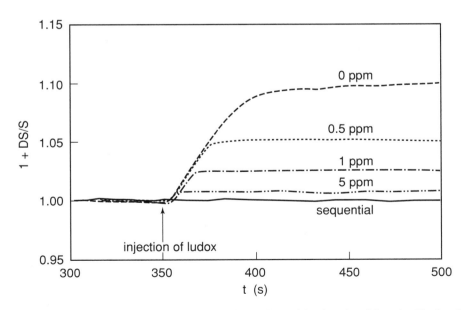

FIGURE 4.5 Effect of free polymer in the suspension on the particle adsorption. Adsorption kinetics of Ludox AS40 in stagnation point flow, studied by reflectometry, on surfaces pretreated with 50% quaternized PVI in the presence of different concentrations of poly(acrylic-acid). The quantity $\Delta S/S$ on the vertical axis is proportional to the adsorbed amount.

containing poly(acrylic acid) (PAA) at different concentrations,[33] at a pH at which no PAA adsorption on silica occurs. Therefore the amount of added polymer equals the amount of free polymer. If reflectometry experiments are performed on surfaces modified with positively charged polymers, a clear effect of the PAA adsorption can be seen, as shown in Figure 4.5. It has been demonstrated that PAA hardly shows a reflectometric signal upon adsorption.[34] Even at very low PAA concentrations the surface is largely blocked for particle adsorption. This can be explained by the much higher diffusion coefficient of the polymer coils in comparison with the particles. More extensive measurements can be found in Böhmer.[33]

4.4   PRETREATMENT OF SURFACE WITH POLYMERS

Polymers can also be used to promote particle adhesion by pretreating the macroscopic surface[23,31,33,35–37] instead of adding polymer to the suspension. While an additional process step is involved, flocculation of the suspension is prevented, as are competition effects between polymer and particles. Efficient particle adsorption is optimized by hit-and-stick conditions, which depend on the choice of the polymer used to pretreat the surface. Polymer pretreatment and the subsequent adsorption of small particles can be conveniently studied by reflectometry in stagnation point flow. Figure 4.6 illustrates the kinetics of adsorption of silica particles (Ludox AS40) on a silica surface pretreated with partially methylated poly(vinyl imidazoles) with degrees

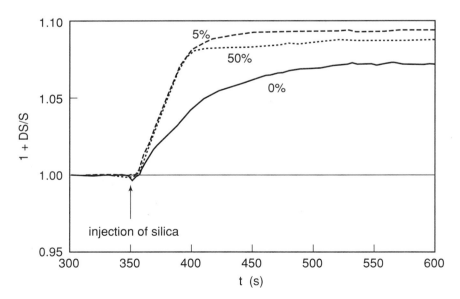

FIGURE 4.6   Adsorption kinetics of silica particles in stagnation point flow on silica surfaces pretreated with PVI of different degrees of quaternization (data taken from Ref. 33). The quantity $\Delta S/S$ on the vertical axis is proportional to the adsorbed amount.

of methylation of 0, 5, and 50% respectively (abbreviated as 0, 5, and 50% mPVI). Details are given in Böhmer et al.,[30] which also deals with the adsorption of PVI onto silica particles and macroscopic surfaces. The initial slopes of the adsorption curves are the same for the surfaces pretreated with 5 and 50% mPVI while the initial slope for 0% mPVI is lower. The zeta potential is still negative if 0% mPVI is adsorbed on different kinds of silica, which results in a repulsive contribution to the interaction between particle and surface. For 5% mPVI adsorbed onto silica the zeta potential is approximately zero, hence the electrostatic repulsion vanishes. The zeta potential is positive for 50% mPVI adsorbed onto silica. The two latter cases give the same initial slope, which stays constant up to high coverages. This is a strong indication for hit-and-stick conditions. The same slope is observed if the surface is pretreated with an amino-silane. It can be concluded that 0% mPVI is less effective in inducing silica particle adsorption.

Pretreatment of surfaces with polymer is a delicate matter, and the effects of polymer concentration and adsorption time[5] have been reported in the literature. We also found that pretreatment of a surface with polymer is very subtle. With dilute polymer solutions, long adsorption times are required to obtain uniform surface properties (e.g., effective immobilization of adsorbed particles across the whole surface). The removal of excess polymer, which may interfere with subsequent particle adsorption, has to be done very thoroughly by repeatedly immersing the substrate in water at the same pH and salt concentration. In stagnation point flow the polymer adsorption is much faster, due to the constant transport toward the surface, and rinsing can be performed very effectively, which makes it a suitable tool for studying polymer pretreatments and subsequent particle adsorption.

4.5 ADHESION STRENGTH OF PARTICLES ADHERING TO POLYMER-PRETREATED SURFACE

Different techniques can be used to assess the strength of the net attractive force that causes a particle to adhere to a surface. These include stagnation point flow[38-40] and parallel plate flow,[41] centrifugal methods,[42-44] and flow in capillary channels.[35,37] In all of these methods, well-defined fluid shear forces are exerted on the adhered particles by means of a controlled flow. More recently the atomic force microscope (AFM), with a colloidal probe attached to the cantilever, has been used to measure directly the adhesion force between a single particle and a surface.

An overview is given by Hartley in Chapter 11, so here we will only consider hydrodynamic techniques. Unlike AFM, these techniques probe the adhesion strength not of single particles but of a large number of particles within a certain area and therefore reflect an average over this number of particles. However, the experiments are easier to perform than colloidal AFM. A limited number of experiments have addressed the influence of adsorbed polymer on the adhesion strength.

We are mainly interested in the adhesion of particles to polymer-pretreated surfaces. The usual procedure is to pretreat the surface with polymer, rinse away the excess polymer, and then apply a particle layer by sedimentation of a suspension. The adhesion strength is then determined by imposing a laminar flow of increasing intensity parallel to the surface, and to monitor at each flow rate the number of particles that remain adhered. The average adhesion strength can then be determined from the flow rate at which a predetermined percentage of the originally adhered particles (typically 50%) is dislodged from the surface. Usually, fairly large particles are used (≥ 1 µm) since the force exerted by the liquid flow approaches zero close to the wall. This has led to the belief that smaller particles adhere more strongly to surfaces than larger ones, however, this is not in general true; they are just more difficult to remove by means of a flowing liquid (e.g., in cleaning a surface by rinsing).

For bare particles adhering to a bare surface, reasonable agreement is found between the average adhesion strength and the theoretically expected critical shear strength determined from a DLVO (Derjaguin, Landau, Vervey, and Overbeek) type of calculation.[44] If the surface is pretreated with polymer, the situation changes markedly. Generally speaking, the adhesion strength increases. Let us consider from now on the case of negatively charged particles, adhering to a negatively charged surface pretreated with either an uncharged or a cationic polymer. There are two possible mechanisms by which the adsorbed polymer can alter the adhesion strength between a particle and the surface. First, the presence of adsorbed polymer changes the DLVO interactions, both van der Waals and electrostatic interactions are affected, with the change in electrostatic interactions usually being most important. Even an uncharged polymer can change the zeta potential markedly,[37] by an outward shift of the plane of shear.[45] If the polymer is a cationic polyelectrolyte, the change in the zeta potential is much stronger, and the zeta potential of the surface may vanish or even change sign. Second, the polymer may primarily exert its influence through the formation of bridges between particle and surface, resulting in an attraction of entropic origin. Hubbe[36] has performed experiments with TiO_2 particles adhering to glass and cellulose surfaces that were either bare or pretreated with different cationic polyelectrolytes. Attempts to explain increased adhesion to the pretreated surfaces solely by means of modified DLVO interactions were unsuccessful, showing that bridging contributes to the observed behavior.

The adhesion strength in the presence of polymer depends on several factors. Of most interest to us are the molecular weight of the polymer and the concentration of the polymer solution used for pretreatment of the surface. The adhesion strength (for surfaces that have reached the adsorption plateau) is found to increase gradually with increasing molecular weight of the polymer. The increase in adhesion strength with concentration of the pretreatment polymer solution is directly related to the adsorbed amount. However, at high concentrations the results depend on the measurement method. With the capillary method the strength decreases at high concentrations,[35] whereas in the centrifugal method[36,42] the strength reaches a plateau value. The most probable explanation is that in the case of the small capillaries the excess of polymer cannot be completely rinsed away after the pretreatment if the concentration is high. Any polymer remaining inside the capillary will then adsorb onto the particles during the sedimentation step. This leads to a decrease in the bridging attraction due to an osmotic repulsion between the polymer layers of the particle and the surface. The centrifugal method may therefore be regarded as more reliable at higher concentrations.

Other parameters that influence the adhesion strength include the pH and ionic strength of the flowing medium, the time allowed between sedimentation and the removal, and the size of the particles under consideration. Experimental data on the influence of these parameters are limited and sometimes conflicting.

4.6 MULTILAYER ADSORPTION

Just as cationic and anionic polymer additions can be alternated to arrive at thicker layers,[34,46–48] polymer pretreatment and particle adsorption steps can also be repeated.[7–9] This is obviously a slow process, but it can be well controlled. It was demonstrated earlier (Fig. 4.1) that the amount of particles adsorbed after a certain time interval depends on particle size and salt concentration, provided hit-and-stick conditions are present. Reflectometry results shown in Figure 4.7 for alternating deposition of PVI and silica reveal that the 5% mPVI yields much higher adsorbed amounts than the 50% mPVI. This accounts for the differences in signal in the polymer adsorption steps. The amount of silica adsorbed after each polymer addition remains approximately constant. This pattern is very similar to that found for multilayers of polyelectrolytes.[34] It can be concluded that the salt concentration and number of adsorption steps can be manipulated such that the amount of particles adsorbed can be very well controlled.

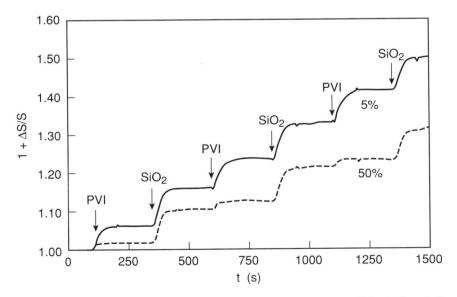

FIGURE 4.7 Buildup of multilayers by alternating supply of polymer (5 or 50% mPVI) and silica particles measured by reflectometry. The quantity $\Delta S/S$ on the vertical axis is proportional to the adsorbed amount.

4.7 Cr_2O_3 COATINGS PREPARED BY PARTICLE ADSORPTION

A thin Cr_2O_3 particle coating can be used as a resistivity coating in flat displays.[3] One way to deposit such a thin coating on a perforated glass substrate is by particle adsorption based on electrostatic attractions between the surface and the particles.[7,8,49–53]

4.8   SUSPENSION STABILITY

A prerequisite for the preparation of high-quality Cr_2O_3 coatings is a stable Cr_2O_3 suspension. Unstable suspensions may result in inhomogeneous layers due to the formation of large aggregates in the suspension. The electroacoustic sonic amplitude (ESA) technique,[54] from which the zeta potential of the particles can be determined, is a powerful method to monitor the electrostatic contribution to the stability of a concentrated suspension. This measurement is usually done as a function of pH. It should be noted that conclusions about the colloidal stability from ESA measurements can only be made if the suspension contains no components that give steric stabilization.

The absolute magnitude of the zeta potential of a Hacros Cr_2O_3 suspension is relatively low over a wide pH range, as shown in Figure 4.8 (open circles). Cr_2O_3 particles, unlike silica, cannot be stabilized by merely adjusting the pH to values well above or below the isoelectric point (iep) (around pH = 4.5). However, addition of a

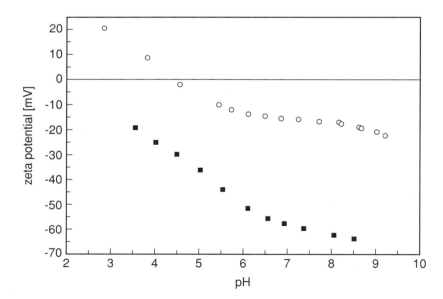

FIGURE 4.8   Zeta potential as a function of pH for a suspension of bare Hacros Cr_2O_3 particles (open symbols) and for the same particles with adsorbed Dispex (closed symbols).

small amount of anionic polyelectrolyte [0.05 wt % Dispex A40 (Allied Colloids)] to the suspension changes the situation dramatically as shown in Figure 4.8 (filled squares). The zeta potential is now large and negative over a wide pH range, and electrosterically stabilized suspensions are obtained.

The addition of polyelectrolyte also causes a complication: the presence of free (nonadsorbed) polyelectrolyte chains that compete with the Cr_2O_3 particles for adsorption on the surface may result in poor particle adsorption. The same effect was discussed earlier for silica (Fig. 4.5).

Rheological measurements and sedimentation tests can also be used to monitor suspension stability. In rheological measurements, Newtonian behavior is expected for stable suspensions whereas unstable suspensions will show a higher viscosity and shear thinning behavior. The latter can be ascribed to the break up of aggregates into smaller units with increasing shear rate. This results in a lower apparent particle volume fraction because the liquid in the pores between the particles of large aggregates also contributes to the apparent particle volume fraction. In a sedimentation test,[55] (sub)micron particles are allowed to settle by gravity in a measuring cylinder. A small, compact sediment volume will be obtained for stable suspensions because the individual particles can rearrange during settling. If the suspension contains a fraction of particles that are too small to settle by gravity, the supernatant will remain turbid. For unstable suspensions, a large, loose sediment volume will be found because no rearrangements of the individual particles are possible during settling of the aggregates. The supernatant will become clear because small particles will aggregate and subsequently settle. Results of rheological and sedimentation experiments on Cr_2O_3 suspensions are summarized in Table 4.1. Sedimentation tests were performed in 10-mL measuring cylinders. The packing density was calculated from the sediment volume, the weight of the Cr_2O_3 particles in the sediment (determined after removing the supernatant), and the density of the Cr_2O_3 particles (5.25 g/cm^3). The rheological measurements were carried out with a Contraves Low Shear LS 30 using a 1–1 geometry and shear rates up to 128 s^{-1}.

It is clear that suspensions are unstable below a critical Dispex concentration (around 0.1 wt %). Below this critical concentration, the amount of adsorbed Dispex is insufficient to obtain stabilization. A Cr_2O_3 suspension with 1.0 wt % Dispex is less

Table 4.1 Rheological and Sedimentation Results of Cr_2O_3 Suspension as Function of Amount of Dispex 40 Added

Dispex A40 (wt %)	Rheological Behavior	Packing Density (%)
0.0001	Shear thinning	6
0.001	Shear thinning	6
0.01	Shear thinning	6
0.1	Newtonian	42
1.0	Shear thinning	36

stable than a Cr_2O_3 suspension with 0.1 wt % Dispex. This effect can be ascribed to the excess of Dispex, which acts as an electrolyte and suppresses the thickness of the double layer that forms the barrier against aggregation.

4.9 FORMATION OF PARTICLE COATINGS: INTRODUCTION AND EXPERIMENTAL ASPECTS

Particle adsorption experiments were carried out on glass substrates precoated with Merquat 550, a commercial high-molecular-weight copolymer of dimethyldiallyl ammonium chloride and acrylamide obtained from Chemviron. The molecular weight of this polymer is about 5×10^6 g/mol with a composition ratio dimethyldiallyl-ammoniumchloride/acrylamide of 50/50. The precoating was carried out by dipping the glass substrates in a 1 wt % Merquat solution for 15 min followed by washing and drying. This results in an attractive electrostatic interaction between the positively charged precoated glass and the negatively charged Dispex-coated Cr_2O_3 particles. The excess suspension was removed after a certain adsorption time by gentle washing. The deposited particle coating was characterized after drying by a number of different methods. In addition to the more common characterization techniques such as scanning electron microscopy (SEM) and layer thickness and surface roughness measurements, UV/VIS (visible) photospectroscopy (Perkin-Elmer PU 8800 UV/VIS spectrophotometer) was used to characterize the thin particle layers in a fast, semiquantitative way. The absorbance at 500 nm, which contains a scattering as well as an absorption contribution, is roughly proportional to the amount of deposited particles. However, no information about the homogeneity of the surface coverage is obtained. The technique is suitable for studying precoat effects for a given suspension. However, if the suspension characteristics are changed, flocculation may severely affect the absorption, which will then no longer be proportional to the surface coverage. In this case SEM pictures should be taken. It should be noted that UV/VIS measurements are limited to relatively thin particle layers.

The surface coverage obtained for particle layers prepared by particle adsorption is affected by parameters such as suspension stability, adsorption time and electrolyte concentration. The influence of these parameters will now be considered for Cr_2O_3 suspensions.

4.10 SUSPENSION STABILITY

The effect of suspension stability on the adsorption of Cr_2O_3 particles onto Merquat pretreated glass substrates was studied for a 5 vol % Cr_2O_3 suspension with an adsorption time of 10 min. SEM measurements showed that at Dispex concentrations up to 0.01 wt %, the precoated glass surface was partly coated with large, compact aggregates of Cr_2O_3 particles. This result points to unstable suspensions, and agrees with the suspension characterization results (see Table 4.1). In the case of the stable Cr_2O_3 suspension with 0.1 wt % Dispex, the surface is covered mainly with single

Cr_2O_3 particles, but the surface coverage is still incomplete. The surface coverage is less homogeneous for a Cr_2O_3 suspension with 1.0 wt % Dispex than for a Cr_2O_3 suspension with 0.1 wt % Dispex. A relatively large amount of (small) aggregates and relatively large, poorly covered regions are found.

4.11 ADSORPTION TIME

The effect of adsorption time on surface coverage for a stable 0.1 wt % Dispex containing Cr_2O_3 suspension is shown by way of the absorbance at 500 nm as a function of adsorption time in Figure 4.9. These results indicate that the surface coverage increases with increasing adsorption time, which was confirmed by SEM. However, even after 1 h of adsorption, the surface coverage is still incomplete (uncoated glass regions remain visible). This is contrary to the results of stagnation point flow experiments on silica particles where the maximum surface coverage is obtained in a much shorter time. Apparently the particle transport to the surface is much faster in the case of stagnation point flow.

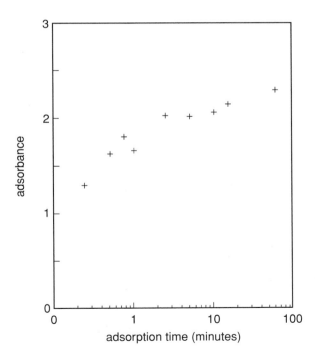

FIGURE 4.9   Absorbance at 500 nm as a function of time for a Cr_2O_3 layer prepared from a 0.1 wt % Dispex containing Hacros Cr_2O_3 suspension.

4.12   ELECTROLYTE CONCENTRATION

A fairly low surface coverage even after 1 h of adsorption can be ascribed to the fact that the negatively charged particles repel each other [see Eq. (4.2)]. This repulsive interaction prevents the adsorption of more particles on a surface that is already partly covered. The repulsion between the negatively charged particles can be suppressed by adding electrolyte to the suspension. Increasing the conductivity of a Cr_2O_3 suspension with 0.1 wt % Dispex from 0.8 to 3.4 mS/cm, by adding some KCl, resulted in an excellent surface coverage after an adsorption time of 10 min as shown in Figure 4.10. Greater electrolyte concentrations result in less homogeneous layers because of flocculation. A marginally stable suspension is needed to obtain an optimal surface coverage, as might be expected. This is quite a technological challenge since small changes in the conditions may cause the suspension to become unstable.

4.13   MULTILAYER PREPARATION

A particle layer prepared by particle adsorption is roughly limited to a layer thickness corresponding to one particle diameter. However, thicker layers can be prepared by

FIGURE 4.10   SEM micrograph showing a Cr_2O_3 layer at high surface coverage obtained by adding electrolyte to a 0.1 wt % Cr_2O_3 suspension.

repeating the precoat step and particle adsorption step a number of times, removing the excess particles and Merquat by careful rinsing each time. The layer thickness and surface roughness are given in Table 4.2 as a function of the number of adsorption steps for a Cr_2O_3 suspension with 0.1 wt % Dispex having a particle volume fraction of 0.05 and a conductivity of 2.5 mS/cm. The particle adsorption time was 10 min. Both layer thickness and surface roughness were measured with a Tencor Alpha Step using a 2-μm radius needle at a force level of 2 mg. The surface roughness values are arithmetic averages using a graphical centerline method. These results show that relatively thick layers can be prepared by repeating the adsorption of polymer and particles a number of times. The amount of deposited particles is roughly constant for each step, as it was for the case of silica particles discussed earlier. The surface roughness increases after each step.

4.14 PHOSPHOR COATINGS: INTRODUCTION AND EXPERIMENTAL ASPECTS

Under certain conditions the presence of polymer in a suspension can lead to unwanted interactions between particles and a surface. In this section some aspects of the role of polymers in the surface–particle interaction will be discussed using the example of phosphor particles/layers. Television screens and computer monitors require phosphor layers with thicknesses of about two particle diameters and a packing density of about 50%. These layers are subsequently exposed through a shadow mask and the nonexposed parts are developed ("washed away"). This should result in a surface with phosphor lines or dots at the exposed areas and empty regions at the nonexposed areas. It is important that all particles are removed from the nonexposed areas, otherwise haze or color contamination occurs. We have studied the effect of unwanted adhesion of phosphor particles to nonexposed areas after rinsing. Phosphor layers were prepared by settling 50 mL of a dilute suspension (phosphor volume fraction about 0.1 vol %) on both bare and polymer-precoated glass substrates. The substrates were placed horizontally in a Petri dish. After the suspension settled, phosphor layers of about two particle diameters thick were obtained. The diameter of

TABLE 4.2 Layer Thickness and Surface Roughness of Multilayer Cr_2O_3 Coatings as Function of Number of Particle Adsorption Steps

Number of Particle Adsorption Steps	Layer Thickness (μm)	Surface Roughness (nm)
1	0.57 ± 0.03	47 ± 8
2	1.02 ± 0.15	160 ± 45
3	1.6 ± 0.2	320 ± 100
4	2.4 ± 0.2	500 ± 110

the phosphor particles was about 5 μm. The wet phosphor layers were subsequently rinsed with excess water (to mimic development). The remaining amount of phosphor particles on both surfaces after drying was determined using SEM, layer thickness measurements, or absorbance measurements.

We consider two effects that polymers may have on haze formation. First, we consider the effect of precoating the surface with uncharged polymers of different molecular weights. Second, we consider the effect of adding a cationic polyelectrolyte to a suspension that is already stabilized by adsorbed anionic polyelectrolyte.

4.15 EFFECT OF MOLECULAR WEIGHT OF POLYMER PRECOAT ON HAZE

It was shown in the previous section that good surface coverages can be obtained with submicron Cr_2O_3 particles when there are attractive electrostatic interactions between the negatively charged particles and the positively charged glass surface. In this case the cationic copolymer Merquat acts as a bridge between the negatively charged glass and Cr_2O_3 particles.

Bridging is not limited to polyelectrolytes. For example, if a nonionic poly(vinyl pyrrolidone) (PVP)-coated glass substrate is used, a considerable amount of phosphor particles remain on the glass after settling and washing. A green Philips ZnS phosphor with a thin homogeneous silica coating (prepared by J. Opitz from our sister lab) was used for this experiment. PVP chains have a high affinity for glass[56] and it is reasonable to expect a similar affinity for silica. Thus strong polymer bridging is expected between the PVP-precoated glass and the silica-coated ZnS phosphor particles. The bridging efficiency increases with increasing molecular weight of the

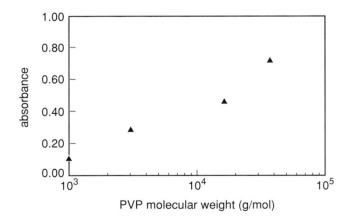

FIGURE 4.11 Absorbance at 500 nm as a function of the molecular weight of the PVP precoat for phosphor layers prepared from silica-coated ZnS phosphor particles.

PVP, as is demonstrated in Figure 4.11. This is in accord with the dependence of adhesion strength on molecular weight discussed previously.

4.16   EFFECT OF FREE POLYMER ON HAZE FORMATION

Haze formation due to polymer bridging can be solved by adding a small amount of a bridging polymer to the phosphor suspension. The deposition of a red Y_2O_2S phosphor (Nichia) will be considered as an example. The Y_2O_2S suspension is stabilized with a negatively charged polyelectrolyte and dispersed in 0.5 wt % poly(vinyl alcohol) (PVA) binder solution. The substrates used are a bare glass plate and a glass plate precoated with Merquat. Small amounts of Merquat are added to the suspension before it is allowed to settle on the substrates. After settling and washing, the absorbance at 500 nm is measured on both substrates. The absorbance values are presented in Figure 4.12. Very fast settling, which can be ascribed to bridging flocculation, occurs at Merquat concentrations between 2.5×10^{-4} and 1.5×10^{-3} wt %. Relatively stable suspensions are obtained at higher Merquat concentrations, most probably because enough Merquat is present to cover all the Y_2O_2S particles, resulting in a charge reversal from negative to positive. It can be seen in Figure 4.12 that no haze is found on the Merquat-precoated substrate at Merquat concentrations greater than about 5×10^{-4} wt %. This can be ascribed to the charge reversal that results in a repulsion between the positively charged particles and the positively charged surface. A repulsive interaction exists on the uncoated substrate if no Merquat is added to the suspension. However, small amounts of Merquat in the suspension result in an

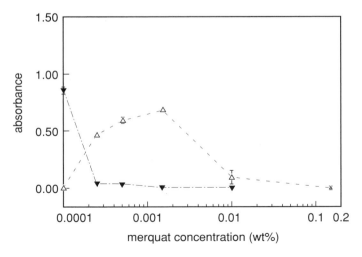

FIGURE 4.12   Absorbance at 500 nm as a function of the concentration of Merquat added to the phosphor suspension for bare glass substrates (open symbols) and Merquat pretreated glass substrates (closed symbols).

attractive interaction between the negatively charged glass and the particles covered with (some) Merquat. The haze is suppressed at high Merquat concentrations because nonadsorbed Merquat chains can compete with the Merquat-coated phosphor particles for adsorption sites on the glass. The same kind of phenomena have been observed by Varennes and van de Ven for both cationic polyelectrolytes and poly(ethyleneoxide).[23,39] It can be concluded from these results that if haze occurs due to polymer bridging, addition of some extra bridge-forming polymer to the suspension will suppress the haze formation. It should be noted that the haze level obtained for the case of negatively charged phosphor particles on Merquat-precoated glass substrates strongly depends on the way the Merquat is applied. For example, in the case of a precoat solution with a low Merquat concentration (0.01 wt %), the haze level strongly increases with the ionic strength of the solution (varied by adding KNO_3). The haze also increases with increasing Merquat concentration of the solution. Corak et al.[57] also found an increase of the particle adsorption with increasing polymer concentration for latex particles on poly(-dimethyl diallylammonium)-covered polyethylene films. Hubbe[36] also found that the adhesion of TiO_2 particles to glass or cellulose increased with increasing concentration of the cationic polymer used to precoat the substrate. Similar effects were found by Varennes and van de Ven[40] for latex particles covered with a cationic polyelectrolyte and deposited on glass.

4.17   THIN PARTICLE LAYERS

If particles stick to a surface upon arrival, then a definite but random configuration will result. In the RSA model the jamming limit is 54.6%. In a gravitational field this value increases to 61%.[17] If the particles sediment, but do not adhere to the surface, a higher packing density is achieved because the particles can rearrange on the surface after arrival. If monodisperse spheres are used, a two-dimensional packing density of 91% can in principle be achieved.

   If more than one particle layer is formed by sedimentation, which is an essential step in making phosphor layers on TV screens, then the spatial distribution of the first particle layer will affect the next layer. We performed sedimentation experiments with 6-μm glass beads from aqueous suspensions onto glass substrates and determined the packing density as a function of the deposited weight. Prior to the experiments, the slides were cleaned. One set of slides was used after a pretreatment with a cationic polymer (Merquat 550), while the other set was not pretreated. Microscopic observation, using an inverted microscope, showed that on the Merquat-pretreated glass hit-and-stick conditions were present. Particles falling on top of a previously deposited particle rolled off and found a position next to it. On the other slides, the particles could easily push each other aside leading to a high packing density with small ordered domains. The experiments were performed as a function of polymer molecular weight. The samples were dried with a small amount of PVA added as a binder to avoid drying artifacts. Then the layer thickness was determined, which together with the weight gives the packing density. The results are shown in Figure

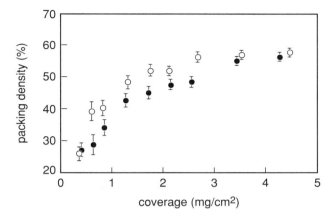

FIGURE 4.13 Packing density of spheres in thin layers with (closed symbols) and without (open symbols) surface pretreatment to fix the first layer.

4.13. The packing densities up to about four equivalent particle layers are slightly lower on pretreated glass slides than on the clean glass slides. A "surface effect" therefore plays a role only for layers thinner than four particle diameters, which is the case for phosphor layers and Cr_2O_3 layers, for example.

4.18   SUMMARY AND CONCLUSION

Polymer adsorption at solid–liquid interfaces can have a strong impact on particle layers created at macroscopic surfaces. If a macroscopic surface is pretreated with a polymer solution prior to exposure to particles, the adsorbed polymer may result in particles binding more strongly to the surface. This phenomenon is closely related to bridging flocculation. Therefore the study of particle adhesion onto a polymer-pretreated surface may provide useful insights in polymer bridging. If attraction between a particle and a surface exists, the surface coverage is mainly controlled by particle size and salt concentration.

If polymer is present in the suspension, however, a more delicate competition between polymer adsorption (on the particles and/or on the macroscopic surface) and particle adsorption may occur. This may even lead to complete absence of particle adsorption. With polymer pretreatments it is difficult to obtain multilayers by particle adsorption. However, by repeated sequential adsorption of polymer and particles, multilayers can be obtained.

Insights in polymer–particle–macroscopic surface interactions are valuable in the application of Cr_2O_3 layers where a stable suspension and fairly high electrolyte concentration are a prerequisite to arrive at dense and smooth particle layers or to prevent the undesired adsorption of phosphor grains on the inside of TV screens.

Often more than one layer of particles is needed. The morphology of the first layer of particles is affected by the interactions between particles and surface. As the morphology of a second particle layer depends on that of the first layer, an effect can be seen on the overall packing density. This is detectable up to four equivalent sedimented particle layers.

REFERENCES

1. D.H. Napper, *Polymeric Stabilization of Colloidal Dispersions*, Academic, London, 1983.

2. E.W.A. Young, A.R. de Wit, A.A.M. van Weert, E. van der Voort, S.M.R. Gelderland, L.G.J. Fokkink, M.H. Blees, and A.R. Balkenende, *Philips J. Res.* **50**, 421 (1996).

3. A.R. Balkenende, G.T. Jaarsma, W. Hoogsteen, H.P. Löbl, and M. Scholten, *Philips J. Res.* **50**, 407 (1996).

4. G. Onoda and P. Somasundaran, *J. Colloid Interface Sci.* **118**, 169 (1987).

5. Z. Adamczyk and P. Warszynski, *Adv. Colloid Interface Sci.* **63**, 41 (1996).

6. C.A. Johnson and A. Lenhoff, *J. Colloid Interface Sci.* **179**, 587 (1996).

7. R.K. Iler, *J. Colloid Interface Sci.* **21**, 569 (1966).

8. J.H. Fendler and F.C. Meldrum, *Adv. Mat.* **7**, 607 (1995).

9. A. Krozer, S.A. Nordin, and B. Kasemo, *J. Colloid Interface Sci.* **176**, 479 (1995).

10. L. Grimm, K.J. Hilke, and E. Scharrer, *J. Electrochem. Soc.: Solid State Sci. Tech.* **130**, 1768 (1983).

11. J.C. Dijt, M.A. Cohen Stuart, J.E. Hofman, and G.J. Fleer, *Colloids Surfaces* **51**, 141 (1990).

12. J.C. Dijt, M.A. Cohen Stuart, and G.J. Fleer, *Adv. Colloid Interface Sci.* **50**, 79 (1994).

13. P.C. Hiemenz, *Principles of Colloid and Surface Chemistry*, 2nd ed., Dekker, New York, 1986.

14. M. Hull and J.A. Kitchener, *Trans. Faraday Soc.* **65**, 3093 (1969).

15. M. Elimelech, J. Gregory, X. Jia, and R. Williams, *Particle Deposition & Aggregation*, Butterworth-Heinemann, Oxford, 1995.

16. P. Wojtaszczyk, P. Schaaf, B. Senger, M. Zembala, and J.C. Voegel, *J. Chem. Phys.* **99**, 7198 (1993).

17. P. Wojtaszczyk, E.K. Mann, B. Senger, J.C. Voegel, and P. Schaaf, *J. Chem. Phys.* **103**, 8285 (1995).

18. M.R. Böhmer, E.A. van der Zeeuw, and G.J.M. Koper, *J. Colloid Interface Sci.* **197**, 242 (1998).

19. E.K. Mann, E.A. van der Zeeuw, G.J.M. Koper, P. Schaaf, and D. Bedeaux, *J. Phys. Chem.* **99**, 790 (1995).

20. M. Semmler, E.K. Mann, J. Ricka, and M. Borkovec, (submitted) *Langmuir* **14**(18), 5127 (1998).

21. T. Dabros and T.G.M. van de Ven, *Colloid Polym. Sci.* **261**, 261 (1983).

22. Z. Adamczyk, B. Siwek, M. Zembala, and P. Warszynski, *J. Colloid Interface Sci.* **130**, 578 (1989).

23. S. Varennes and T.G.M. van de Ven, *PCH PhysicoChemical Hydrodynamics* **10**, 229 (1988).

24. N. Merston and B. Vincent, *Langmuir* **13**, 14 (1997).

25. M. Polvevari and T.G.M. van de Ven, *Langmuir* **11**, 1870 (1995).

26. J.M. Meinders and H.J. Busscher, *Colloid Polym. Sci.* **272**, 478 (1994).

27. N.D. Denkov, O.D. Velev, P.A. Krachevski, I.B. Ivanov, H. Yoshimura, and K. Nagayama, *Langmuir* **8**, 3183 (1992).

28. P.A. Krachevski and K. Nagayama, *Langmuir* **10**, 23 (1994).

29. G.J. Fleer, M.A. Cohen Stuart, J.M.H.M. Scheutjens, T. Cosgrove, and B. Vincent, *Polymers at Interfaces*, Chapman and Hall, London, 1993.

30. M.R. Böhmer, W.H.A. Heesterbeek, A. Deratani, and E. Renard, *Colloids Surfaces A: Physicochem. Engr. Asp.* **53**, 99 (1995).

31. T.G.M. van de Ven and S.J. Kelemen, *J. Colloid Interface Sci.* **181**, 118 (1996).

32. M.Y. Boluk and T.G.M. van de Ven, *Colloids Surfaces* **46**, 157 (1990).

33. M.R. Böhmer, *J. Colloid Interface Sci.* **197**, 251 (1998).

34. N.G. Hoogeveen, M.A. Cohen Stuart, G.J. Fleer, and M.R. Böhmer, *Langmuir* **12**, 3675 (1996).

35. R.H. Pelton and L.H. Allen, *J. Colloid Interface Sci.* **99**, 387 (1984).

36. M.A. Hubbe, *Colloids Surfaces* **25**, 325 (1987).

37. M.L. Janex, V. Chaplain, J.L. Counord, and R. Audebert, *Colloid Polym. Sci.* **275**, 352 (1997).

38. S. Varennes and T.G.M van de Ven, *PCH PhysicoChemical Hydrodynamics* **9**, 537 (1987).

39. S. Varennes and T.G.M. van de Ven, *PCH PhysicoChemical Hydrodynamics* **10**, 415 (1988).

40. S. Varennes and T.G.M. van de Ven, *Colloids Surfaces* **33**, 63 (1988).

41. J. Sjollema, H.J. Busscher, and A.H. Weerkamp, *J. Microbiol. Methods* **9**, 73 (1989).

42. K. Mühle, *Colloid Polym. Sci.* **263**, 660 (1985).

43. J. Visser, *J. Colloid Interface Sci.* **34**, 26 (1970).

44. M.A. Hubbe, *Colloids Surfaces* **25**, 311 (1987).

45. M.A.C. Stuart and J.W. Mulder, *Colloids Surfaces* **15**, 49 (1985).

46. Y. Lvov, F. Essler, and G. Decher, *J. Phys. Chem.* **97**, 13773 (1993).

47. J.D. Hong, K. Lowack, and G. Decher, *Progr. Colloid Polym. Sci.* **93**, 98 (1993).

48. M. Ferreira, J.H. Cheung, and M.F. Rubner, *Thin Solid Films* **244**, 806 (1994).

49. V. Priman, H.L. Frisch, N.R. Ryde, and E. Matijevic, *J. Chem. Soc. Faraday Trans.* **87**, 1371 (1991).

50. N. Ryde, N. Kallay, and E. Matijevic, *J. Chem. Soc. Faraday Trans.* **87**, 1377 (1991).

51. Y.M. Lvov and G. Decher, *Crystallography Rep.* **39**, 628 (1994).

52. Y. Lvov, K. Ariga, I. Ichinose, and T. Kunitake, *Langmuir* **12**, 3038 (1996).

53. Y. Lvov, K. Ariga, I. Ichinose, and T. Kunitake, *J. Am. Chem. Soc.* **117**, 6117 (1995).

54. R.W. O'Brien, D.W. Cannon, and W.N. Rowlands, *J. Colloid Interface Sci.* **173**, 406 (1995).

55. J. Lyklema, *Fundamentals of Interface and Colloid Science*, Part I, Academic, London, 1991.
56. M.A. Cohen Stuart, G.J. Fleer, and B.H. Bijsterbosch, *J. Colloid Interface Sci.* **90**, 310 (1982).
57. M. Corak, M. van de Wiele, R. Pelton, and A. Hrymak, *Colloids Surfaces* **90**, 203 (1994).

PART II
Fundamentals of Colloid–Polymer Interaction

5 Diffusion-Controlled Phenomena in Adsorbed Polymer Dynamics

MARIA M. SANTORE

Department of Chemical Engineering, Lehigh University, Bethlehem, Pennsylvania 18015

5.1 INTRODUCTION

Kinetic behavior represents the new frontier in polymer adsorption and is motivated by the need to better control processing behavior and performance of complex fluids. Until the last 10 years or so, interest had focused primarily on structural aspects of adsorbed polymer layers. With the advent of experimental methods capable of temporal resolution, the study of polymer adsorption, desorption, and exchange was made possible.

Polymer adsorption consists of three primary steps: diffusion of chains from solution to an interfacial region, adhesion to the surface, and subsequent interfacial relaxation and interfacial rearrangement. If the adsorbing sample is polydisperse or contains different species, the process becomes more complicated. Multiple species arriving at an interface may all, to some extent, adsorb. It is rare, however, that the equilibrium surface composition will be identical to the bulk solution. As a result of the surface selectivity, exchange between the surface and the solution becomes an additional step in adsorption.[1-5]

Experiments have proven that polymer adsorption is generally rapid. In contrast, adsorbed polymer layers resist washing from a surface in flowing solvent. Desorption is, therefore, slow as a result of the large energy needed to remove a chain from a surface.[6] With about 100 segment-surface contacts for a high-molecular-weight chain, with each contact contributing on the order of a kT, desorption energies can approach 100 kT even though a clean surface is the state preferred at equilibrium. Despite the apparent irreversible nature of polymer adsorption, radiotracer studies indicate that adsorbed polymer layers are indeed dynamic: There is a continuous exchange of

Colloid–Polymer Interactions: From Fundamentals to Practice, Edited by Raymond S. Farinato and Paul L. Dubin
ISBN 0-471-24316-7 © 1999 John Wiley & Sons, Inc.

chains between an adsorbed layer and the polymer solution that it contacts.[7] Such exchange occurs much more rapidly than desorption because less energy is involved. There is no net energy gain for self-exchange and if it occurs one or two segments at a time, the energy cost is just a few kT.

Although initial work has identified the approximate timescales[7-9] of macromolecular processes at interfaces, ultimate quantification of interfacial polymer kinetics must be done in carefully controlled conditions. Studies employing colloids (e.g., a serum replacement cell), or a stagnant adsorption cell, are limited in their ability to probe interfacial dynamics. The former has a poorly defined flow field near the surface of the particles, as this varies considerably over the flow cell. In a quiescent cell, interfacial concentration gradients continually evolve with time, making all but the slowest interfacial processes difficult to measure, and confounding the determination of intermediate rate behavior. Because the mass transfer of chains from solution to the surface (or vice versa) is a key mechanism in adsorption, efforts must be taken in its quantification.

Close scrutinization of the contribution of diffusion to polymer adsorption kinetics is relatively new. While recent works identify mass-transport-limited behavior for polymer adsorption,[10-12] limited efforts address mass-transport issues in desorption[13] and competitive adsorption situations.[3-5] These works have been ground-breaking in that they have dispelled the notion that adsorbed polymer layers are always nonequilibrium entities whose kinetics are difficult to anticipate. A significant number of systems can be described by diffusion-controlled models.

The purpose of this chapter is, therefore, to provide a background on the development of mass-transport models of polymer adsorption. Interfacial thermodynamics and a crude treatment of intrinsic interfacial kinetics are included. A number of experimental cases are then examined to illustrate the strengths of these models in predicting adsorption, desorption, exchange, and competitive adsorption behaviors.

5.2 DEVELOPMENT OF MODEL

5.2.1 Step 1: Diffusion

The first step of adsorption is diffusion of the adsorbing species from solution to the interface. The diffusive flux J toward the surface can be described in a general treatment, accounting for multiple species i, according to

$$J_i = M_i (C_i - C_i^*) \tag{5.1}$$

where C_i is the bulk solution concentration of species i, C_i^* is the concentration of species i in the fluid nearest the surface, and M_i is the mass-transfer coefficient for i. While C_i is usually held constant during a portion of an experiment, C_i^* typically varies (and is unknown) as adsorption proceeds; M_i is constant and depends on the flow geometry.

Equation (5.1) is completely general: For adsorption, J_i is positive with the bulk concentration C_i exceeding that near the surface, C_i^*. In the case of desorption, pure flowing solvent maintains $C_i = 0$, but a finite concentration persists for C_i^*. In this case the flux is negative indicating diffusion away from the surface. When C_i^* corresponds exactly to the concentration on the x axis of an isotherm (see Fig. 5.1) for a particular adsorbed amount, Γ_i, then *local equilibrium* is achieved. Note that in general the bulk solution concentration C_i might not be equilibrated with the adsorbed amount on the y axis except at long times. Then $C_i = C_i^*$, giving no net flux at equilibrium.

A special case of Eq. (5.1) gives rise to what is typically called *mass-transfer-limited adsorption*. When the adsorption and relaxation steps are very rapid relative to the diffusion of an adsorbing species from solution to the region nearest the surface, and this condition persists for some time, then all species reaching the interface are adsorbed as rapidly as they arrive. As a result, the local concentration of adsorbers in the fluid near the surface is vanishingly small, that is, $C_i^* = 0$. This gives

$$J_i = M_i C_i \qquad (5.2)$$

This expression for mass-transport-limited adsorption yields the maximum rate that can be achieved for a particular concentration of adsorbers, under the particular flow rate and geometry. Each molecule adsorbs as fast as it can diffuse to the surface. While Eq. (5.2) describes mass-transport-limited adsorption, it should be noted that application of Eq. (5.1) for conditions of local equilibrium (requiring reversible adsorption) is similar in principle since both cases involve fast surface processes relative to diffusion. The difference between these two scenarios will become clear below.

Before delving further into the development of the kinetic treatment, a discussion of the origin of the mass-transfer coefficients is necessary. The starting point for the calculation of a mass-transfer coefficient is a mass balance on an element of fluid in the flow cell of interest. For flow in a slit shearing cell (see Fig. 5.2) with wall shear rate γ, the appropriate form of the convection–diffusion equation is

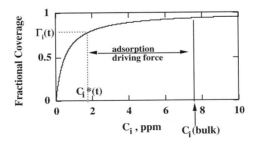

FIGURE 5.1   Illustration of $C_i^*(t)$ at local equilibrium for a Langmuir isotherm with $K_i = 2 \text{ ppm}^{-1}$.

FIGURE 5.2   Adsorption in a slit-shear cell.

$$\frac{\partial C_i}{\partial t} + \gamma y \left(1 - \frac{y}{b} \right) \frac{\partial C_i}{\partial x} = D_i \frac{\partial^2 C_i}{\partial y^2} \tag{5.3}$$

Here y is the direction normal to the surface with $y = 0$ at the plane of adsorption; b is the thickness of the flow chamber; D_i is the bulk solution diffusion coefficient of species i, assumed to be concentration independent (a reasonable assumption when solutions are dilute); x is the direction of flow, with $x = 0$ at the entrance to the cell. Equation (5.3) presumes fully developed laminar flow. This partial differential equation cannot be solved analytically; however, several numerical treatments have been presented.[14–17] The numerical solution applied to the following boundary and initial conditions is especially enlightening: Initially the cell contains only flowing solvent so $C_i = 0$ everywhere. At time zero, a valve is switched and a solution of absorbing species flows through the cell with concentration C_{i0} at the centerline. Diffusion and convection proceed via Eq. (5.3). Diffusers reaching the surface adsorb as they arrive, maintaining $C_i^* = 0$ at $y = 0$.

Figure 5.3 illustrates the development of the concentration profile for a single species of dimensionless concentration C_i/C_{i0}, which is unity at the centerline of the cell. Dimensionless distance x along the surface gives $x/L_\infty = 1$ at the point of

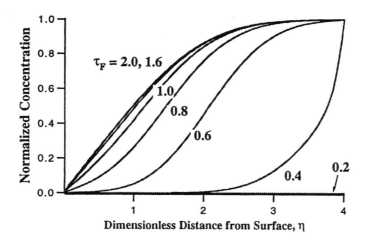

FIGURE 5.3   Evolving concentration profile near adsorbing surface of infinite capacity. Pe = 1290.

observation, L_∞. It is also convenient to represent the dimensionless distance normal to the surface,

$$\eta_i \equiv Pe_i^{1/3} y/b \qquad (5.4)$$

and dimensionless time, τ_F indicating F for flow,

$$\tau_{iF} \equiv \frac{t}{L_\infty Pe_i^{1/3}/(\gamma b)} \qquad (5.5)$$

with Pe, the Peclet number, given by

$$Pe_i \equiv \frac{\gamma b^3}{D_i L_\infty} \qquad (5.6)$$

In Figure 5.3, once the cell's contents switch from pure solvent to a solution of the adsorbing species, the concentration increases toward the surface as diffusion proceeds. Ultimately, however, a pseudo-steady-state concentration gradient near the surface is established and remains for the rest of time because the concentration in the fluid near the surface is always zero. This pseudo-steady-state concentration profile is key to mass-transport-limited adsorption because diffusion of adsorbers down this gradient gives the constant (maximum) adsorption rate, in Eq. (5.2).

According to Figure 5.3, there is time needed for the pseudo-steady-state concentration gradient to be established, at $\tau_F \sim 1$. Prior to this time, the local gradient near the surface is less than the ultimate value. This means that the adsorption rate increases with time and reaches a constant (maximum) value when the pseudo-steady-state condition is attained. This adsorption rate will then persist as long as the surface has sufficient capacity to maintain fast adsorption, in Eq. (5.2). It is important to understand, therefore, that Eq. (5.2) is the mass-transport-limiting rate and requires, in addition to fast adsorption, a surface of relatively large capacity. In reality, adsorption will not persist indefinitely per Eq. (5.2). Rather, equilibrium will limit coverage per Eq. (5.1).

Leveque solved the diffusion convection equation in Eq. (5.3), dropping the time-dependent term to give the pseudo-steady-state solution for interfacial flux, or the ultimate adsorption rate:

$$J_i = \frac{1}{\Gamma\left(\frac{4}{3}\right) 9^{1/3}} \left(\frac{\gamma}{D_i L_\infty}\right)^{1/3} D_i C_i \qquad (5.7)$$

A comparison of Eqs. (5.2) and (5.7) reveals that the mass-transfer coefficient for a slit-shear cell is, then,

$$M_i = \frac{1}{\Gamma\left(\frac{4}{3}\right) 9^{1/3}} \left(\frac{\gamma}{D_i L_\infty}\right)^{1/3} D_i \tag{5.8}$$

In Eq. (5.8) only, Γ is the gamma function. (Elsewhere Γ is adsorbed amount.) A similar development using a diffusion–convection equation in polar coordinates gives a mass-transfer coefficient for an impinging jet geometry.[18] A variety of mass-transfer coefficients may be used in Eq. (5.1), depending on the flow geometry of interest. In our adsorption studies, the surface is of sufficient capacity that the time of adsorption or desorption is much greater than the time needed to established the interfacial concentration gradient. Therefore, the use of Eqs. (5.1) and (5.2) are appropriate, and solution of the full diffusion convection Eq. (5.3) is not necessary.

5.2.2   Surface Processes Using a Langmuir Model

As discussed above, Eq. (5.2) describes the maximum adsorption rate for a particular species adsorbing in a particular flow field. The drawback of Eq. (5.2) is that it predicts adsorption to continue indefinitely, that is, the surface will never saturate. Treatment of surface saturation requires some information about the surface thermodynamics.

The Langmuir model is one of the simplest treatments possible, and despite the fact that no interfacial polymer physics are included, its use provides a useful perspective for polymer adsorption, desorption, and exchange kinetics. The Langmuir model can be written for one component, or generalized for multiple species, i. It presumes a limited number of surface sites $C_{\theta T}$ (T is for total), with those that are empty and available for adsorption represented as θ, and those occupied by species i, θ_i. Adsorption occurs reversibly:

$$i + \theta \underset{k_i'}{\overset{k_i}{\rightleftharpoons}} \theta_i \tag{5.9}$$

with equilibrium constant K_i given by

$$K_i = \frac{C_{\theta_i}}{C_i C_\theta} \tag{5.10}$$

Inversion of Eq. (5.10) yields the familiar isotherm in Figure 5.1:

$$C_{\theta_i} = \frac{K_i C_i}{1 + K_i C_i} \tag{5.11}$$

In Eqs. (5.10) and (5.11), C_i represents the free solution concentration of species i, as though no gradients exist in solution. If only local equilibrium occurs, then C_i^* should be used in Eqs. (5.10) and (5.11), and use of C_i reserved for equilibration at long times when $C_i = C_i^*$.

When multiple species i adsorb, the conservation of surface sites, $C_{\theta T}$, applies:

$$C_{\theta T} = C_\theta + \sum_i C_{\theta_i} \tag{5.12}$$

Equation (5.9) suggests the following kinetics for the net adsorption rate of i:

$$R_i = k_i C_i^* C_\theta - k_i' C_{\theta_i} \tag{5.13}$$

Here k_i and k_i' are the fundamental forward and reverse rate constants. In Eq. (5.13), C_i^* reminds us that the fundamental surface kinetics are influenced only by the concentration in the fluid nearest the surface. At equilibrium, the forward and reverse rates in Eq. (5.13) cancel, and there are no gradients in free solution. As a result Eq. (5.10) is recovered, with $K_i = k_i/k_i'$.

Combination of Eqs. (5.1) and (5.13) to eliminate C_i^* yields

$$R_i = J_i = \frac{M_i (C_i C_\theta - C_{\theta_i}/K_i)}{M_i/k_i + C_\theta} \tag{5.14}$$

Equation (5.14) can be solved analytically for one specie. If multiple species adsorb, Eq. (5.14) represents a set of coupled differential equations. In either case, the solution requires conservation of surface sites in Eq. (5.12). Depending on the initial conditions and C_i, adsorption, desorption, or exchange can be treated. Equation (5.14) contains three main parameters: the mass-transfer coefficient, the equilibrium constant, and the ratio of the mass-transfer coefficient to the intrinsic rate constant. The latter vanishes for diffusion-controlled adsorption (with local equilibrium) when the intrinsic surface kinetics are relatively rapid compared with the rate of chain arrival at the interface.

5.2.3 Local Equilibrium at the Surface from Self-Consistent Mean Field Theory

The Langmuir treatment of adsorption kinetics and equilibrium in the previous section does not address adsorption features that can directly be attributed to the polymeric nature of the adsorbate. To go to such a level of detail may not be necessary; however, it may be useful to examine expectations for polymer adsorption, where the physics of chain connectivity and multiple segment-surface contacts have been taken into account.

The self-consistent mean field theory (SCMFT) of Scheutjens and Fleer[19-22] captures much polymer detail and makes predictions for equilibrium isotherms that can be incorporated into kinetic treatments. Adsorption is often predicted to be extremely high affinity, with saturated layers in equilibrium with extremely dilute polymer solutions on the order of a fraction of a part per million. The predicted extension of the isotherm plateau down to the very dilute region is approximated by the following isotherm shape[13]:

$$C = C_r 10^{(\Gamma - \Gamma_r)/p}$$

$$(5.15)$$

Here, Γ_r and C_r refer to a reference adsorbed amount and reference solution concentration and p is a parameter, the slope of the isotherm on a semilog plot. During the modeling of adsorption, it is useful to set C_r to the bulk solution concentration and Γ_r to the equilibrium adsorbed amount. For desorption kinetics, it is useful to choose C_r to correspond to the bulk solution concentration in equilibrium with the adsorbed layer, Γ_r, at time zero. (The bulk solution concentration is then changed to zero *during* desorption.)

For a kinetic treatment in which local equilibrium is maintained at all times between the adsorbed layer and the nearest elements of polymer solution, the isotherm in Eq. (5.15) is solved for C, which is then used as C^* in Eq. (5.1):

$$\frac{d\Gamma}{dt} = M_i(C_i - C_r 10^{(\Gamma - \Gamma_r)/p})$$

$$(5.16)$$

For desorption, $C_i = 0$ in Eq. (5.16), which represents a kinetic treatment including a "polymeric" perspective on the local equilibrium, which is maintained at all times.[13]

5.3   EXPERIMENTAL

All experiments were conducted in a slit-shear cell per the design of Shibata and Lenhoff.[15] Adsorption, desorption, and exchange studies employed poly-(ethyleneoxide) (PEO) molecular weight standards from Polymer Labs, some of which were fluorescently tagged to facilitate studies using total internal reflectance fluorescence (TIRF).[23,24] TIRF only sees the adsorbed amount of fluorescently labeled species (see Chap. 14). Other measurements of the total evolving interfacial mass were made with near-Brewster optical reflectivity.[10] (see also Chap. 13). The substrate is silica,[10,25] obtained by controlled exposure of a microscope slide to sulfuric acid. All the experimental methods have been described in detail elsewhere.

5.4   RESULTS

5.4.1   Equilibrium Isotherms

Figure 5.4 compares Langmuir and SCMFT isotherms with data for PEO (33,000 molecular weight) on silica in deionized (DI) water, measured by optical reflectivity.[10] In Figure 5.4a, the free solution concentration is on a log axis to highlight the extension of the plateau down to very dilute concentrations. The Langmuir K_i value must be high to give a reasonable fit to this data, and several K_i values are included to illustrate its influence. Also in Figure 5.4a, the fractional surface coverage in terms of C_θ only loosely translates to the measured adsorbed amount. It is more important to choose K_i such that all the data for the plateau coincide with the plateau region of a Langmuir isotherm. The Langmuir model also shows a dilute region where the coverage drops toward zero: This behavior is not observed experimentally. Therefore, the comparison of theory to data defines a minimum Langmuir K_i value that should be used in the kinetic treatment of Eq. (5.14). Though the Langmuir model is minimally appropriate for features of the experimental polymer isotherm, its utility for kinetic predictions will become apparent.

Figure 5.4b compares the same PEO adsorption data and the SCMFT; $\Gamma_r = 1$ is chosen at $C_r = 100$ ppm, and several values of p are included to illustrate its effect. For

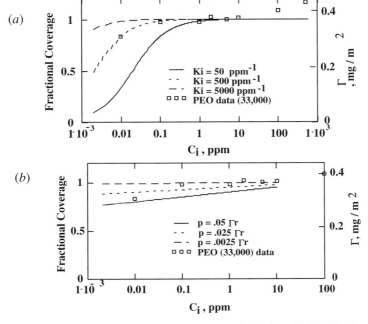

FIGURE 5.4 (*a*) Langmuir adsorption isotherms compared with data. (*b*) SCMFT adsorption isotherms compared with data.

PEO adsorbing on silica, a value of $p = 0.025\Gamma_r$ is in reasonable agreement with the data. The SCMFT captures the essential experimentally observed features of PEO adsorption: a plateau that slopes very gradually down at low coverages.

5.4.2   Adsorption Kinetics

Figure 5.5 illustrates the adsorption kinetics predicted by Eqs. (5.14) and (5.16) and makes a comparison to experiment in Figure 5.6 for PEO adsorption on silica. Adsorption is modeled via the Langmuir approach by integrating Eq. (5.14) analytically for one species, with an initial condition of zero coverage. Figure 5.5a shows fractional coverage as a function of dimensionless time, $\tau = tC_iM_i/C_{\theta T}$ at fixed K_i value of $50/C_i$. The other free parameter in Eq. (5.14) is M_i/k_i. Its dimensionless equivalent, $M_i/(k_iC_{\theta T})$, is varied in Figure 5.5a. At large values adsorption is slow; however, when this parameter becomes small, transport-limited adsorption is achieved. For the mass-transfer-limited adsorption, coverage rises nearly linearly until the equilibrium is approached, near $\tau = 1$.

In Figure 5.5b, the diffusion-controlled limit of the Langmuir model is further explored by fixing $M_i/(k_iC_{\theta T}) = 0$, and varying K_i. Small K_i values yield rounder adsorption traces toward saturation. It is incorrect to mistake this rounded shape as the onset of surface-dominated kinetics. Though the approach to the final equilibrium is more gradual for weak adsorption, the calculations in Figure 5.5b have maintained $M_i/(k_iC_{\theta T}) = 0$, and therefore local equilibrium between the adsorbed layer and the nearby polymer solution has been maintained. When this condition holds, saturation is approached near $\tau = 1$. Figure 5.5b highlights the difference between the *mass transport-limited* maximum adsorption rate achieved for surfaces of high affinity and capacity and the more general case of *local equilibrium*. In both cases, adsorption is fast and diffusion through solution dominates the kinetics.

Figure 5.5c illustrates the adsorption kinetics from Eq. (5.16) for SCMFT. The bulk solution concentration is C_r, such that the ultimate long-time surface coverage will be Γ_r. In Figure 5.5c, adsorption traces for different values of the parameter p are illustrated. Like the transport-limited kinetics for Figures 5.5a and 5.5b, in 5.5c there is a nearly linear rise in coverage that extends almost to saturation. Larger values of p tend to give a more gradual approach to saturation, though local equilibrium is maintained at all times.

Adsorption kinetics for PEO on silica are shown in Figure 5.6 for $C_i = 5$ ppm and varied flow through a slit cell.[10] The mass-transfer coefficients have been calculated and the time axis represented in dimensionless units, $\tau = M_iC_it/C_{\theta T}$. The superposition of three runs at different wall shear rates confirms transport-limited adsorption. The nearly linear rise of the adsorption trace to its final coverage indicates that a high-affinity Langmuir or SCMFT is an adequate representation of the adsorption equilibrium. A comparison of Figures 5.5b and 5.6 demonstrates that even though the Langmuir treatment in Figure 5.4a is a marginal representation of the equilibrium, as long as a large enough K_i value is chosen (high-affinity adsorption), transport-limited adsorption kinetics can be modeled accurately with the Langmuir treatment.

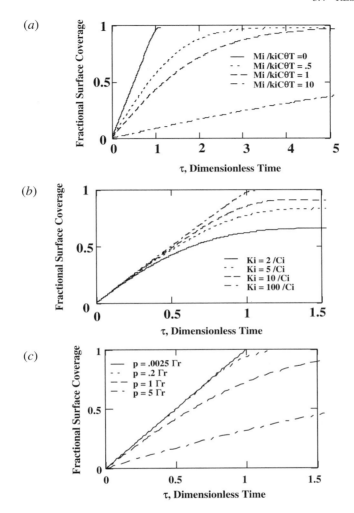

FIGURE 5.5 (*a*) Langmuir model predictions for adsorption with varying degrees of surface rate influence with $K_i = 50/C_i$. (*b*) Langmuir model predictions for diffusion-controlled adsorption with varying adsorption strengths. (*c*) SCMFT model predictions for diffusion-controlled adsorption with varying adsorption strengths.

5.4.3 Desorption Kinetics

Diffusion-controlled desorption kinetics anticipated by the Langmuir and SCMF theories are shown in Figure 5.7. In Figure 5.7a, Eq. (5.14) was solved with an initial condition of equilibrium surface coverage for a solution of concentration C_i (which is part of the dimensionless time $\tau = M_i C_i t / C_{\theta T}$.) The diffusion-limited decay rates (maintaining local equilibrium) were calculated by analytical integration of Eq.

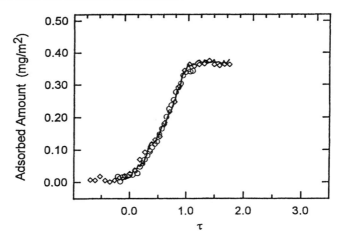

FIGURE 5.6   PEO (33,000 molecular weight) adsorption for a bulk solution concentration C_i of 5 ppm. Experiments at different wall shear rates: 22.6 s^{-1} (○), 4.52 s^{-1} (—), and 1.13 s^{-1} (◇). Dimensionless time $\tau = C_i M_i t / C_{\theta T}$.

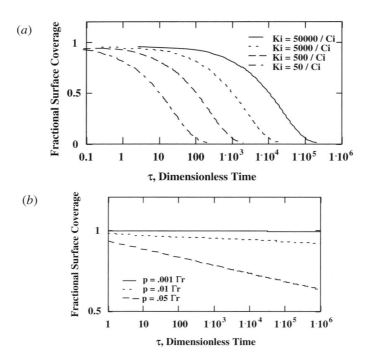

FIGURE 5.7   (a) Langmuir treatment of diffusion-controlled desorption with variations in K_i. (b) SCMFT treatment of diffusion-controlled desorption with variations in p.

(5.14), with the bulk solution concentration set to zero, the rate parameter $M_i/k_iC_{\theta T} = 0$, and different values of K_i. Because the concentration prior to desorption, C_i, is included in the dimensionless time, different values for K_i give slightly different initial surface coverages. In Figure 5.5b, the adsorption time was always near $\tau = 1$, regardless of K_i. In contrast, K_i has a significant influence on the desorption time, in Figure 5.7a, even when $M_i/k_iC_{\theta T} = 0$. The Langmuir kinetic treatment predicts slower desorption when the isotherm is higher affinity. In all cases, the desorption time scale is much slower than the adsorption time scale for the same physical parameters.

The desorption kinetics for the SCMFT show a trend similar to the Langmuir-based calculations, though the shape of the desorption curves in Figure 5.7b are linear on a log τ axis. Like the Langmuir approach, the SCMFT predicts desorption times that are much greater than the initial adsorption times, and which decrease for stronger adsorption. For both models, the shapes of the desorption kinetics show a strong similarity with the isotherm shapes, though the two representations are inverted left-to-right. During desorption, one essentially traverses backward down the isotherm.[13]

We have not conducted extensive studies on polymer desorption kinetics because the desorption has always been observed to be longer than the baseline stability of our instrumentation. For PEO desorption measured by reflectivity,[10] we observe that adsorbed layers are stable in flowing solvents for at least 10 h, following an initial drop in coverage that is fairly rapid and accounts for about 5% of the signal. We attribute this initial drop to loosely bound chains entangled with those adsorbed onto the silica substrate. Dijt et al.[13] found the behavior in Figure 5.7b in good agreement with their experiments.

5.4.4 Adsorption of Multiple Species

One of the strengths of the Langmuir model is that it is readily extended to competitive adsorption kinetics. The SCMFT can also be extended to treat multiple species[2]; however, it becomes much more involved because of the number of parameters and computational effort involved in calculating competitive equilibria.[3,4] Equation (5.14) already contains everything needed to predict competitive kinetics via the Langmuir model.

The simplest experiment requiring a multiple-species approach is self-exchange. Here an adsorbed layer is equilibrated and exposed to flowing solvent, during which time there is minimal desorption. The layer is then exposed to a solution of the labeled version of the original molecules and exchange takes place. At equilibrium, the surface will contain an adsorbed layer of only labeled molecules because the free solution is continually replenished and contains only the labeled species. At intermediate times, however, the interface will contain labeled and unlabeled molecules. The main premise of the self-exchange experiment is that labeling does not influence the affinity of the polymer for the surface. This limit has been approached experimentally for some systems.[26]

Figure 5.8 illustrates the Langmuir predictions for self-exchange. Here Eq. (5.14) has been numerically integrated for two species. Initially, the surface contains an adsorbed amount of species 1, which has previously equilibrated with a bulk solution of C_1. At time zero, the surface is exposed to a bulk solution of 2 at the same concentration, C_2. In Figure 5.8, $M_i/k_iC_{\theta T} = 0$ for $i = 1, 2$ such that local equilibrium is maintained between the surface and the nearby polymer solution. Dimensionless time employs the C_2 during the exchange. In Figure 5.8, $K_1 = K_2$ and $M_1 = M_2$, to simulate the case where a label distinguishing the incoming and desorbing polymer does not influence behavior.

In Figure 5.8, several families of curves track the surface populations during self-exchange, and $K_1 = K_2$ is varied to show the influence of adsorption affinity. Variations in surface affinity cause the initial coverage of species 1 to be slightly different for each case. In all cases, though, the ultimate coverage achieved by species 2 is equal to the initial coverage of species 1, and exchange takes place at constant total surface mass. Most of the self-exchange occurs, independent of affinity, by $\tau = 2$, which is slower than the initial adsorption of species 1, also shown for one case.

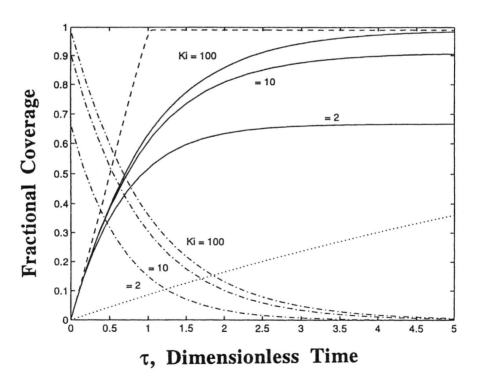

τ, Dimensionless Time

FIGURE 5.8 Influence of adsorption strength on self-exchange kinetics (at diffusion-controlled local equilibrium) following preadsorption of species 1. During self-exchange, species 2 (—) replaces species 1 (-·-·-). Initial transport-limited rate of species 1 (- - -) shown for comparison. Also, for comparison is self-exchange with significant surface influence: $K_i = 100$ and $M_i/k_iC_{\theta T} = 0.1$ ($\cdots\cdots$).

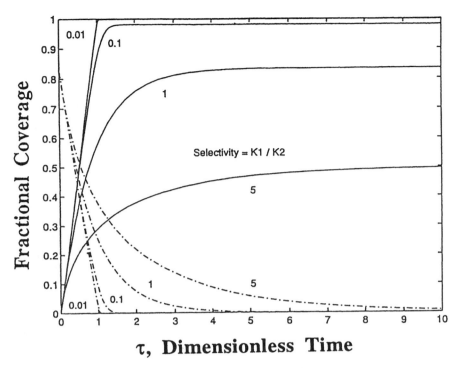

τ, Dimensionless Time

FIGURE 5.9 Diffusion-controlled replacement of preadsorbed species 1 (·····, $K_1 = 5$) by species 2 of the same bulk solution concentration (——, $K_2 = 500, 50, 5, 1$).

Self-exchange processes may also proceed more slowly than $\tau = 2$, as illustrated for $K_i = 100$, and a value of $M_i/k_iC_{\theta T} = 0.1$, which includes the influence of surface kinetics.

Figure 5.9 illustrates a competitive adsorption "challenge" experiment, which is similar to self-exchange, except that the two species now have different K_i's, but the same M_i's. In Figure 5.9, the species initially adsorbed has a moderate surface affinity, with $K_1 = 5$. The various curves show the evolution of surface composition when this initial layer is challenged by a second species, whose affinity for the surface is varied from $K_2 = 1$ to 500. When the initial species has the greater surface affinity, the exchange proceeds slowly. Even though equilibrium ultimately favors only species 2 on the surface (because the bulk solution contains only species 2), locally both species are present and the surface prefers the first species. When the challenging species is preferred on the surface, the exchange proceeds more rapidly. When the selectivity of the surface is orders of magnitude in favor of the second species, the preadsorbed layer of species 1 offers no resistance and species 2 adsorbs at a rate approaching the mass-transport limit.

Figure 5.10 illustrates a self-exchange experiment using PEO [120,000 narrow molecular weight (MW) standard]. Initially, a coumarin-tagged version of the

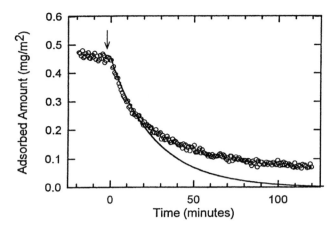

FIGURE 5.10 Experimental results for displacement of coumarin-tagged PEO (120,000 molecular weight) with untagged PEO of same molecular weight and bulk solution concentration. Line shows predictions for $K_{\text{C-PEO}}/K_{\text{PEO}} = 1.2$, $M_i k_i C_{\theta T} = 0.013$.

molecule has been preadsorbed, and, after equilibrium has been achieved, the adsorbed layer is challenged with a PEO solution of the same concentration but without the fluorescent label. (In this experiment, we have independently established that coumarin is minimally invasive such that the ratio of the K_i values for the labeled/unlabeled pair is 1.2.) Figure 5.10 shows a fluorescence decrease that results from the displacement of the tagged PEO by the native PEO. Also shown is a best fit to the data, employing Eq. (5.14). In this case, the exchange was not quite transport limited, with a value of $M_i/k_i C_{\theta T} = 0.013$. Also, at long times, not all the coumarin-tagged PEO desorbed, a result of kinetic trapping on the surface (motivating future investigation).

Multiple species adsorption occurs when a bare surface is exposed to a binary solution and both species adsorb simultaneously. This situation can be modeled by numerical integration of Eq. (5.14) with an initial condition of zero surface coverage, and the various parameters chosen to reflect the properties of the two species. Here we consider the case where both have the same diffusivity, giving $M_1 = M_2$, and local equilibrium is maintained, $M_1/k_1 C_{\theta T} = M_2/k_2 C_{\theta T} = 0$. Experimentally, this corresponds to a binary mixture of PEO chains (33,000 MW), where both native and fluorescein end-tagged chains are present. While native PEO gives a classical homopolymer-adsorbed layer, the fluorescein end tags (containing two negative charges) on some of the chain ends are repelled from the negatively charged silica surface, making these chains less favored at equilibrium. Of particular interest is that the interaction of the fluorescein-tagged chains with the surface is tunable through ionic strength, allowing the user to tune selectivity.

Figure 5.11 illustrates binary mixture (50% fluorescein-tagged PEO plus 50% native PEO) adsorption experiments compared with the predictions of Eq. (5.14). The data for the surface behavior of fluorescein-tagged PEO have been measured using

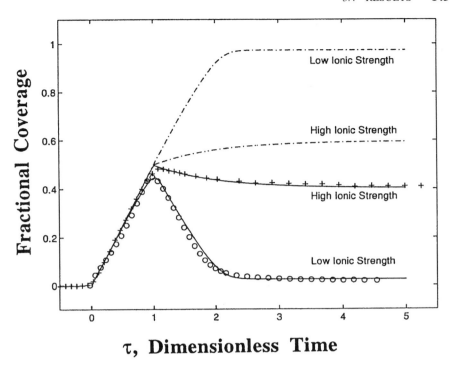

FIGURE 5.11 Competitive adsorption of end-charged chains in 50–50 mixture of native and end-charged PEO. (+, o) are data for end-charged chains at two ionic strengths. Native PEO is invisible. Predictions from diffusion-controlled model for end-charged (—) and native (-.-.-..) PEO.

TIRF, wherein the adsorbed amount of the native PEO is not detected. In Figure 5.11, the total concentration is 5 ppm. Two experiments for NaCl concentrations of 0 and 0.1 M are superposed. Also in Figure 5.11 are the predictions from Eq. (5.14), with $K_{native}/K_{tagged} = 40$ at low ionic strength and $= 1.5$ at high ionic strength. In Figure 5.11, time is made dimensionless according to $\tau = M(C_{native} + C_{tagged})t$.

In Figure 5.11, both native and tagged PEO initially adsorb at their mass-transport-limited rates until surface saturation. Then the native PEO, which is preferred on the surface, continues to adsorb and displace some of the labeled PEO, giving an overshoot in the TIRF data, sensitive only to labeled PEO. The shape of the overshoot depends on the selectivity of the surface. At low ionic strengths, the repulsion between the fluorescein tag and the surface is pronounced, giving a surface selectivity of $K_{native}/K_{tagged} = 40$. At higher ionic strengths, the electrostatic repulsion in screened, minimizing the selectivity.

In Figure 5.11, the TIRF and reflectivity data are in excellent agreement with the diffusion-controlled calculations in Eq. (5.14). It is worth pointing out that there are essentially no free parameters in this treatment, because the M_i's and K_i's have been independently determined, in separate studies.

5.5 SUMMARY

This chapter has illustrated the power of diffusion-controlled (local equilibrium) adsorption treatments in interpreting a wide variety of polymer adsorption phenomena. It was shown that both Langmuir and SCMFT treatments adequately represent much kinetic data, even though the former takes into account no polymer physics. Langmuir and SCMFT treatments are in good quantitative agreement with PEO adsorption on silica. Both treatments also explain a polymer layer's resistance to desorption as a consequence of high affinity adsorption, not requiring nonequilibrium surface phenomena in the model. The Langmuir kinetic model is also a powerful tool for predicting the behavior of multiple species in competitive adsorption. This chapter presented a general formulation for competitive adsorption of multiple species and provided examples of competitive PEO adsorption experiments for comparison. It was found that the model's predictions were in excellent quantitative agreement with experimental data for the evolution of interfacial populations.

Acknowledgments   This work was made possible by grants from the National Science Foundation (CTS 9209290 and CTS 9310932) and The Whitaker Foundation (RG 94-0355). Special thanks go to Z. Fu and E. Mubarekyan for their help with figures and especially H.S. Caram for insightful discussions of kinetic models.

REFERENCES

1. I. Kolthoff and R. Gutmacher, *J. Phys. Chem.* **56**, 740 (1952).

2. M. Kawaguchi, K. Maeda, T. Kato, and A. Takahashi, *Macromolecules* **17**, 1666 (1984).

3. J. Dijt, M.A. Cohen Stuart, and G. Fleer, *Macromolecules* **27**, 3219 (1994).

4. J. Dijt, M.A. Cohen Stuart, and G. Fleer, *Macromolecules* **27**, 3229 (1994).

5. M. Santore and Z. Fu, *Macromolecules* **30**(26), 8516 (1997).

6. P.G. deGennes, *Adv. Colloid Interface Sci.* **27**, 190 (1991).

7. E. Pfefferkorn, A. Carroy, and R. Varoqui, *J. Polym. Sci. (Phys.)* **23**, 1997 (1985).

8. E. Enriquez, H. Schneider, and S. Granick, *J. Polym. Sci. B. (Phys.)* **33**, 2429 (1995).

9. P. Frantz and S. Granick, *Macromolecules* **27**, 2553 (1994).

10. Z. Fu and M. Santore, *Colloids Surfaces A: Physiochem. Eng. Aspects* **135**(1–3) 63 (1998).

11. J. Dijt, M.A. Cohen Stuart, J. Hofman, and G. Fleer, *Colloids Surfaces* **51**, 141 (1990).

12. V. Rebar and M. Santore, *Macromolecules* **29**, 6273 (1996).

13. J. Dijt, M.A. Cohen Stuart, and G. Fleer, *Macromolecules* **25**, 5416 (1992).

14. B. Lock, Y. Cheng, and C. Robertson, *J. Colloid Interface Sci.* **91**, 104 (1983).

15. C. Shibata and A. Lenhoff, *J. Colloid Interface Sci.* **148**, 485 (1992).

16. M.S. Kelly, Master's Thesis, Lehigh University, 1993.

17. M. Santore, M.S. Kelly, E. Mubarekyan, and V. Rebar, in Sharma, R., ed., *Surfactant Adsorption and Surface Solubilization*, ACS Symposium Series 615, American Chemical Society, Washington, DC, 1995, p. 183.

18. T. Dabros and T. van ve Ven, *Colloid Polymer Sci.* **261**, 694 (1983).

19. G. Fleer, M.A. Cohen Stuart, J.M.H.M. Scheutjens, T. Cosgrove, and B. Vincent, *Polymers at Interfaces*, Chapman and Hall, London, 1993.

20. J.M.H.M. Scheutjens and G. Fleer, *J. Phys. Chem.* **83**, 1619 (1979).

21. J.M.H.M. Scheutjens and G. Fleer, *J. Phys. Chem.* **84**, 178 (1980).

22. M.A. Cohen Stuart, J.M.H.M. Scheutjens, and G. Fleer, *J. Polym. Sci. (Phys.)* **18**, 559 (1980).

23. M. Kelly and M. Santore, *Colloids Surfaces A: Physiochem. Eng. Aspects* **6**, 199 (1995).

24. V. Rebar and M. Santore, *Macromolecules* **29**, 6263 (1996).

25. V. Rebar and M. Santore, *J. Colloid Interface Sci.* **178**, 29 (1996).

26. Z. Fu and M. Santore *Macromolecules* **31**(20), 7014 (1998).

6 Depletion-Induced Aggregation and Colloidal Phase Separation

ANDREW MILLING
IRC in Polymer Science, Department of Chemistry, University of Durham, Durham DH1 3LE, United Kingdom

BRIAN VINCENT
School of Chemistry, University of Bristol, Bristol BS8 1TS, United Kingdom

6.1 INTRODUCTION

It is now well-established that there are two primary mechanisms for inducing aggregation in initially stable colloidal dispersions by the addition of macromolecules: bridging and depletion. The former situation arises when the added polymer (or polyelectrolyte) adsorbs onto the particles, but, initially at least, the polymer concentration is sufficiently low that the rate of particle collisions is significantly faster than the rate of (further) polymer adsorption onto the particles. Under such conditions adsorbed polymer chains can become "coadsorbed" onto more than one particle, forming polymer "bridges." For maximum effect, it is usually a necessary condition that the "final" adsorbed amount of polymer on the colloidal particles is much less than full coverage. If not, bridges formed initially may be disrupted as polymer adsorption continues. Bridging flocculation occurs on a reasonably rapid time scale, dependant on the initial particle concentration and, in general, leads to "open" (i.e., low fractal number), reasonably strong flocs.

Depletion flocculation is due to a very different effect. If we consider two flat plates immersed in a solution of *nonadsorbing* polymer molecules, then for large interplate separations the polymer concentration in the region between the plates will essentially be the same as for the bulk solution. However, if the interplate separation is reduced to a value less than the "effective" dimensions of the polymer molecules, then any

Colloid–Polymer Interactions: From Fundamentals to Practice, Edited by Raymond S. Farinato and Paul L. Dubin
ISBN 0-471-24316-7   © 1999 John Wiley & Sons, Inc.

polymer molecules entering the space between the two plates will have a reduced configurational entropy, since the total number of possible conformations of such chains is now reduced. As a result, the mean polymer segment concentration in the gap is reduced, compared to the bulk, and so the chemical potential of the solvent is higher in the region between the plates than in the bulk solution. Solvent molecules will tend to move out of the gap region into the bulk, effectively "pulling" the two plates together, giving rise to the so-called depletion attraction effect. The magnitude of this depletion attraction energy is proportional to the osmotic pressure difference (Π) between the interplate region and the bulk solution.

A similar attraction exists between colloidal particles dispersed in a solution of a nonadsorbing polymer. Scheutjens et al.[1] have derived the following analytical equation for the depletion attraction potential (V_{dep}) between two spheres of radius a, as a function of their separation h:

$$V_{dep} = -(2\pi/3)\Pi(2\delta - h)^2(3a + 2\delta + h/2)$$

(6.1)

where δ is the depletion layer thickness for a nonadsorbing polymer at a *single* surface, that is, 2δ is the effective diameter of a polymer coil in solution ($\sim 2R_g$, for dilute solutions, where R_g is the radius of gyration of the coil). Note that $V_{dep} = 0$ for $h > 2\delta$. When $h = 0$, V_{dep} has its maximum value, $V_{dep}(0)$. Note also that δ decreases with increasing polymer concentration over the semidilute and concentrated polymer concentration regions.

For colloidal particles, flocculation occurs when the polymer concentration in the continuous phase of the dispersion, and, hence, $V_{dep}(0)$, both reach critical values. Although, as discussed above, depletion flocculation is more likely to be observed in systems where the polymer concerned does not actually adsorb onto the particles in the dispersion, under certain circumstances, it may also be observed when the particles concerned carry either a layer of pregrafted polymer chains or a saturated layer of physically adsorbed polymer chains, and the excess added polymer builds up in solution to the critical concentration referred to. Depletion flocculation usually manifests itself as a colloidal phase separation phenomenon, rather than the classical formation of fractal flocs, per se. This is because, in general, depletion interactions are significantly weaker than bridging interactions, such that the particle aggregation process is a *reversible* phenomenon. Under these circumstances the aggregation process has many features in common with *molecular* condensation from the vapor to the liquid or solid states, that is, it is essentially a nucleation and growth process rather than a *continuous* (diffusion- or rate-limited) aggregation process. For these reasons, depletion-induced colloidal phase separation may take typically several hours to come to "equilibrium."

Depletion flocculation may arise, therefore, in any practical formulation where the system concerned contains colloidal particles (or indeed droplets or bubbles) and a sufficiently high concentration of polymer in the continuous phase. There are many industrially important products or processes where this situation pertains. Depletion flocculation may be something to be avoided in the formulation of stable products (e.g., cosmetics, pharmaceuticals) but may be a useful tool in helping to bring about a state of controlled, weak, reversible aggregation, in order to impart, for example, desirable rheological properties (e.g., paints) or "anticaking" properties (e.g., agrochemical products). By analogy with depletion-induced particle–particle aggregation in dispersions, one may expect to see depletion-induced adhesion (adsorption) of particles at *macroscopic surfaces*, when polymer is present in solution. There have been few studies of this effect.[2] One situation, for example, where such an effect might play a role is in the adhesion of bacteria to solid substrates, particularly where the bacteria themselves are known to generate extraneous polysaccharide-type polymers in situ in the vicinity of an adsorbing wall. Although classically ascribed to polymer bridging, it may be that polymer depletion may also play a role in some cases.

In this review we will first outline methods for characterizing polymer depletion layers at isolated surfaces and then discuss both theoretical and experimental aspects of depletion-induced pair potentials. Finally, we report on theoretical and experimental aspects of colloidal phase separation in these types of systems. There have been two major recent reviews of polymer depletion phenomena of which we are aware: Fleer et al.[3] and Jenkins and Snowden.[4] In order to save space and to avoid duplication, previous work discussed in either or both of those articles will not be reviewed at length here. It is suggested that the reader who wishes to have a complete overview of this subject should first read those two articles prior to this chapter. A further useful article, containing some material of a review nature is that by Seebergh and Berg.[5]

6.2 POLYMER DEPLETION AT SINGLE INTERFACES

The various theoretical models (numerical–Monte Carlo, self-consistent field, scaling, and "pragmatic") describing this situation have been extensively reviewed in the articles by Fleer et al.[3] and by Jenkins and Snowden.[4]

As far as we are aware, all attempts to characterize depletion layers at interfaces experimentally have been with regard to the solid–polymer solution interface, at which the polymer is "nonadsorbing," or, more precisely, in thermodynamic terms, the solvent is adsorbed in excess.

Although studies of depletion layers at the air–liquid interface have been made (see later in this section) there have been no attempts to monitor polymer depletion at a liquid–liquid interface. Nor have there been any experimental studies made of the depletion layer at a "soft" (i.e., the partially penetrable) interface between a polymer solution and a solid surface carrying a pregrafted or a saturated, physically adsorbed layer of polymer chains. Similarly, there have been few, if any, direct experimental studies of depletion at interfaces with solutions of *mixed* polymers: different types, or just different molecular weights of the same type, which leads on to considerations of polydispersity effects.

Most of the body of experimental work has concentrated on developing methods for studying depletion layers of monodisperse (assumed if not actual!) polymers at hard, solid surfaces. Several major techniques have emerged over the past few years.

Ideally one would like to determine the segment–concentration (density) profile, both normal $[\rho(z)]$ and tangential $[\rho(x, y)]$ to the surface; but in practice no techniques have emerged yet for the latter case. Moreover, even in the case of the segment distribution normal to the interface, many techniques only yield some moment(s) or other parameter characteristic of that distribution, typically an effective thickness (Δ) of the depletion layer. For polyelectrolytes one would like to know the associated counterion distribution.

In principle, small-angle neutron scattering (SANS) ought to yield the $\rho(z)$ profile for polymer chains at the interface between spherical colloidal particles and a *nonadsorbing* polymer solution, in a similar manner to that successfully employed for *adsorbing* systems, but attempts to date have failed. This is largely because of the much smaller difference in concentration between the depletion layer (some average value) and the bulk solution, compared to that between an adsorbed polymer layer and the bulk solution. This leads to sensitivity problems. However, neutron reflectivity has been successfully used to determine Δ for polystyrene at the air–toluene solution interface.[6] Other methods that lead to values for Δ for polymers at the solid–solution interface include evanescent wave fluorescence,[7-9] capillary flow studies,[10] nuclear magnetic resonance (NMR) relaxation time (T_2) studies on particulate dispersions,[11] and electrophoretic mobility experiments on dispersions of *charged* particles by Donath's group (see, e.g., Donath et al.[12]). Fuller details of the results obtained using these methods are given in the two review articles mentioned in the beginning of this section. More recently, Donath has suggested[13] that dynamic light scattering may also be used to determine Δ for particulate dispersions. Comparisons of theoretical predictions with experimental values are presented for liposomes in aqueous solutions of dextran. Evidence is also presented for the first time of depletion layer relaxation effects.

6.3 THEORETICAL ASPECTS OF THE DEPLETION PAIR POTENTIAL

Again, most of the theoretical developments in this subject (up to ~ 1995) have been adequately described in the two articles referred to previously. The most recent developments have been in the excluded volume theories of the depletion interaction, where the depleting species are treated as hard, immutable spheres. Both Walz and Sharma[14] and Mao et al.[15] have extended the basic excluded-volume model of the depletion interaction to include effects due to nonideal (higher order) behavior of the depleting solute particles. Walz and Sharma[14] used a force–balance approach in which interplate solute distributions were calculated using Boltzmann statistics. Access to higher-order terms were facilitated by using a virial expansion. The energetic term in the Boltzmann distribution was calculated for both hard-sphere and electrostatic interactions (truncated at the second-order with respect to solute concentration). The most interesting (and pertinent) part of this work is where the hard-sphere potentials are replaced with screened electrostatic potentials, typical of colloidal dispersions. First-order solutions showed that the depletion interaction begins at a far greater surface separation than for uncharged particles, and the attractive force is up to three orders of magnitude greater. The inclusion of solute–solute electrostatic interactions to the solute partition function leads to the formation of a secondary maximum in the particle pair-potential curve, the height of this maximum rising to very high values (>100 kT) at high solute concentration; this is illustrated in Figure 6.1. The effect of adding an interparticle DVLO (after Derjaguin, Verwey, Landau, and Overbeek)[16,17] interaction to the second-order depletion potential produced the pair-potential reproduced as Figure 6.2. However, it should be noted that calculations involving the addition of DLVO and depletion energies are quite parameter sensitive, and subtle changes in input data may drastically effect the depth and width of the attractive well, and the height of the secondary maximum. Mao et al.[15] also examined higher-order effects upon the depletion interaction in a system comprising large spheres dispersed in a solution of smaller spheres. Analytical solutions for depletion potentials were derived for up to second-order effects, while for third-order effects a nomograph of expansion coefficients (as a function of separation) facilitates calculation of the third-order depletion pair-potential. The work clearly demonstrates that for mutually repulsive, hard solute particles higher-order solute correlations give rise to an oscillatory (as a function of particle separation) potential, the final oscillation being the repulsive secondary maximum that may lead to "depletion stabilization." However, considering that neutral homopolymer solutions do not exhibit interchain structure factors, higher-order effects may be markedly reduced, and sample polydispersity (encountered in the majority of polymer samples) may destroy these subtle higher-order effects. The predicted secondary maximum of these hard-sphere theories [and also the theory of Feigin and Napper[18]] is far in excess of that predicted by the Scheutjens and Fleer[19,20]

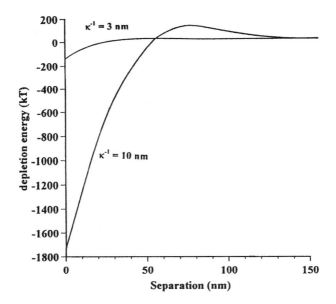

FIGURE 6.1 Second-order excluded volume theory depletion potential for the depletion of electrostatically charged solute spheres (radius 5 nm, 1% volume fraction) from between two spheres of 10 μm diameter. The potential of all surfaces was assumed to be -50 mV. The inverse Debye lengths are as indicated in the diagram. Adapted from Waltz and Sharma.[14]

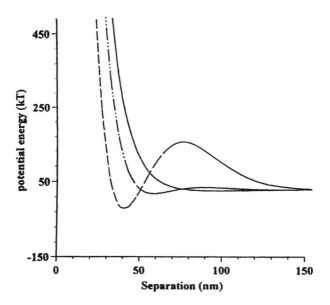

FIGURE 6.2   Total potential energy (depletion + electrostatic + van der Waals energies) for the depletion of electrostatically charged solute spheres (radius 5 nm) from between two polystyrene spheres (10 μm diameter) at various solute volume fractions (1% broken line; 0.4% dot-dashed line; 0% solid line). The potential of all the surfaces was assumed to be -50 mV, and the inverse Debye lengths were all assumed to be 10 nm. Adapted from Waltz and Sharma.[14]

self-consistent mean-field theory for polymers at interfaces, which also includes a second-order excluded volume term (in the Flory–Huggins χ parameter[21]).

Walz[22] has recently considered the effect of polydispersity on the depletion interaction between two parallel plates and between two spherical, colloidal particles in solutions of (assumed) hard-sphere macromolecules. A general force–balance approach is used. A normal distribution is assumed for the macromolecules, the polydispersity in diameter being characterized by the coefficient of variation (σ). Another assumption is that the concentration of macromolecules is low enough for higher-order effects arising from interactions between the macromolecules to be negligible. It is found that the interaction between parallel plates may be represented in terms of a complimentary error function of a dimensionless separation distance. The Derjaguin approximation is used to obtain similar results for spherical particles. For dispersions of spherical particles in solutions of macromolecules, at a fixed *number concentration*, it is predicted that polydispersity increases both the range and magnitude of the depletion force. The effect is relatively small, however, unless the polydispersity is substantial (typically, $\sigma > 50\%$). Note that at constant polymer *volume fraction*, it is predicted that the depletion energy actually decreases with increasing polydispersity. In related studies, Chu and co-workers[23,24] have investigated the behavior of particles constrained within thin liquid films. Monte Carlo simulations for repulsive hard spheres were used to calculate the population distribution of particles between two plates immersed in a solute reservoir. Again, these authors demonstrated that the hard-sphere depletion pair-potential is a damped oscillatory function at high solute volume fraction. The use of Monte Carlo methods allowed ready examination of solute polydispersity, the main effect of which was to reduce the magnitude of the potential oscillations, which is not really surprising as increasing polydispersity will intrinsically decrease any interparticle structural correlations. A common feature of the theories discussed in this section is that the extent of the higher-order effects is strongly dependent on the relative sizes of the solute (r) and the colloidal particles (a). The magnitude of the potential oscillations increases with increasing (a/r). Thus, while large particles ($>$ ca. 2 to 5 μm) [such as those used in atomic force microscopy (AFM) and total internal reflectance microscopy (TIRM) experiments] may sample an oscillatory force/potential, particles of submicron size may perhaps only experience a depletion potential with reduced higher-order effects.

Dahlgren and Leermakers[25] have used the Scheutjens–Fleer, self-consistent field theory to study depletion effects in polyelectrolyte solutions. The $\rho(z)$ profiles were found to be independent of the chain length. It would seem that in polyelectrolyte solutions, at low ionic strengths, a depletion zone develops at adsorbing, as well as nonadsorbing surfaces. Interactions can be fully repulsive; on the other hand, interactions may be repulsive at some surface separations and attractive at others, for example, for very low chain charge densities or at intermediate ionic strengths. At very high ionic strengths (and very low charge densities) the situation essentially reverts to that for neutral polymers, leading to net attraction between the surfaces. Effects of chain-length polydispersity on $\rho(z)$ were found to be small. There is only a minor effect on the depletion region, even at low ionic strengths, provided there is not a large fraction of very short chains present. Only depletion zones near adsorbing surfaces tend to be preferentially populated by short chains. The behavior of weak polyelectrolyte molecules at like-charged interfaces was investigated by Chatellier and Joanny,[26] who derived polymer–polymer and polymer–surface correlation functions for weakly ionizing polyelectrolyte molecules (coiled as opposed to rodlike), interacting via mean-field electrostatic forces (Debye–Hückel theory). The main finding of this work was that, at low ionic strength, there is damped oscillatory polymer structuring at the interface. These oscillatory functions give rise to oscillatory pair potentials between the surfaces and are present even when there are weakly adsorbed polymer layers. The main theoretical predictions seem to be in good qualitative agreement with the recent experimental work of Milling and Vincent,[27] as discussed later.

6.4 EXPERIMENTAL STUDIES OF THE DEPLETION INTERACTION

6.4.1 Introduction and Experimental Techniques

Direct experimental measurements of the depletion interaction have been achieved in recent years using AFM, the surface forces apparatus (SFA), and TIRM. This work is described in some detail below.

One of the first measurements of the depletion force was in 1988 when Evans and Needham[28] measured the adhesion force between two lipid membranes. They observed that the addition of polydextran to the solution increased the adhesive force between the membranes, the adhesion increasing with increasing polymer concentration and molecular weight. Unfortunately, the technique was only capable of measuring the actual adhesive force between the lipid membranes and details such as the force–distance relationship were inaccessible.

Atomic force microscopy employs a device capable of measuring the deflection of a probe (normally a micron-sized sphere) mounted upon the tip of cantilever spring as a surface is scanned toward or away from it. Spring deflection and scanner displacement data may be readily transformed using the analysis of Ducker et al.[29] to yield quantitative force–distance data, with an absolute force sensitivity approaching the order of piconewtons. The use of AFM in measuring surface forces is reviewed in Chapter 11. Atomic force microscopy enjoys some advantages over the use of SFA in measuring depletion forces. For example, the probe and surface materials can be readily varied. Second, the effective surface overlap area is considerably smaller. Hence, the diffusive pathlength from the center of surface overlap to the bulk solution is typically a factor of 30 to 60 less (this factor is approximately proportional to the square root of the ratio of the radius of curvature of the AFM probe to that of the SFA cylinders). Thus, hydrodynamic drainage[30] and solution relaxation problems are less likely to be encountered in AFM measurements. Additionally, some earlier investigations into the forces between mica surfaces in toluene solutions of (nonadsorbing) polystyrene led the experimenters to question whether (their particular) SFA was sufficiently sensitive to measure depletion forces.[31]

6.4.2 Electrically Neutral Polymer Depletion

An AFM has been used to measure the force–distance relationship between a silica sphere and an optically polished silica surface (both bearing a grafted sheath of n-octadecyl chains), in the presence of solutions of polydimethylsiloxane (PDMS) in cyclohexane.[32] The authors reported that in pure cyclohexane there was negligible (long-range) interaction between the surfaces. Upon the addition of PDMS, however, there was an attractive force between the surface, commencing at a separation of ca. 20 nm (see Figure 6.3). Using just one independent fitting parameter (Δ) to calculate the depletion force, the experimental data were shown to be in very good agreement with the mean-field theory of Scheutjens et al.[1] There was very little evidence of a repulsive maximum, prior to entering the attractive depletion well, as the surfaces approached, as suggested by Feigin and Napper[18] and Napper.[33] This could be due to the high polydispersity of the PDMS sample used, or, although less likely (the measurements were made in the semidilute regime), to the fact that the polymer concentration was too low for this proposed effect to be seen.

The SFA has been used to measure the forces between crossed mica cylinders, each bearing an adsorbed lipid bilayer, in solutions of polyethylene glycols (PEG) of various molecular weights.[34] Motivated by the use of PEG solutions in cell fusion

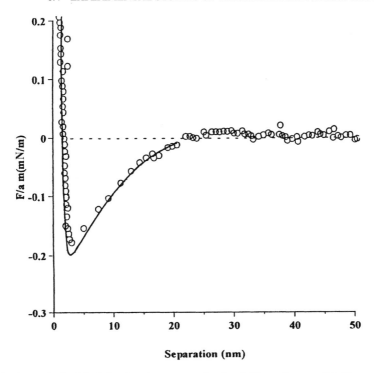

FIGURE 6.3 Normalized depletion force between octadecylated silica surfaces mediated by cyclohexane solutions of PDMS (volume fraction 0.06%). The solid line represents a theoretical fit comprising a depletion term[1] with $\Delta = 10$ nm, and a short-range steric component.[52] Adapted from Milling and Biggs.[32]

applications, the authors observed (for PEGs of molecular weight 8000 and 10,000 g mol^{-1}) an attractive "jump-in" (which occurs when the attractive force gradient is greater than the restoring spring stiffness), as the surfaces approached, and also an increase in the force required to separate the surfaces. The surface separation at the jump-in point decreased, and the strength of the adhesion increased, with increasing polymer concentration. At surface separations greater than the jump-in distance a slowly decaying repulsion was observed. This repulsion does not correspond to a typical electrical-double-layer (EDL) interaction and is most probably a nonequilibrium hydrodynamic drainage effect. Indeed, at surface separations greater than 2Δ the surfaces would be separated by a viscous polymer solution, whereas upon closer approach the viscosity will decrease as polymer vacates the intersurface region. With PEG of molecular weight ca. 1000 g mol^{-1} no depletion interaction was observed, probably because the radius of gyration of this polymer is too small to develop sufficient depletion force. On the other hand, experiments performed using a PEG of molecular weight ca. 20000 g mol^{-1} showed a repulsive force as the surfaces approached, which was attributed to polymer *adsorption* of this higher-molecular-weight sample. The theoretical analysis employed in this work, which was based upon the theory of Vrij,[35] may be criticized in two ways. First, the depleting polymer coils were treated as hard spheres (implying an invariant value for Δ). Second, the osmotic pressure (Π) of the polymer solution was calculated assuming an *ideal* solution (for which Π is proportional to the number concentration of the polymer molecules), whereas perhaps the use of a more sophisticated polymer solution theory (e.g., Flory–Huggins theory[21]) to calculate Π may have led to better agreement between theory and experiment.

6.4.3   Polyelectrolyte Depletion

Both AFM and SFA have recently been used to measure the depletion interaction and also to probe local ordering effects near the interfaces concerned, in systems where both the depleted "species" and the interfaces are electrically charged. In this Chapter we focus on depletion effects with polyelectrolyte molecules, although the attention of the interested reader is drawn to reports of depletion and ordering effects in the following systems: (i) SFA measurements involving micellar solutions[36–38] and wormlike phases[39] with ionic surfactants; (ii) TIRM[40] studies, where the depleting species were (small) colloidal particles[41] or micelles[42]; and (iii) light-scattering studies of depletion effects again involving solutions of ionic surfactants.[43] Both surfactant and particle depletion/ordering phenomena have been investigated at the gas–liquid interface, in studies concerned with thin-film stability.[23,24]

Although SFA work has mainly been concerned with surfactant-mediated interactions and has not yet been used to directly measure depletion forces due to polyelectrolytes, Marra and Hair[44] indirectly inferred depletion of sodium (polystyrene sulfonate) (NaPSS) from the water–mica interface, although the high surface potential of the mica interface entailed that EDL forces dominated and any attractive depletion potential was effectively swamped.

Atomic force microscopy has been used to determine force–distance relationships for silica surfaces in the presence of polyelectrolytes. Polyelectrolytes are often classified in two classes: (i) "weak," where the degree of ionization/charge separation is determined by solution conditions such as ionic strength and pH (typically these involve polymer chains bearing either weakly basic (e.g., 1° to 3° amines) or weakly acidic groups (e.g., carboxylic functionalities) and (ii) "strong," where the degree of ionization is only very weakly affected by ionic strength or pH (e.g., 4° amines and sulfonic acid groups). This work is described below.

Strong Polyelectrolytes   Intersurface forces between silica, in the presence of sodium poly(styrene sulfonate) (NaPSS), have been investigated, using AFM, as functions of the polyelectrolyte molecular weight and concentration, the background electrolyte concentration, and the solution pH, by Milling.[45] Figure 6.4 illustrates typical force curves for various NaPSS concentrations, in the *absence* of added electrolyte. The interactions between the surfaces exhibit a damped oscillatory behavior, suggestive of the development of partial structure factors between the polymer molecules in the intersurface region.

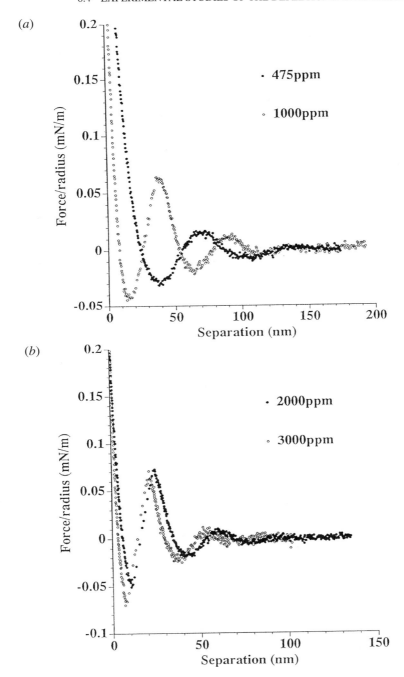

FIGURE 6.4 Sphere-plate interactions between silica surfaces in the presence of NaPSS (M_n 200,000 g mol^{-1}) at various polymer concentrations and intake absence of added electrolyte. (a) (●) 475 ppm, (○) 1000 ppm, and (b) (●) 2000 ppm, (○) 3000 ppm). Adapted from Milling.[45]

The main features of the force curves are: (1) With increasing polyelectrolyte concentration, both the oscillatory period and the depletion thickness (equivalent to half the surface separation of the secondary maximum) decreased, and the oscillatory force amplitude initially increased and, thereafter, slowly decreased. The oscillatory period was found to scale as $c_p^{-1/2}$. Both the depletion layer and the oscillations eventually vanished at high polymer concentrations and the interaction became monotonically repulsive. (2) Varying the polyelectrolyte molecular weight (at fixed monomer unit concentration) had a small effect upon the periodicity of the force oscillations [for a limited data set of four NaPSS samples (M_n 23,000 to 200,000 g mol^{-1}) these scaled as (approximately) $M_n^{-1/8}$]. However, the magnitude of the force oscillations increased with increasing polyelectrolyte molecular weight. (3) The addition of salt decreased the magnitude of the force oscillations (through reduction of the osmotic pressure of the polyelectrolyte solutions). (4) At higher polyelectrolyte concentrations (> 1000 ppm) the depletion thickness increased with increasing bulk solution pH (eventually reaching a plateau value at a pH value of ca. 7), while the oscillation periodicity was unaffected.

Only a limited theoretical analysis, based on the Scheutjens, Fleer, and Vincent (SFV) analysis[1] was presented since the presence of oscillatory forces suggests that multibody polymer–polymer interactions, which are beyond the applicability of theories such as those of Vrij[35] or SFV,[1] are appreciable in this system. Additionally, at close surface separations, theoretical modeling is further complicated by the presence of EDL interactions. Enumeration of the EDL contribution to the total force is difficult because in the presence of polyelectrolytes the surface potential, local pH, and ionic strength are independently intractable quantities. The presence of the force oscillations provides direct corroboration to the observation that there indeed may be an increase in the kinetic stability of colloids in the presence of strong polyelectrolyte molecules, as observed in dispersion stability studies.[46] Recently, similar structural oscillations were seen in thin films of (semidilute) polyelectrolyte solutions.[47] The exact physical interpretation of these structural force oscillations is a moot point.

Polyelectrolyte scaling theory[48,49] provides some insight and suggests the following possible interpretations: (1) An electrostatic molecular crystal is formed [comprising of either "coils" (semidilute) or smectic rods (dilute regime)] and (2) development of an entangled isotropic mesh in the semidilute regime. Unfortunately, in these models electrostatic polymer correlation lengths all scale with $c^{-1/2}$, and thus further comment would be a matter of conjecture. Figure 6.5 illustrates the experimentally measured scaling relationships for polymer solutions between silica surfaces. A recent work by Walz et al.[50] also reports upon the depletion and structural effects of nonadsorbing NaPSS between a large (15-µm diameter) latex sphere and a glass plate, using the TIRM technique. The experimental findings of this study are essentially the same as for Milling,[45] although it should be remembered that that raw TIRM data is deconvoluted to give pair potential energy curves and raw AFM data is transformed to give force–separation curves. The TIRM technique has the added drawback that attractive regions of the potential energy curve may not be properly sampled, the manifestation being discontinuities in the attractive regimes of potential energy curve.

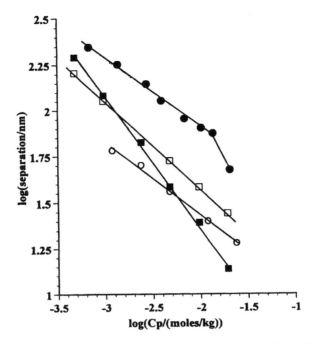

FIGURE 6.5 Scaling relationships between polymer concentration and (1) position of secondary maximum: PAA M_n 112,000 g mol^{-1} (filled circles); NaPSS 200,000 g mol^{-1} (filled squares) and (2) force oscillation period: NaPSS 23,000 g mol^{-1} (open circles) and 200,000 g mol^{-1} (filled circles).

However, with an AFM, choice of cantilever stiffness can be used to overcome some of these instability problems. Nevertheless, the corroboration between these AFM and TIRM studies is indeed satisfying.

Weak Polyelectrolytes The behavior of weak polyelectrolyte molecules at metal oxide–water interfaces is of importance in fields such as ceramic processing and water purification. The forces between silica surfaces in the presence of poly(acrylic acid) (PAA) have been measured using an AFM.[27] Drawing on the earlier work of Macmillan,[51] who investigated the rich phase behavior of silica sols in the presence of PAA, the authors reported first the development, and then the disappearance, of a long-range attractive force between a silica sphere and a silica flat surface, as they increased the PAA concentration from 50 to 2000 ppm. Figure 6.6 illustrates these findings. Upon close approach of the surfaces a repulsion was observed, and while this resembles a typical EDL force curve, both the Donnan equilibrium established between the surfaces (the intersurface region being analogous to a semipermeable membrane) and the presence of an adsorbed PAA layer entail that local properties such as surface potential, pH, and ionic strength are independently intractable. Additionally, a small repulsive maximum was noted prior to entrance into the depletion well, although the integrated height of this force barrier would be less than

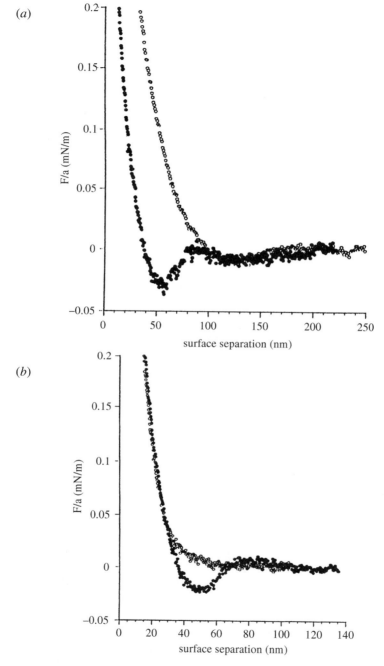

FIGURE 6.6 Normalized forces between silica surfaces in the presence of PAA; M_n 112,000 g mol^{-1} (no added electrolyte). (a) (•) 100 ppm, (○) 500 ppm and (b) (•) 2000 ppm, (○) 2500 ppm. Adapted from Milling and Vincent.[27]

the several kT required to "kinetically stabilize" a colloidal dispersion. Under similar solution conditions a silica sol was shown to be unstable over polymer concentrations (c_p) in the range $1100 < c_p < 1600$ ppm. Although Macmillan[51] had shown that at the prevalent pH values (< 7) examined, PAA weakly adsorbs at the silica–water interface, there is a large excess of PAA present in solution and the attractive force is due to a depletion mechanism and not polymer bridging. The polymer is thought to adsorb via a hydrogen bonding mechanism. The sol stability experiments demonstrate a nonequilibrium effect, namely that a colloidal dispersion may undergo depletion flocculation while bearing an adsorbed layer of the same polymer. We shall return to this theme later. In the absence of added electrolyte the data were successfully fitted to a theoretical expression, comprising the summation of depletion[1] and EDL[52] forces. At higher PAA concentrations (2500 ppm) the authors explored the effect of neutralizing the PAA solutions. The data presented showed that, despite there being no depletion minimum, as the solution pH was raised (thus increasing the charge density of the interface and the polymer molecules), there was evidence of polymer desorption at close approach and also some extremely weak PAA structural forces.

6.4.4 Summary

In summary, the force measurements performed with SFA and AFM have clearly demonstrated the existence of depletion forces in neutral polymer and both strong and weak polyelectrolyte-containing systems. The principle differences are: (1) For neutral polymers there is no evidence of depletion stabilization, and the depletion thickness appears to be of similar dimensions to the radius of gyration of the depleting polymer molecules. (2) For strong polyelectrolyte molecules there are dampened structural forces present (the depletion force effectively being the final structural oscillation). For low background electrolyte strength the structural oscillations obey a periodicity that scales according to the inverse square root of the polymer concentration (similar to studies of polyelectrolyte solution properties in the semidilute regime[49]). There seems to be no simple scaling exponent for the depletion layer thickness with respect to polymer concentration, the scaling exponent varying according to a complexity of factors including pH, ionic strength, and free polymer molecular weight. The thickness of polyelectrolyte depletion layers are apparently determined by two factors: an entropic contribution, as with neutral polymers, and an electrostatic contribution (due to competing surface–polymer and polymer–polymer interactions). The latter contribution would seem to be dominant and the ensuing depletion layer may be quite thick (tens to hundreds of nanometers) at low polymer concentrations, but readily collapses under the osmotic pressure of the bulk solution as the polymer concentration increases. Figure 6.5 presents data for the depletion layer thickness (both PAA and NaPSS solutions) and the force–oscillation correlation lengths (NaPSS solution). Addition of background electrolyte suppresses the depletion force. (3) Based upon the limited evidence provided by the PAA study, weak polyelectrolytes seem to exhibit properties of both neutral and strong polyelectrolyte-containing systems. These are very weak kinetic barriers, potentially thick depletion layers, salt/pH sensitivity, and weak, damped structural oscillations as the polymer molecules acquire a higher degree of charge.

6.5   PHASE BEHAVIOR OF COLLOIDAL DISPERSIONS

The fact that colloidal phase separation into coexisting phases may occur on the addition of sufficient "free" polymer to a dispersion has been referred to in Section 6.1. A detailed overview of the important contributions to this subject (up to ~ 1995) is given in the review by Jenkins and Snowden.[4] Therefore, only some general points will be made here and some more recent work outlined.

6.5.1 Effect of Particle Radius and Depletion Layer Thickness

Generally, two coexisting phases are observed; the molecular analogs are vapor + liquid, vapor + solid, and liquid + solid.[52] If the range of the depletion interaction (d) is very much smaller than the particle radius (a), then only the vapor–solid transition is seen. Effectively one is dealing with dispersions of "sticky spheres," such that the particles are held at close separations in the dense phase. Equilibrium phase separation would give rise to a crystal structure, but frequently, for kinetic reasons, amorphous (random close-packed) structures are observed, particularly if the depletion interaction is relatively strong, and the phase separation (aided in general by gravity) is relatively fast. Essentially the superposition of an interparticle depletion attraction widens the order–disorder (Kirkwood–Alder) coexistence region over that for true hard-sphere dispersions, which spans the particle volume fraction (ϕ) range: $0.494 < \phi < 0.545$.

   If d/a is ~ 1, then a liquid (i.e., less dense) separated phase may be sustained. This has been observed by many authors, perhaps first by DeHek and Vrij.[54] Vincent et al.[55] demonstrated the existence of all three types of phase coexistence, referred to above, for the one system: hydrophobed silica particles (i.e., silica particles with a close-packed layer of terminally grafted n-octadecyl chains: SiO_2-g-C_{18}) dispersed in cyclohexane solutions of PDMS, in which two sets of silica particles were used, $a = 134$ nm ($d/a \ll 1$) and $a = 8.3$ nm (d/a ~ 1). With increasing polymer concentration, into the semidilute region, most of the theories predict that the depletion interaction should pass through a maximum. The result is that a restabilization effect should occur at sufficiently high free polymer concentrations. However, at these concentrations the viscosity of the polymer solution is usually sufficiently high as to make phase separation studies intractable. However, Vincent[56] reported such a restabilization effect for SiO_2-g-C_{18} particles ($a = 134$ nm) dispersed in bromocyclohexane solutions of PDMS at 30 °C . This corresponds to theta conditions, where the depletion interaction is weak compared to good solvent conditions, and, hence, the restabilization region occurs at lower polymer concentrations. Other examples of the restabilization effect at higher polymer concentrations are given in the review by Jenkins and Snowden.[4]

Ternary Component Systems Perhaps the most recent, complete statistical mechanical analysis of the phase behavior of colloidal dispersions in the presence of monodisperse, nonadsorbing homopolymers is the one by Lekkerkerker et al.[57] This work is important in that it "corrects" the implicit assumption, made in previous theoretical approaches, that such systems can be treated effectively as *two*-component systems: particles plus polymer solution. The fact that the polymer may partition between coexisting phases was neglected. Such systems should, therefore, strictly be treated as *three*-component ones. This leads to the prediction, under certain conditions, that a *three-phase* coexistence region should exist.

6.5.2 Experimental Studies of Depletion-Induced Colloid Phase Separation

Electrically Neutral Polymers Leal Calderon et al.[58] and Faers and Luckham[59] have reported the first observations of three-phase coexistence, for the systems: polystyrene latex particles + water + hydroxyethylcellulose (HEC). In the latter case prevention of adsorption of HEC onto the latex particles was prevented by preadsorbing a layer of a particular triblock copolymer PEO–PPO–PEO [poly(ethyleneoxide-b-propylene-oxide-b-ethyleneoxide)] (Synperonics).

Poon et al.[60] have carried out a direct test of the theoretical work of Lekkerkerker et al.,[57] finding good agreement, using as a model, a supposedly hard-sphere colloid: core–shell particles (radius 217 nm) having a poly(methylmethacrylate) core and thin shell of terminally grafted poly(12-hydroxystearic acid) chains. The polymer was polystyrene (hydrodynamic radius 12.8 nm) and the solvent was *cis*-decalin. For this system d/a is therefore $<< 1$, and only vapor–solid type of transitions are expected. The main effect of adding polymer to the dispersion was indeed, as discussed above, to widen the fluid–crystal, two-phase coexistence region. The separated dilute and dense colloidal phases showed marked polymer partitioning, giving rise to a strong osmotic compression of the colloidal crystal phase. At sufficiently high polymer concentrations, kinetic restrictions set in and a gel phase was observed, rather than the crystalline (equilibrium) phase.

Bean et al.[61] have recently studied the effect of adding an AB block copolymer (polystyrene-*b*-polyisoprene) (PS-*b*-PIP) to dispersions of SiO_2-*g*-C_{18} particles in (benzene + *n*-heptane) solvent mixtures. Benzene is a common solvent for both parts of the block copolymer, whereas *n*-heptane is a "hemi-solvent" for the PS block. In pure *n*-heptane and in the solvent mixtures, up to ~ 0.1 mole fraction benzene, the polymer forms micelles (the critical micelle concentration increasing with benzene mole fraction) and adsorbs onto the particles, thereby imparting steric stabilization. At benzene mole fractions higher than ~ 0.1, there is no evidence of micelles, and depletion-induced phase separation is observed. The minimum polymer concentration required to achieve this was found to decrease with increasing benzene mole fraction in the solvent mixture, reflecting the increase in net solvency for the block polymer chains in solution.

Vincent's group (see review by Jenkins and Snowden[4]) has made extensive studies (both experimental and theoretical) of so-called soft spheres, based on silica particles as cores that carry shells of terminally grafted polymer chains, such as PS, PDMS, and poly(methylmethacrylate) (PMMA). In this case free polymer chains may partially penetrate the solvent-swollen grafted polymer shells, thereby weakening the depletion interaction, compared to hard spheres, under equivalent conditions. Thus, higher concentrations of added free polymer are required to achieve phase separation for the soft spheres.[62] One interesting finding by Cawdery et al.,[63] for the SiO_2-g-PMMA systems, was that the required concentration of free polymer for phase separation passes through a *maximum* with increasing grafted amount; the point is that as the grafted chain coverage increases, the particle periphery becomes first "softer," then "harder" (i.e., less penetrable) again.

The combination of electrostatic forces (arising from surface charge on the particles) and depletion forces has been well-studied, in particular, for dispersions in aqueous media (see, e.g., Seebergh and Berg[5] and Milling and Vincent[27] for details of recent studies). One study of such systems in *nonaqueous* media is that reported by Cawdery et al.,[63] for SiO_2-g-PMMA particles in a variety of solvents [e.g., dioxan, acetone, tetrahydrofuran (THF), toluene, chloroform, methylmethacrylate monomer]. In all of these media (except dioxan) the SiO_2-g-PMMA particles had a net negative electrophoretic mobility, as determined using the phase analysis light-scattering technique, developed by Miller et al.[64] The surface charge is thought to be associated with surface—OH groups on the core silica particles undergoing proton exchange with water adsorbed at this interface. In this regard it is noted that water has a low solubility in all the solvents mentioned (except dioxan). It was found that, for dispersions of the SiO_2-g-PMMA particles in THF, the addition of the THF-miscible electrolyte, tetrabutylammonium acetate, reduced the electrophoretic mobility of the particles, and also, therefore, the interparticle electrostatic repulsion. There was a commensurate reduction in the minimum concentration of free, homopolymer PMMA required to achieve colloidal phase separation.

Polyelectrolytes   The use of polyelectrolytes (again mostly in aqueous media) to achieve depletion-induced colloidal phase separation has been increasingly studied in recent years (see Jenkins and Snowden[4]). Almost invariably, the particles have an associated surface charge of the same sign as the polyelectrolyte. This leads again to an electrostatic repulsion contribution to the overall pair-potential, which offsets the depletion attraction term. On the other hand, the depletion interaction is in general stronger in a polyelectrolyte solution compared to a neutral polymer under otherwise equivalent conditions. This is because of the counterion contribution to the osmotic pressure. Sharma et al.[65] have recently used small (Ludox) silica spheres (5 to 7 nm diameter) as model "polyelectrolytes" to study depletion-induced phase separation in aqueous dispersions of much larger polystyrene latex particles (0.5 to 1 μm diameter). Both sets of particles were negatively charged. Two critical silica particle concentrations were noted, the first corresponding to the onset of depletion flocculation, and the second (higher) concentration corresponding to restabilization of the system. Both these critical silica particle concentrations were found to decrease with increasing size of the latex particles, but with decreasing size of the silica particles themselves. This use of *particles*, rather than (flexible) macromolecules as such to induce depletion phase separation, reflects a growing interest in this subject; but this topic is beyond the scope of this review.

Simultaneous Polymer Adsorption and Depletion Effects   As mentioned earlier, polymer *depletion* attraction is mainly observed when the polymer does not adsorb onto the particles, whereas polymer *bridging* attraction occurs under conditions where polymer adsorption does occur. It has been of interest, therefore, to see whether, by switching conditions in a dispersion to which polymer is added, from adsorbing to nonadsorbing, a change in flocculation mechanism may be observed. One way of achieving this would be to reduce the mutual electrostatic repulsion between the added polyelectrolyte and particles of the same charge sign simply by adding electrolyte, thus allowing adsorption to occur. Snowden et al.[66] first reported such a situation for the system of silica particles in water, plus hydroxyethylcellulose. A more subtle, nonaqueous system had been previously studied by Jones and Vincent.[67] This comprised SiO_2-g-PS particles, dispersed in toluene, to which polyvinylmethylether (PVME) was added. The PS–PVME–toluene phase triangular diagram has central "loop" region where coacervation occurs, that is, the two polymers phase separate *together*, indicating a strong mutual attraction (*negative* Flory χ parameter) between them. Thus, when PVME is added at *low* concentrations to the SiO_2-g-PS particles in toluene, bridging flocculation is observed, resulting from some (limited) "adsorption" of the PVME chains onto (and maybe, partially, *into*) the sheath of grafted PS chains. At higher concentrations, full coverage of adsorbed PVME chains around the SiO_2-g-PS particles is achieved, so interparticle bridging by these chains is now prevented, and the dispersions were stable. As the PVME concentration was increased further still, until the concentration of excess, free PVME in solution exceeded some critical value, then depletion-type phase separation was observed. One suspects that this is a further example of depletion flocculation occurring in a polymer *adsorption* situation (cf. the earlier example quoted from the work of MacMillan[51]). A more recent example, again with a more conventional aqueous system, is referred to in the work by Smith and Williams.[68] They showed that HEC adsorbs onto PS latex particles, even in the absence of added electrolyte. At sufficiently high HEC concentrations in the latex, much greater than those corresponding to "plateau" adsorption, flocculation occurred; it would seem this must be depletion flocculation. This situation must often be reached in many systems of practical interest. However, *equilibrium* theories of polymers at interfaces, such as the Scheutjens–Fleer, self-consistent mean-field theory[3], imply that, with neutral polymers, one ought *not* to see depletion attraction in a system where the polymer is physically (and reversibly) adsorbed onto the particles. The results referred to here would seem to imply perhaps that *nonequilibrium* conditions apply in many cases of depletion flocculation. We shall return to this theme later.

Another example where polymer bridging and depletion have been seen in essentially the same system, but with a change in conditions, was reported by Cawdery et al.[63] Their system was PS latex particles, carrying terminally grafted layers of poly(ethylene oxide) (PEO), that is, PS-g-PEO particles, dispersed in water to which poly(acrylic acid) (PAA) was added at a controlled background ionic strength and pH.

At low pH values (< ~ 4), PAA is largely undissociated and is known to coacervate with PEO in solution, presumably through some form of H-bonding association between the two polymers (the —COOH group of PAA with the ether —O— of PEO?); at higher pH values (> ~ 5) the polyacrylate salt is formed, and coacervation with PEO no longer occurs. Thus, it is perhaps not surprising that the addition of PAA to the PS-g-PEO particles led to bridging flocculation at low pH values, but depletion phase separation at higher pH values. This work has been extended recently by Allan et al.[69] who made rheological studies (shear-wave propagation and controlled stress rheometry) on the corresponding depletion-flocculated systems at pH 9. These authors showed that in order to fully account for the rheological parameters measured, in terms of theoretical models, then due account had to be taken of the viscoelastic nature of the polyelectrolyte solutions themselves. This reflects the weak nature of depletion flocs.

Rheological Studies of Depletion Phase-Separated Dispersions   Earlier rheological studies of aqueous dispersions of PS-g-PEO particles, undergoing depletion flocculation on adding PEO or PAA, by Tadros's group have been reviewed by Jenkins and Snowden.[4] The rheological properties of the continuous phase was not explicitly considered in these earlier studies. Faers and Luckham[59] have also reported rheological measurements on their system (see earlier). Otsubo[70] has presented a comparison of the viscosity/shear rate dependence for (similar) dispersions undergoing either bridging or depletion flocculation. Basically, he has shown that the systems showing bridging flocculation exhibit a yield stress, but no yield stress was observable (over the shear-rate and stress ranges employed) for those systems showing depletion flocculation, even though they were shear thinning. This is broadly in line with the comments about the relative strengths of the two modes of polymer-induced flocculation made in Section 6.1.

Neutron Scattering Studies   Small-angle neutron scattering has recently been used to investigate the development of structure factors for the colloid-rich phase of a (depletion) flocculated dispersion.[71] The experimental system studied was comprised of small (2-nm radius), sterically stabilized, calcium carbonate particles dispersed in decane, to which nonadsorbing hydrogenated poly(isoprene) was added. Model calculations were performed by obtaining the Fourier transform $[C(Q)]$ of a calculated radial correlation function $[C(r)]$, comprising an attractive depletion and a short-range steric repulsion $[C_{hs}(r)]$ term:

$$C(r) = \begin{cases} C_{hs}(r) & r \leq d \\ -U(r)/k_b T & r > d \end{cases} \tag{6.2}$$

where d is the particle diameter, k_B is the Boltzmann constant, and the depletion potential $U(r)$ was calculated using the expression of Vrij.[35] The structure factor $[S(Q)]$ was then calculated using

$$S_c(Q) = 1/[1 - \rho_c C(Q)] \tag{6.3}$$

where ρ_c is the colloid number density. Fig. 6.7 reproduces both experimental and theoretically modeled $S_c(Q)$ functions at various polymer concentrations. There is excellent agreement between the two respective data sets, although the theoretical generation of the structure factors may be relatively insensitive to subtle differences in the various models available to calculate $U(r)$. Thus, SANS studies, while clearly demonstrating the development of structure factors, may not uniquely test theories of the depletion interaction (for further discussion of deconvolution problems concerning scattering experiments see, e.g., Rajagopalan[72]). Ye et al.[71] considered the depletion thickness (Δ) to be invariant with respect to polymer concentration; this is an oversimplification as some of the data is for polymer concentrations above the critical coil overlap concentration, beyond which Δ decreases with increasing polymer concentration. The osmotic pressure is the only adjustable fitting parameter used, and in calculating the actual free polymer concentration the authors accounted for the excluded volumes of the particles and their attendant depletion layers. Neglecting the concentration dependence of Δ led the authors to the conclusion that the depletion interaction was less than anticipated at the higher polymer concentrations, because an increase in the polymer–polymer interactions requires the system to do more work to expel interstitial polymer molecules. It is not clear from this conclusion whether the authors are supportive of the depletion stabilization theory of Feigin and Napper,[18] and Napper,[33] which should be treated with some caution for neutral polymer molecules

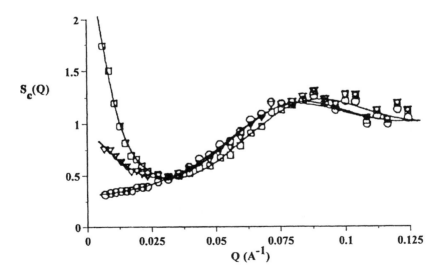

FIGURE   6.7 Measured colloid structure factors [$S_c(Q)$] for the colloid-rich phase of depletion-flocculated sterically stabilized calcium carbonate particles (volume fraction 0.146) in solutions of hydrogenated poly(isoprene) at various polymer concentrations. The solid curves represent fitted values (see text) and the polymer concentrations (g cm^{-3}) are as follows: ($\circ$), 0.0039; ($\triangle$) 0.0165; ($\square$), 0.0308. Adapted from Ye et al.[71]

where interpolymer structure factors are very weak, or it is an implicit acknowledgment of some other phenomenon (a likely candidate may be the erroneous invocation of "higher-order" effects found in excluded volume theories for hard immutable spheres at high solute volume fraction).

Polymer Polydispersity Effects   We return now to the discussion of polydispersity effects. In many of the (particle + polymer + solvent) systems studied in academic laboratories, stress is placed on the monodispersity of not only the particles but also the polymer samples used. However, in many practical situations polydisperse systems have to be tolerated. Polydispersity effects in polymer adsorption are well described; higher molecular mass fractions are preferentially adsorbed—at least at low, overall polymer concentrations.[3] Reference has already been made earlier to the theoretical considerations by Walz[22] with regard to the effect of the polydispersity of an added polymer on the depletion pair-potential. Sear and Frenkel[73] have analyzed the effect of polymer polydispersity on the phase separation behavior of colloidal dispersions, using essentially an extension of the theory by Lekkerkerker et al.,[57] referred to earlier. They show that the greater the polydispersity, the wider the fluid–fluid coexistence region, implying flocculation is seen at lower polymer concentrations, for a given particle number concentration (and vice versa). This is in accord with the prediction of Walz[22] that polydispersity increases the depletion interaction (see earlier). Sear and Frenkel[73] have also considered possible fractionation of the polymer sample between the two coexisting colloidal phases. They showed that the concentration ratio of smaller polymer chains to larger polymer chains is larger in the colloid-rich phase than in the colloid-poor phase. That is, the colloid-rich phase contains predominantly smaller chains. Warren[74] carried out a similar analysis to Sear and Frenkel, but for the case of *binary mixtures*, rather than for polydisperse polymers, per se. The general conclusions reached are similar. Moreover, Warren[74] directly compared his theoretical predictions with the experimental data of Jenkins and Vincent,[75] who studied the phase separation behavior of dispersions of SiO_2-g-C_{18} particles in (i) toluene solutions of binary mixtures of PS or (ii) cyclohexane solutions of binary mixtures of PDMS. A typical data set, for the latter case, is shown in Figure 6.8. The M_n values of the two PDMS polymers used in the case were 53,000 and 580,000; ϕ_2^+ is the total volume fraction of polymer required to induce depletion flocculation; W_L is the weight fraction of the larger polymer in the binary mixture. To Jenkins and Vincent, the shape of the experimental curve (i.e., convex to the W_L axis) seemed intuitively wrong; they expected to find a curve concave to the W_L axis, as illustrated by the dashed line in the figure. Put simply, one would expect the smaller chains to partially fill any depletion layer formed by the larger chains, so the smaller chains should control the depletion flocculation behavior, at least until the value of W_L approaches unity. Jenkins and Vincent backed up this conjecture by some theoretical calculations based on estimating second virial coefficients in the dispersions, and, hence, predicting the (spinodal) phase boundary conditions. However, the calculations by Warren[74] suggest that the observed *shape* of the curve is indeed correct, even if the theoretical curve he finds for ϕ_2^+ against W_L lies considerably above the experimental one shown in Figure 6.8. The situation is made more complex by the

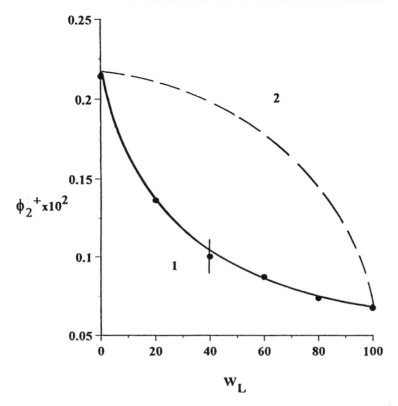

FIGURE 6.8   (O) Experimental flocculation boundary curve: *total* polymer volume fraction (ϕ_2^+) versus weight fraction (W_L) of the *larger* polymer, for SiO_2-g-C_{18} particles dispersed in cyclohexane, in the presence of mixtures of PDMS (M_n 53,000) and PDMS (M_n 580,000); dotted line: the expected stability boundary. Adapted from Jenkins and Vincent.[75]

fact that Jenkins and Vincent[75] were able to show that the shape of the ϕ_2^+ versus W_L curve actually depends on the order of mixing of the system. If a small amount of the smaller polymer is added first, but insufficient to reach ϕ_2^+, and then sufficient of the larger polymer added until the flocculation boundary is reached, one may calculate the effective W_L value, corresponding to that value of ϕ_2^+. In that case, the compiled experimental ϕ_2^+ versus W_L curve follows exactly the curve shown in Figure 6.8 for the premixed systems. However, if a similar set of experiments is carried out in which the *larger* polymer is added first, then a curve having the shape shown by the dashed line in Figure 6.8 is found! Jenkins and Vincent[75] offer a nonequilibrium explanation of these effects. If the smaller polymer is added first, then relatively "close-packed" *transient* aggregates are formed in the dispersion below the phase separation boundary (cf. short-lived, transient clusters in molecular condensation processes—e.g., vapor to liquid). They are close-packed because the range of the depletion interaction is small when the smaller polymer is added. The larger polymer chains are subsequently effectively excluded from penetrating into these transient clusters. This has two consequences: (i) the clusters act effectively as much larger particles; and (ii) the effective volume fraction of the larger polymer chains in free solution is increased. Both of these effects lead to a reduction in ϕ_2^+, below the equilibrium value and hence the shape of the experimental ϕ_2^+ versus W_L curve is as shown in Figure 6.8. This would represent *nonequilibrium* behavior. The expected equilibrium behavior (the dashed curve) is found in the case that the *larger* polymer is added first, since then the transient clusters formed are much more open (the range or the depletion interaction is longer), and the smaller chains, when added subsequently, *are* able to penetrate these clusters. Clearly, this issue awaits further clarification.

The theme of nonequilibrium behavior in depletion flocculation systems has been taken up by Allan et al.[69] They report studies on the addition of sodium polyacrylate to dispersions of Ludox silica particles in aqueous electrolyte solutions. They have shown that the method of achieving the flocculation boundary not only affects the position of that boundary but completely changes the structure of the flocs formed. Two methods of mixing were employed: conventional, straight addition of the polyelectrolyte to particulate dispersion, slowly increasing the concentration of the polyelectrolyte until phase separation is first observed and (ii) mixing the particles and polyelectrolyte at concentrations (for both) *below* the flocculation phase boundary, and then slowly evaporating the mixture until the boundary is reached, and beyond. This process could be conveniently monitored in this work using turbidiometry, by having a control without any polyelectrolyte present. Careful monitoring of the water weight loss with time was made, so that the particle and polyelectrolyte concentrations were always known at any time. Using method (i) led to the formation of flocs that were irregular in shape, size, and density (and hence fractal dimension). Using method (ii), however, led to reasonably monodisperse, pseudo-spherical flocs, which were close packed, and whose average size increased as the evaporation process continued, that is, as the polyelectrolyte and particle concentrations both increased. It is suggested that the results from method (ii) lie much closer to the expected equilibrium behavior.

It was not possible to determine whether the close-packed flocs found using method (ii) were actually crystalline or not because the constituent particles were just too small for diffraction studies. However, Hemsley et al.[76] have carried out similar slow evaporation experiments on systems with much larger particles, namely aqueous polystyrene latex dispersions, to which carboxymethylcellulose (CMC) was added. Evaporation was continued over a time scale of several weeks. In this case, after a certain period, depletion-induced phase separation set in. The dense colloid phase that separated was iridescent, and therefore crystalline in nature. These experiments were actually carried out to mimic the formation of colloidal crystals found in nature in the walls of certain spores. In that case the colloidal particles are biologically formed latex particles (probably based on *p*-coumaric acid) and the depletion-induced phase separation (colloid crystallization) is caused by the presence of natural polysaccharides in the aqueous solution present, along with the latex particles, inside the "sac" containing the spores, before it breaks and the spores are released.[77]

REFERENCES

1. J.M.H.M. Scheutjens, G.J. Fleer and B. Vincent, *Polymer Adsorption and Dispersion Stability*, ACS Symp. Ser., **240**, American Chemical Society, Washington, D.C. (1984), p. 245.

2. S. Rawson, K. Ryan, and B. Vincent, *Colloids Surfaces*, **34**, 89 (1988).

3. G.J. Fleer, M.A. Cohen Stuart, J.M.H.M. Scheutjens, T. Cosgrove, and B. Vincent, *Polymers at Interfaces*, Chapman Hall, London, 1993.

4. P.D. Jenkins, and M.J. Snowden, *Adv. Colloid Interface Sci.* **68**, 57 (1996).

5. J.E. Seebergh, and J.C. Berg, *Langmuir* **10**, 454 (1994).

6. L.T. Lee, O. Guiselin, A. Lapp, B. Farnoux, and J. Penfold, *Phys. Rev. Lett.* **67**, 2838 (1991).

7. C. Alain, D. Ausserre, and F. Rondelez, *Phys. Rev. Lett.* **49**, 1694 (1982).

8. D. Aussere, H. Hervet, and F. Rondelez, *Phys. Rev. Lett.* **54**, 1948 (1985).

9. D. Aussere, H. Hervet, and F. Rondelez, *Macromolecules* **19**, 85 (1986).

10. G. Chauveteau, M. Tirrel, and A. Omari, *J. Colloid Interface Sci.* **100**, 41 (1984).

11. T. Cosgrove, T.M. Obey, and K. Ryan, *Colloids Surfaces* **65**, 1 (1992).

12. E. Donath, A. Krabi, G.C. Allan, and B. Vincent, *Langmuir* **12**, 3425 (1996).

13. E. Donath, A. Krabi, M. Nirschl, V.M. Shilov, M.I. Zharkikh and B. Vincent, *J. Chem. Soc., Faraday Trans.* **93**, 115 (1997).

14. J.Y. Walz, and A. Sharma, *J. Colloid Interface Sci.* **168**, 485 (1994).

15. Y. Mao, M.E. Cates, and H.N.W. Lekkerkerker, *Physica A* **222**, 10 (1995).

16. B.V. Derjaguin, and L. Landau., *Acta Physiochim., URSS* **14**, 633 (1941).

17. E.J. Verwey, and J.Th.G. Overbeek, *Theory of Stability of Lyophobic Colloids*, Elsevier, Amsterdam, 1948.

18. R.I. Feigin, and D.H. Napper, *J. Colloid Interface Sci.* **7**, 117 (1979).

19. J.M.H.M. Scheutjens, and G.J. Fleer, *J. Phys. Chem.* **83**, 1619 (1979).

20. J.M.H.M. Scheutjens, and G.J. Fleer, *J. Phys. Chem.* **84**, 178 (1980).

21. P.J. Flory, *Principles of Polymer Chemistry*, Cornell University Press, Ithaca, NY, 1953.

22. J.Y. Walz, *J. Colloid Interface Sci.* **178**, 505 (1996).

23. X.L. Chu, A.D. Nikolov, and D.T. Wasan, *Langmuir* **10**, 4403 (1994).

24. X.L. Chu, A.D. Nikolov, and D.T. Wasan, *J. Chem. Phys.* **103**, 6653 (1995).

25. M.A.G. Dahlgren, and F.A.M. Leermakers, *Langmuir* **11**, 2996 (1995).

26. X. Chatellier, and J.F. Joanny, *J. Phys. France II* **6**, 1669 (1996).

27. A.J. Milling, and B. Vincent, *J. Chem. Soc., Faraday Trans.* **93**, 3179 (1997).

28. E. Evans, and D. Needham, *Macromolecules* **21**, 1822 (1988).

29. W.A. Ducker, T.J. Senden, and R.M. Pashley, *Nature* **353**, 239 (1991).

30. D.Y.C. Chan, and R.G. Horn, *J. Chem. Phys.* **83**, 5311 (1985).

31. P.F. Luckham, and J. Klein, *Macromolecules* **18**, 721 (1985).

32. A.J. Milling, and S. Biggs, *J. Colloid Interface Sci.* **170**, 604 (1994).

33. D.H. Napper, *Polymeric Stabilisation of Colloidal Dispersions*, Academic, London, 1983.

34. T. Kuhl, Y. Guo, J.Y. Alderfer, A.D. Berman, D. Leckband, J. Israelachvili, and S.W. Hui, *Langmuir* **12**, 3003 (1996).

35. A. Vrij, *Pure Appl. Chem.* **48**, 471 (1976).

36. P. Richetti, and P. Kekicheff, *Phys. Rev. Lett.* **68**, 1951 (1992).

37. J.L. Parker, P. Richetti, P. Kekicheff, and S. Sarman, *Phys. Rev. Lett.* **68**, 1955 (1992),

38. P. Kekicheff, and P. Richetti, *Prog. Colloid Polym Sci.* **88**, 8 (1992).

39. P. Kekicheff, F. Nalett, and P. Richetti, *J. Phys. II France* **4**, 735 (1994).

40. D.C. Prieve, and N.A. Frej, *Langmuir* **6**, 396 (1990).

41. A. Sharma, and J.Y. Walz, *J. Chem. Soc., Farday Trans.* **92**, 4997 (1996).

42. D.L. Sober, and J.Y. Waltz, *Langmuir* **11**, 2352 (1995).

43. O. Mondain-Monval, F. Leal-Calderon, and J. Bibette, *J. Phys. II France* **6**, 1313 (1996).

44. J. Marra, and M.L. Hair, *J. Colloid Interface Sci.* **128**, 511 (1989).

45. A.J. Milling, *J. Phys. Chem.* **100**, 8986 (1996).

46. D. Sohn, P.S. Russo, A. Davila, D.S. Posche, and M.L. McLaughlin, *J. Colloid Interface Sci.* **177**, 31 (1996).

47. A. Asnacios, A. Espert, A. Colin, and D. Langevin, *Phys. Rev. Lett.* **78**, 4974 (1997).

48. P.G. De Gennes, P. Pincus, R.M. Velasco, and F. Brouchard, *J. Phys. France*, **37**, 1461 (1976).

49. A.V. Dobrynin, R.H. Colby, and R.H. Rubenstein, *Macromolecules* **28**, 1859 (1995).

50. A. Sharma, S.N. Tan, and J.Y. Walz, J. Colloid Interface Sci. **191**(1), 236 (1997).

51. R. Macmillan, Ph.D. Thesis, University of Bristol, 1989.

52. J. Israelachvili, *Intermolecular and Surface Forces*, 2nd ed., Academic, San Diego, 1992.

53. S. Emmett, and B. Vincent, *Phase Transitions* **49**, 121 (1990).

54. H. DeHek, and A. Vrij, *J. Colloid Interface Sci.* **70**, 592 (1979).

55. B. Vincent, J. Edwards, S. Emmett and R. Croot, *Colloids Surfaces*, **31**, 267 (1988).

56. B. Vincent, *Colloids Surfaces*, **24**, 269 (1987).

57. H.N.W. Lekkerkereker, W.C. Poon, P.N. Pusey, A. Stroobants, and P.B. Warren *Europhys. Lett.* **20**, 559 (1992).

58. F. Leal Calderon, J. Bibette, and J. Biais, *Eur. Phys. Lett.* **23**, 157 (1993).

59. M.A. Faers, and P.F. Luckham, *Langmuir* **13**, 2922 (1997).

60. W.C.K. Poon, J.S. Selfe, M.B. Robertson, S.M. Illett, A.D. Pirie, and P.N. Pusey, *J. Phys II France* **3**, 1075 (1993).

61. K. Bean, P. Jenkins, and B. Vincent, in *Conference on Colloid Chemistry, Proc. 7th* (1996), Z. Horvoelgyi, Zs. Nemeth, and I. Puszli, eds., Hungarian Chemical Society, Budapest, Hungary (1997), p. 21; K. Bean, Ph.D. thesis Bristol, 1997.

62. A.J. Milling, B. Vincent, S. Emmett, and D.A.R. Jones, *Colloids Surfaces* **57**, 185 (1991).

63. N. Cawdery, A.J. Milling, and B. Vincent, *Colloids Surfaces* **86**, 239 (1994).

64. J.F. Miller, K. Schätzel and B. Vincent, *J. Colloid Interface Sci.* **143**, 532 (1991).

65. A. Sharma, S.N. Tan, and J.Y. Walz, *J. Colloid Interface Sci.* **191**, 236 (1997).

66. M.J. Snowden, M.S. Clegg, P.A. Williams, and I.D. Robb, *J. Chem. Soc. Faraday Trans.* **87**, 230 (1991).

67. A. Jones, and B. Vincent, *Colloids Surf.* **42**(1–2), 113 (1989).

68. N.J. Smith, and P.A. Williams, *J. Chem. Soc., Faraday Trans.* **91**, 1483 (1995).

69. G.C. Allan, M.J. Garvey, J.W. Goodwin, R.W. Hughes, R. MacMillan, and B. Vincent, in R.H. Ottewill and A.R. Rennie, eds., in *Modern Aspects of Colloidal Dispersions*, Kluwer, Dordrecht, Netherlands 1998.

70. Y. Otsubo, *Heterogeneous Chem. Revs.* **3**, 327 (1996).

71. X. Ye, T. Narayanan, P. Tong, and J.S. Huang, *Phys. Rev. Lett.* **76**, 4640 (1996).

72. R. Rajagopalan, in A.K. Arora and B.V.R. Tata, eds., in *Ordering and Phase Transition in Charged Colloids*, VCH, New York, 1996, Chap. 13.

73. R.P. Sear, and D. Frenkel, *Phys. Rev. E* **55**(2), 1677 (1997).

74. P.B. Warren, private communication; *Langmuir* **13**(17), 4588 (1997).

75. P.D. Jenkins, and B. Vincent, *Langmuir* **12**, 3107 (1996).

76. A.R. Hemsley, P.D. Jenkins, M.E. Collinson, and B. Vincent, *Bot. J. Linnean Soc.* **121**, 177 (1996).

77. A.R. Hemsley, M.E. Collinson, W.L. Kovach, B. Vincent, and T. Williams, *Phil. Trans. Royal. Soc. (London) B* **345**, 163 (1994).

7 Polyelectrolyte Adsorption: Theory and Simulation

M. MUTHUKUMAR

Polymer Science and Engineering Department, and Materials Research Science and Engineering Center, University of Massachusetts, Amherst, Massachusetts 01003

7.1 INTRODUCTION

Adsorption by polyelectrolytes to charged surfaces and complexation between oppositely charged polyelectrolytes constitute a vast and rich research area with a wide range of conceptual puzzles and diverse set of industrial applications. Complexation between macromolecules and surfaces having specific chemical patterns leads to an infinitely diverse set of structures, dynamics, and functions of these resulting complexes, as evidenced overwhelmingly in biological contexts.[1,2] One of the most fundamental examples is the delicate complexation between chromosomes and the inner membrane of the nuclear envelope of an eukaryotic cell. The inner membrane contains many proteins, assembled in specific patterns, as binding sites for the chromosome, which in itself is a complex of DNA with hundreds of proteins. These complexation reactions regulate the cell cycle and are vital to the life and death of a cell. Although the chemical natures of the complementary molecules in such complex formations are known in many biological examples, the general priniciples behind pattern recognition in macromolecular self-assembly are not yet understood.

In addition, there is an increasing interest to assemble desired nanoscale structures[3,4] at interfaces and control the lifetimes of such structures. One underlying hope is to fabricate artificial enzymes by polymer synthesis and to manipulate their interfacial behavior. One method for exploring these issues is to pattern a surface in a specific way and then determine the specific complementary sequences of chemical groups needed for macromolecular recognition of such a patterned surface. This issue

Colloid–Polymer Interactions: From Fundamentals to Practice, Edited by Raymond S. Farinato and Paul L. Dubin
ISBN 0-471-24316-7   © 1999 John Wiley & Sons, Inc.

175

appears repeatedly in various biological contexts. Using this approach, it is also possible to explore novel separation protocols.[5]

Although the most straightforward approach to these problems is to directly solve a specific example with appropriate microscopic details, we focus here on uncovering fundamental principles of self-assembly of polyelectrolytes at patterned surfaces.

When the loss in configurational entropy of a polymer due to the presence of a surface is compensated by the accompanying gain in energy of the polymer–surface interaction, the polymer adsorbs[6] to the surface. For charge-bearing polymers, there are no simple arguments to estimate the entropic loss due to the proximity of a surface and to intrachain electrostatic repulsion. The situation becomes more complicated when the adsorbing surface is not uniformly charged or itself is another polyelectrolyte. Under these circumstances, complete adsorption is expected to be frustrated due to the complexation among a few pairs of charged groups with the formation of many loops. Some of these pairs must dissociate and then reassociate differently in a cooperative way to proceed with adsorption so as to achieve the final minimum free energy structure.

In general the adsorbing polyelectrolyte may contain one kind of charge group distributed along the chain backbone in a particular sequence. The chain may also have both kinds of charges, thus constituting a polyampholyte. There are many features that control the nature of a surface. These include the charge density and the sign of the surface charge (in comparison to that of the polyelectrolyte), heterogeneity of the surface charge distribution, surface curvature, and whether the interface is a solid–liquid interface or a liquid–liquid interface.

We illustrate below the key principles by first considering the simplest situation of a uniformly charged polyelectrolyte adsorbing to an oppositely charged planar surface with a uniform charge density and in a medium containing a known amount of added salt and counterions. This problem has been studied using experimental techniques,[6-9] analytical methods,[10-14] numerical methods,[15-18] and Monte Carlo simulations.[19,20] The roles of kinetics[6,21,22] and confinement[23] have also been addressed. In an effort to understand the fundamental aspects of colloidal stability influenced by polyelectrolytes adsorbing to colloidal particles, there have recently been experimental investigations of isolated polyelectrolyte chains complexing with latex spheres,[24] proteins,[25-27] micelles,[28] vesicles,[29] dendrimers,[30-32] and microgel particles.[33] Adsorption of a uniformly charged polyelectrolyte to a uniformly charged sphere has been investigated both analytically[34] and using Monte Carlo simulations.[35-37]

Next a simple model is described depicting the adsorption of a uniformly charged polyelectrolyte to an oppositely charged surface. The surface is taken to be a chemically heterogeneous surface with surface patterns made of charges. We will then summarize only the final results for the critical condition necessary for adsorption of a polyelectrolyte to a planar surface, a curved surface, and a patterned surface. After briefly considering complexation between polyions, the kinetics of adsorption and the role of charge sequences are presented. Finally we will comment on the complicated situation of assemblies formed simultaneously by many chains.

7.2 MODEL

The basic model is sketched in Figure 7.1. We adopt the Kuhn model to represent the connectivity of a polymer chain. The actual chain of contour length L is imagined to consist of N sequentially connected segments with each segment containing several monomers. The Kuhn model chain of N segments obeys random-walk statistics with step length l such that $L = Nl$. The values of the number of Kuhn segments N and the Kuhn step length l are determined empirically for a given polymer backbone structure. The average charge density on the polymer is q. The sequence of charges on the polymer is designated as S_g. The surface contains the pattern made up of opposite charges to that of the polymer. Let the typical size and the average charge density of the pattern be Λ and σ, respectively. The sequence of charges on the pattern will be designated as S_h.

Let us assume that the solution containing the polyelectrolyte chain can be treated as a continuum with an effective dielectric constant ε at a given temperature T. We further assume that the charge density is weak enough so that charges interact among themselves with screened Coulomb interactions of the Debye–Hückel type. Thus, two charges of magnitude e separated by a distance r interact with potential energy,

$$V(r) = \frac{l_B}{r} e^{-\kappa r} \tag{7.1}$$

where the Bjerrum length l_B is

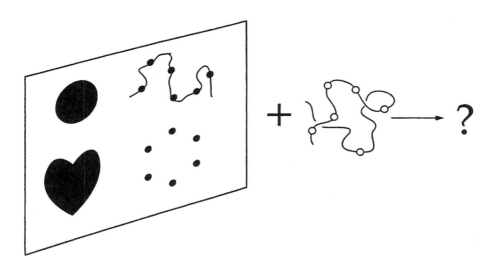

FIGURE 7.1 Sketch of the model problem where a polymer adsorbs to a patterned surface.

$$l_B = \frac{e^2}{4\pi\varepsilon_0 \varepsilon k_B T} \tag{7.2}$$

where ε_0 is the permittivity of vacuum and k_B is the Boltzmann constant; κ is the inverse Debye length, determining the range of the electrostatic interaction. Within the assumed Debye–Hückel approximation, κ is determined by the concentrations of various mobile ions in the system except those on the polyelectrolyte and the surface. It is given by

$$\kappa^2 = 4\pi l_B \sum_i c_i z_i^2 \tag{7.3}$$

where c_i and z_i are the concentration and valence, respectively, of the ith mobile ion. Therefore the strength and range of the electrostatic interaction are l_B/l and κl in dimensionless units.

The polymer is represented as a continuous flexible polyelectrolyte with a uniform charge distribution. There are q charges per Kuhn length. The potential energy associated with intrachain segment–segment interactions is given by

$$\frac{V}{k_B T} = \frac{1}{2} \sum_{i=1}^{N} \sum_{j=1}^{N} \left[\left(\frac{1}{2} - \chi \right) l^3 \delta(r_{ij}) + \frac{l_B q^2 l^2}{r_{ij}} \exp(-\kappa r_{ij}) \right] \tag{7.4}$$

where r_{ij} is the separation between the i and j segments; χ is the usual Flory excluded volume parameter arising from the short-range interaction; and δ is the Dirac delta function. The second term in the square bracket denotes the electrostatic part. Here κ is determined[38] only by the counterions dissociated from the polyelectrolyte and the ions from the added salt.

The potential energy V_s associated with the electrostatic interaction between the polymer and the patterned surface is given by

$$V_s = \sum_{i=1}^{N} \sum_{s=1}^{N_s} \frac{k_B T l_B q l \sigma_s}{r_{is}} \exp(-\kappa r_{is}) \tag{7.5}$$

where the pattern consists of N_s charged ions with σ_s being the charge of the sth ion; r_{is} is the distance of separation between the sth ion of the pattern and the ith segment of the polyelectrolyte. The pattern is assumed to be on an infinite planar surface situated at $z = 0$. The polyelectrolyte solution is in the space $z > 0$. The space $z < 0$ is not accessible to the polyelectrolyte. This condition is accounted for by choosing the appropriate boundary condition in the calculation.[11,12]

7.3   RESULTS

7.3.1   Isolated Polyelectrolyte Chain

Before studying the interaction of the polyelectrolyte with the patterned surface, consider a polyelectrolyte chain in a dilute solution. Even the most fundamental questions about the radius of gyration and charge distribution of an isolated polyelectrolyte chain at different ionic strengths are of much debate, as reviewed in Beer et al.[39] However, a simple result for the mean-square radius of gyration R_g^2 of the chain is obtained by assuming uniform chain expansion due to excluded volume and electrostatic interactions:

$$R_g^2 = \frac{Nl^2}{6}\left(\frac{l_1}{l}\right)$$

(7.6)

where the swelling ratio l_1/l is given by[12]

$$\left(\frac{l_1}{l}\right)^{3/2} - \left(\frac{l_1}{l}\right)^{1/2} = \zeta_{ev} + \zeta_c,$$

$$\zeta_{ev} = \frac{4}{3}\left(\frac{3}{2\pi}\right)^{3/2}\left(\frac{1}{2}-\chi\right)N^{1/2}\left(\frac{l}{l_1}\right)$$

$$\zeta_c = \frac{4}{45}\left(\frac{6}{\pi}\right)^{1/2}\left(\frac{l_B}{l}\right)(ql)^2 N^{3/2}\left\{\frac{15\sqrt{\pi}\,e^a}{2a^{5/2}}(a^2-4a+6)\right.$$

$$\left.\times \text{erfc}\sqrt{a}+\frac{15}{\sqrt{\pi}}\left(-\frac{3\pi}{a^{5/2}}+\frac{\pi}{a^{3/2}}+\frac{6\sqrt{\pi}}{a^2}\right)\right\}$$

(7.7)

with $a = \kappa^2 R_g^2$. From Eq. (7.7) we can show that l_1 is proportional to L and $L^{1/5}\,\kappa^{-4/5}$ in the low (small κl) and high (large κl) salt concentration limits, respectively. In these extremes R_g is proportional to L (rodlike but not a rod) and $L^{3/5}$ (good solution behavior), respectively. The above crossover formula predicts correctly the asymptotic limits for low- and high-salt concentrations and is in good agreement[39] with experimental data on quarternized poly(vinyl pyridines) in water containing added salt.

7.3.2   Adsorption to a Planar Surface

Now we consider the interaction of the uniformly expanded polyelectrolyte chain with a uniformly charged surface. Let the surface charge be opposite to that of the polyelectrolyte and σ be the charge density on the planar surface. The variables σ, q, l_B, κ and the expansion factor l_1/l determine whether the polyelectrolyte chain is adsorbed to the surface or not. The critical condition for adsorption is obtained theoretically[12] as

$$\frac{\sigma q l_B}{\kappa^3 l_1} \geq \frac{0.12}{\pi} \qquad (7.8)$$

As mentioned, l_1 is proportional to L and $L^{1/5}\kappa^{-4/5}$ in the low- (small κl) and high- (large κl) salt concentration limits, respectively. Substituting the limiting values of l_1 into Eq. (7.8) provides limiting laws of adsorption critical conditions. Using the surface charge density σ to tune the adsorption, for example, the chain is adsorbed at $\sigma > \sigma_c$, where

$$\sigma_c \sim \begin{cases} \kappa^3, & \text{low salt} \\ \kappa^{11/5}, & \text{high salt} \end{cases} \qquad (7.9)$$

Using the definition of l_B and the full expression for l_1, the condition for absorption is given in Figure 7.2 where $4\pi\varepsilon\varepsilon_0 k_B T/\sigma q e^2 l^2$ is plotted against κl for $\chi = 0.5$ and $N = 60$. In the region below the theoretical curve, there is adsorption. There is no

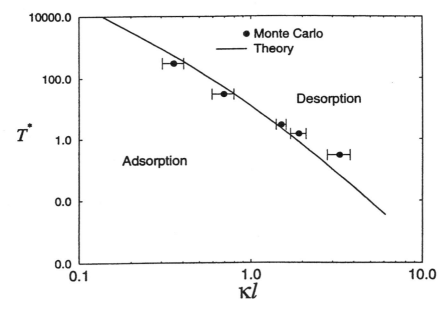

FIGURE 7.2   Comparison between Eq. 7.8 and Monte Carlo simulation data for adsorption of a uniformly charged polyelectrolyte ($\chi = 0.5$, $N = 60$) to a uniformly charged surface. Ordinate is the reduced temperature $T^* = 4\pi\varepsilon\varepsilon_0 k_B T/\sigma q e^2 l^2$.

adsorption in the parameter range above the curve. Also adsorption can be induced by increasing σ, q, or l_B or by decreasing κ. The data points given in the figure are from Monte Carlo simulations[20] for $\chi = 0.5$ and $N = 60$, which verify the validity of the theoretical calculation.

7.3.3 Adsorption to a Sphere

If the adsorbing surface is a sphere of radius R and uniform charge density σ, the critical condition for adsorption is[34]

$$\frac{\sigma q l_B (1 - e^{-2\kappa R})}{\kappa^3 l_1} \geq \frac{0.12}{\pi} \tag{7.10}$$

In addition to the variables σ, q, l_B, l, and κ, the sphere radius R can also be tuned to control the adsorption. The dependence of the critical charge density on κ is now modified by the sphere radius. If R is sufficiently large, Eq. 7.10 reduces to Eqs. 7.8 and 7.9. If R is small, but still satisfying the inequality of Eq. 7.10, the critical surface charge density becomes

$$\sigma_c \sim \begin{cases} \kappa^2, & \text{low salt} \\ \kappa^{6/5}, & \text{high salt} \end{cases} \tag{7.11}$$

The high-salt limit has been observed experimentally.[28–30] For example, if $\sigma q l_B l^2 = 1$ and $N = 60$, the critical radius R_c necessary for adsorption is given in Figure 7.3 as a function of κl. In the region above the curve, there is adsorption. The curve is in reasonable agreement with the Monte Carlo data[20] obtained for the same model. It is clear from Figure 7.3 that as the salt concentration increases, increasingly larger spheres desorb their polymer. Adsorption conditions similar to Eqs. 7.10 and 7.11 were also derived for a uniformly charged cylinder.[34]

The formula of Eq. 7.10 shows that by fixing other variables at suitable values, the critical radius required for complexation increases monotonically with salt concentration. This criterion is in qualitative agreement with the observed depletion of increasingly larger histones from chromatin fiber[1,2] upon a continuous increase in salt concentration. One manifestation of this effect might enable an efficient protein separation method based on eluting them from their complexes with polyelectrolytes by using different ionic strengths. Mattison et al.[25] have investigated complexation between the globular protein bovine serum albumin (BSA) and a synthetic polyelectrolyte. Using turbidimetric pH titrations, they determined the specific pH values where soluble complex formation is initiated (pH_c) and where phase separation occurs (pH_ϕ). These values, collected at different ionic strengths, can be presented as phase boundaries that may be used in the selection of optimal pH and ionic strength for maximum yield and purity in protein separations.

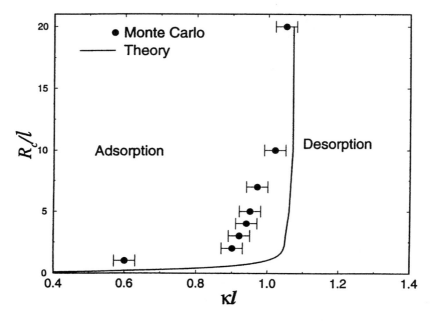

FIGURE 7.3   Dependence of critical radius on κ. The curve is Eq. 7.10 and the data are from Monte Carlo simulations.

7.3.4   Adsorption to a Patterned Surface

Let a pattern of circular symmetry of radius Λ with charge density σ be imprinted on a planar surface to which an oppositely charged polyelectrolyte adsorbs. Now the critical condition for the binding of the polymer to a pattern of size Λ is given by[40]

$$\frac{\sigma q l_B [1 - (\kappa\Lambda + 1)e^{-\kappa\Lambda}]}{\kappa^3 l_1} \geq \frac{0.12}{\pi} \tag{7.12}$$

This result was also verified by performing Monte Carlo simulations.[40] Recently Mattison et al.[27] have used light scattering and pH titration to study the binding of synthetic polyelectrolytes to proteins. The critical protein charge required to induce protein–polyelectrolyte complexation was found to vary directly with κ. Similar results have been observed[28-32] for complexation between polyelectrolytes and micelles or dendrimers. By substituting the experimental values for the parameters of Eq. 7.12, Mattison et al.[41] have found an agreement between their observations and the predicted critical condition for complexation. However, much more experimental work is required in this crucial area of research.

7.3.5   Complexation Between Polyions

The same calculational method used in obtaining the critical condition for binding of a polyelectrolyte to a charged surface can be used to obtain the condition for complexation between a polyanion and a polycation. Two possible situations may be identified. In one case, let a mobile polyion with N_2 segments and charge density q_2 complex with a frozen polyion of opposite charge having N_1 segments and charge density q_1. In the second case, let both polyions be mobile. In the first case, the condition for binding at the high-salt limit turns out to be

$$\frac{q_1 q_2^{1/5}}{\kappa^{6/5}} \left(\frac{l_B}{l}\right)^{3/5} \frac{N_1}{N_2^{1/5}} \geq A \tag{7.13}$$

where A is a constant. It is to be noted that the exponents of N_1 and N_2 are not the same. Similarly the exponents of q_1 and q_2 are different. Thus the frozen host dominates the critical condition for binding over the mobile guest polyion. In the low-salt condition, the above condition changes to

$$\left(\frac{q_1^2 l_B l}{q_2}\right)^{1/3} \frac{N_1^3}{N_2} \geq A_1 \tag{7.14}$$

where A_1 is a constant. Again, the frozen host dominates over the mobile guest. In the second case, where both polyions are mobile, the condition for complexation is determined equally by both chains. For example, the radius of gyration of the complex is comparable to that of either chain and is determined by the loop length,

$$R_g \sim R_{g1} \sim R_{g2} \sim \left(\frac{N}{m}\right)^{1/2} \tag{7.15}$$

where R_{gi} is the radius of gyration of the ith chain, N is the number of segments in either the polycation or polyanion, and m is the number of pairs that are formed in the complex.

7.3.6   Kinetics

So far we considered only the critical condition for the binding of a polyelectrolyte to an oppositely charged surface (planar or spherical) containing a pattern or another polymer. For the case of polyelectrolytes, the recognition of a specific pattern by a polyelectrolyte with a specific sequence is extremely slow as seen in Monte Carlo simulations. In fact the complexation between two oppositely charged polymers has been seen[41] in Monte Carlo simulations to proceed in two steps. The first step is a relatively fast complexation without registry of the pattern followed by a very slow process of registry.

The inability of a chain to recognize a pattern perfectly in a reasonable time can be understood by the following simple argument. Consider only three units (labeled $i =$ 1, 2, 3) making up a linear pattern with a regular spacing b in three-dimensional space. Let us proceed to monitor the binding of a portion of a chain with three special segments (labeled $i_p = 1, 2, 3$) complementary to those making up the pattern. Let there be a spacer of $(\mu - 1)$ neutral segments between any two consecutive special segments. Also, let the first special segment be anchored to the end unit of the pattern as shown in Figure 7.4a. Assuming that the chain obeys Gaussian statistics and the gain in energy for contact between any special segment of the polymer and any unit of the pattern is ε, the free energy corresponding to the configuration of Figure 7.4a is $-\varepsilon$. The second contact between the polymer and the pattern can take place in four possible ways. These are indicated in Figure 7.4b to 7.4e; $i_p = 2$ can be paired either with $i = 2$ or 3; $i_p = 3$ can be paired either with $i = 2$ or 3. When $i_p = 2$ is paired with $i = 2$, the end-to-end distance of the polymer spacer with μ segments is b with an

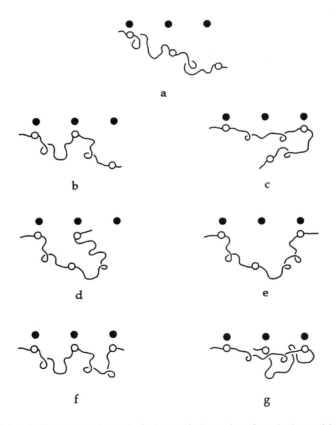

FIGURE 7.4   Different possible topological states during registry for only three pairing events.

accompanying entropic penalty of $s \equiv 3\kappa_B b^2 / 2\mu l^2$. Therefore the free energy of the complex depicted as the graph of Figure 7.4b is $-2\varepsilon + s$ apart from a constant term. Similarly the complexes with contacts $(i_p = 2; i = 3)$, $(i_p = 3; i = 2)$, and $(i_p = 3; i = 3)$ have free energies $-2\varepsilon + 4s$, $-2\varepsilon + s/2$, and $-2\varepsilon + 2s$, respectively. It is to be noted that the complex of Figure 7.4d from $(i_p = 3; i = 2)$ is the most stable complex as far as two-contact complexes are considered and for $b^2/\mu > 2 \ln 2$.

Proceeding now to consider the three-contact complexes, there are only two such complexes, as shown in Figures 7.4f and 7.4g, with respective free energies $-3\varepsilon + 2s$ and $-3\varepsilon + 5s$. Assuming sequential pairing, the complex with full registry given in Figure 7.4f can arise only from Figures 7.4b and 7.4e. The complex of Figure 7.4g can arise from Figures 7.4c or 7.4d. While the most stable complex with only two contacts is that of Figure 7.4d, its further evolution to the three-contact state is to a state with higher free energy and not to the lowest free energy state. It is obvious from the configurations of the complexes shown in Figures 7.4f and 7.4g, that these configurations cannot be directly converted to each other. The only way the complex of Figure 7.4g can relax to that of Figure 7.4f is to trace its trajectory backwards to its original state and try again to avoid the free energy minimum state at an intermediate time. Thus any effort to minimize the free energy of the system at intermediate times can take the trajectory further away from the proper trajectory leading to the fully registered state of the complex.

The above simple argument illustrates the necessity of optimized temporal correlations for full pattern recognition. It also indicates that any computational algorithm that minimizes the free energy at every time step can substantially prolong the time needed for full recognition. The same qualitative arguments are valid for the recognition process involving patterns even within a single polymer chain.

The restrictions on the pathways connecting different topological states have two consequences. If k_{ij} is the rate constant for the "reaction" of going to the jth state from the ith state, the time dependence of the concentration of the system in the ith state is given by

$$\rho(t) = \sum_{\mu=1}^{7} \exp\left(-\frac{t}{\tau\mu}\right) \tag{7.16}$$

where τ_μ is the relaxation time for the μth normal mode. This result can be equally fitted with a stretched exponential (a common feature of relaxations in glassy materials):

$$\rho(t) \simeq \exp\left[-\left(\frac{t}{\tau}\right)^{\beta}\right] \tag{7.17}$$

where τ and β are parameters. Since the free energies of the various states are known, the temperature dependence of the entropy S of the system can be calculated. A typical

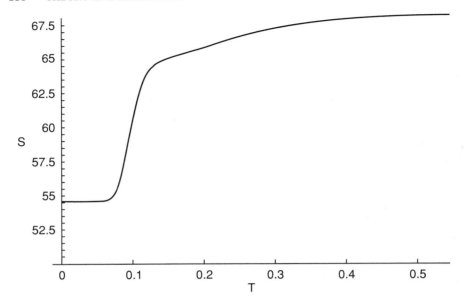

FIGURE 7.5   Temperature dependence of the entropy for the ensemble shown in Figure 7.4. Entropy is in units of the Boltzmann constant and T is the absolute temperature for the particular case discussed in the text.

result is given in Figure 7.5 (see Wang and Audebert[8] for more details). The entropy gradually decreases as the temperature is reduced and then smoothly crosses over to that of a fully registered state without any catastrophy. The stretched exponential nature of time evolution of population and the above-mentioned behavior of entropy with respect to temperature are typical properties[42] of glasses. Therefore these glassy features are borne out in the pattern recognition process.

7.3.7   Effect of Sequence

In the process of recognition, the sequence of the polymer naturally plays a crucial role. As an example, Monte Carlo results[43] for the self-aggregation of a polyampholyte chain are available. The probability distribution of R_g^2, $n(R_g^2)$, is plotted in Figure 7.6 against R_g^2 for a chain of $N = 50$ containing five positive charges and five negative charges at three different reduced temperatures. Similarly, the distribution $n(E)$ of intrachain energy E at these three different temperatures in shown in Figure 7.7. Figures 7.6a and 7.7a correspond to the sequence A,

$$O_6PNO_2NPO_6NOPO_5PO_8NO_2NO_8PO_2$$

and Figures 7.6b and 7.7b correspond to the sequence B,

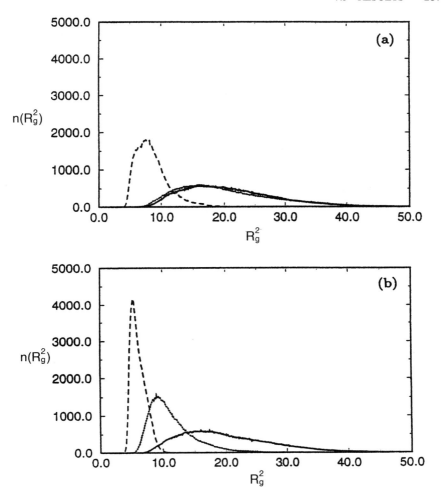

FIGURE 7.6 Distribution of mean square radius of gyration (in units of Kuhn length) at three reduced temperatures [from left to right, 0.078 (- - - -), 0.625 (· · · ·), and 2.5]. (*a*) and (*b*) correspond, respectively, to the sequences *A* and *B*.

$$O_5POPO_{11}NOP_2O_3PO_7NO_5NO_3NONO_3$$

Here the symbols *O*, *P*, and *N* denote, respectively, neutral, positively charged, and negatively charged segments. It is to be noted that sequence *A* contains charges randomly distributed and sequence *B* is close to a diblock case. Sequence *B* has a broad distribution of radii at high temperatures and a narrow distribution at low temperatures. Sequence *A*, on the other hand, has a broad distribution of radii even at

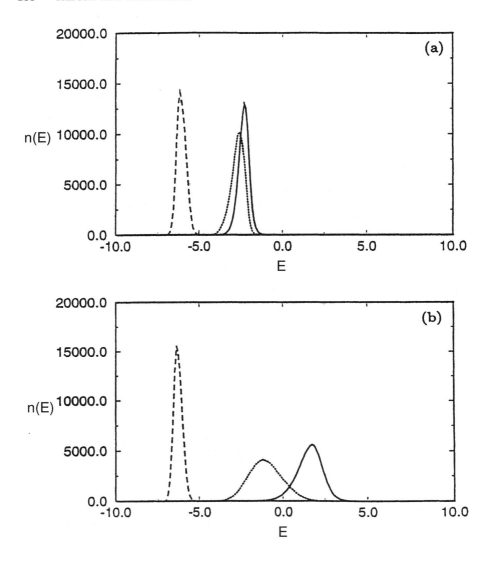

FIGURE 7.7   Distribution of reduced intrachain energy (in dimensionless units) at the same three reduced temperatures as in Figure 7.6 (*a*) and (*b*) correspond, respectively, to the sequences *A* and *B*.

low temperatures. It is also obvious from these figures that knowledge of merely the energy landscape is insufficient to describe the conformational properties of polyampholytes. The entropy associated with the formation of frustrated conformations needs to be accounted for in understanding the conformational properties of heteropolymers with specific sequences.

7.4 CONCLUSIONS

We have reviewed the theoretical and Monte Carlo simulation results for adsorption of isolated polyelectrolyte chains to an oppositely charged surface, whether planar, spherical, cylindrical, patterned, or another polyelectrolyte. Preliminary experimental data compare well with theoretical predictions as outlined above.

While this is a necessary step in the fundamental understanding of adsorption of polyelectrolytes, experimentation on the complexation by an isolated molecule is a nontrivial task. Typically many chains are involved in forming aggregates. The experimental conditions required for determining the complexation by single molecules have not yet been achieved. The complexation between polycations and polyanions in nondilute solutions has been investigated extensively.[5–7] The polyelectrolytes in these studies included both natural and synthetic polymers. Depending on the systems and conditions, the complexes were formed by an aggregation process where many chains were involved. Different states of aggregation were observed. Two extremes were macroscopically homogeneous systems consisting of soluble complexes and the precipitation of "polysalts" by phase separation. The complexes made from polycations and polyanions have many potential areas of applications.[5] While there is an accumulation of data on such multichain complexes, there has not yet been any predictive models to understand these data. We hope to see developments of the above ideas for higher concentrations of adsorbing polyelectrolytes.

When there are many similarly charged polyelectrolytes adsorbing to an oppositely charged surface, a layer is formed with charge reversal. This layer then can be used to adsorb polyelectrolytes of the same sign as the surface. In this way multilayers can be constructed[44] by a consecutive adsorption of polyanions and polycations. These are very stable layered nanoassemblies whose thickness can be tuned by limiting the number of deposited alternating layers. Although a fundamental understanding of the roles played by the various controling factors is not available yet, fabrication of multilayers using oppositely charged molecules (flexible polyelectrolytes, DNA, proteins, surfactants, lipids, clay and other colloids) holds a tremendous opportunity for the synthesis of new materials.

Acknowledgment   Acknowledgment is made to the NSF grant No. DMR-9625485 and the Materials Research Science and Engineering Center at the University of Massachusetts.

REFERENCES

1. B. Alberts, D. Bray, J. Lewis, M. Raff, K. Roberts, and J.D. Watson, *Molecular Biology of the Cell*, Garland, New York, 1994.

2. H. Lodish, D. Baltimore, A. Berk, S.L. Zipursky, P. Matsudaira, and J. Darnell, *Molecular Cell Biology*, W.H. Freeman, New York, 1997.

3. M. Muthukumar, C.K. Ober, and E.L. Thomas, *Science* **277**, 1225 (1997).

4. M. Muthukumar, *Curr. Opinion Colloid Interface Sci.* **3**, 48 (1998).

5. B. Philipp, H. Dautzenberg, K-J. Linow, J. Koetz, and W. Dawydoff, *Prog. Polym. Sci.* **14**, 91 (1989).

6. G.J. Fleer, M.A. Stuart Cohen, J.M.H.M. Scheutjen, T. Cosgrove, and B. Vincent, *Polymers at Interfaces*, Chapman & Hall, London, 1993.

7. H. Dautzenberg, W. Jaeger, J. Koetz, B. Philipp, Ch. Seidel, and D. Stscherbina, *Polyelectrolytes: Formation, Characterization and Application*, Hansen/Gardner, Cincinnati, 1994.

8. T.K. Wang and R. Audebert, *J. Colloid Interface Sci.* **121**, 32 (1988).

9. V. Shubin and P. Linse, *J. Phys. Chem.* **99**, 1285 (1995).

10. F.T.J. Hesselink, *J. Colloid Interface Sci.* **60**, 448 (1977).

11. F.W. Wiegel, *J. Phys. A* **10**, 299 (1977).

12. Muthukumar, M. *J. Chem. Phys.* **86**, 7230 (1987).

13. R. Varoqui, in K.S. Schmitz, ed., *Macro-Ion Characterization: From Dilute Solutions to Complex Fluids*, ACS, Washington, DC, 1994.

14. X. Chatellier and J.P. Joanny, *J. Phys. II France*, **6**, 1669 (1996).

15. H.A. Van der Schee and J. Lyklema, *J. Phys. Chem.* **88**, 6661 (1984).

16. O.A. Evers, G.J. Fleer, J.M.H.M. Scheutjens, and J. Lyklema, *J. Colloid Interface Sci.* **446**, 111 (1986)

17. M.R. Boehmer, O.A. Evers, and J.M.H.M. Scheutjens, *Macromolecules* **23**, 2288 (1990).

18. H.G.M. Van der Steeg, M.A. Cohen Stuart, A. de Keizer, and B.H. Bijsterbosch, *Langmuir* **8**, 2538 (1992).

19. S. Beltran, H.H. Hooper, H.W. Blanch, and J.M. Prausnitz, *Macromolecules* **24**, 3178 (1991).

20. C.Y. Kong and M. Muthukumar, *J. Chem. Phys.* **109**(4), 1522 (1998).

21. T.G.M. van de Ven, *Adv. Colloid Interface Sci.* **48**, 121 (1994).

22. E. Pefferkorn and A. Elaissari, *J. Colloid Interface Sci.* **138**, 187 (1990).

23. I. Borukhov, D. Andelman, and H. Orland, *Europhysics* **32**, 499 (1995).

24. Y. Adachi and T. Matsumoto, *Colloids Surf. A: Physicochem. Eng. Aspects* **113**, 229 (1996).

25. K.W. Mattison, I.J. Brittain, and P.L. Dubin, *Biotechnol. Prog.* **11**, 632 (1995).

26. A. Tsuboi, T. Izumi, M. Hirata, J. Xia, P.L. Dubin, and E. Kokufuta, *Langmuir* **12**, 6295 (1995).

27. K.W. Mattison, P.L. Dubin, and J.I. Brittain, *J. Phys. Chem.* **B102**, 3830 (1998).

28. D.W. McQuigg, J.I. Kaplan, and P.L. Dubin, *J. Phys. Chem.* **96**, 1973 (1992).

29. J.L. Thomas, K.A. Borden, and D.A. Tirrell, *Macromolecules* **29**, 2570 (1996).

30. Y. Li, P.L. Dubin, R. Spindler, and D.A. Tomalia, *Macromolecules* **28**, 8426 (1995).

31. G. Shah, P.L. Dubin, J. Kaplan, G.R. Newkome, C.N. Moorefield, and G.R. Baker, *J. Colloid Interface Sci.* **183**(2), 397 (1996).

32. H. Zhang, P.L. Dubin, J. Kaplan, C.N. Moorefield, and G.R. Newkome, *J. Phys. Chem. B* **101**(18), 3494 (1997).

33. P. Haronska, T.A. Vilgis, R. Grottenmueller, and M. Schmidt, *Macromol Theory Simul.* **107**, 9640 (1997).

34. F. von Goeler and M. Muthukumar, *J. Chem. Phys.* **100**, 7796 (1994).

35. T. Wallin and P. Linse, *Langmuir* **12**, 305 (1996).

36. T. Wallin and P. Linse, *J. Phys. Chem.* **100**, 17873 (1996).

37. T. Wallin and P. Linse, *J. Phys. Chem. B* **101**, 5506 (1997).

38. M. Muthukumar, *J. Chem. Phys.* **105**, 5183 (1995).

39. M. Beer, M. Schmidt, and M. Muthukumar, *Macromolecules* **30**, 8375 (1997).

40. M. Muthukumar, *J. Chem. Phys.* **103**, 4723 (1995).

41. D. Srivastava and M. Muthukumar, *Macromolecules* **29**, 1461 (1994).

42. C.A. Angell, *Science* **267**, 1924 (1995).

43. D. Srivastava and M. Muthukumar, *Macromolecules* **29**, 2324 (1996).

44. G. Decher, *Science* **277**, 1232 (1997).

8 Small-Angle Neutron Methods in Polymer Adsorption Studies

TERENCE COSGROVE
University of Bristol, School of Chemistry, Bristol BS8 1TS United Kingdom

STEPHEN M. KING
Large-Scale Structures Group, ISIS Facility, Rutherford Appleton Laboratory, Chilton, Didcot, Oxon OX11 0QX United Kingdom

PETER C. GRIFFITHS
Department of Chemistry, University of Wales, Cardiff, CF1 3TB United Kingdom

8.1 INTRODUCTION

The adsorption of macromolecules at interfaces has been known since the first graffiti appeared on the walls of caves in which early humans lived. The attraction and, interestingly, also repulsion of polymers at interfaces gives rise to a whole range of phenomena including bacterial adhesion, crystal habit modification, depletion flocculation, enhanced hydrodynamic flow in oil recovery, and many others. In this chapter we shall describe how small-angle neutron scattering methods can be used to probe the interfacial structure of polymers and give examples from a range of practically important systems.

8.2 COLLOID STABILITY

The stability of colloidal particles is generally determined by a balance between attractive van der Waals forces and electrostatic repulsion. In situations where the ionic strength of an aqueous dispersion cannot be controlled or for nonaqueous

Colloid–Polymer Interactions: From Fundamentals to Practice, Edited by Raymond S. Farinato and Paul L. Dubin
ISBN 0-471-24316-7 © 1999 John Wiley & Sons, Inc.

systems, an alternative stabilization mechanism is often necessary. In these situations an adsorbed polymer can be very effective. The basis for this so-called steric interaction is the observation that polymer chains effectively repel each other in a good solvent environment. Two very distinct cases emerge depending on the value of the Flory parameter χ.[1] In a good solvent ($\chi < 0.5$) chains repel each other, whereas when $\chi > 0.5$, then the chains attract, enhancing flocculation. The surface coverage of the particles by polymer is also very important, and at low coverage the steric "force" cannot protect against aggregation, and the adsorbed chains may even enhance flocculation by a bridging mechanism. At high polymer coverages the steric barrier can enhance the stability of a dispersion. The effect of increasing the molecular weight (at fixed coverage) is also important, as a thicker layer will extend the steric barrier beyond the effective range of the van der Waals attraction. Many of these effects can be determined if the volume fraction profile of the adsorbed layer is known.[2]

8.3   VOLUME FRACTION PROFILE

A useful way to describe the structure of the zone next to an interface is through the volume fraction profile of the adsorbed polymer, $\varphi_a(z)$, where z is the distance normal to the interface. Once the volume fraction profile is known, other parameters of interest such as the adsorbed amount Γ, the bound fraction p, and the layer thickness δ can be immediately calculated. It is possible to calculate volume fraction profiles theoretically, by generating all possible conformations of all molecules in the ensemble comprising the surface and the solution. Several approaches have been used to this end, including mean-field, scaling, and Monte Carlo methods. A detailed review of these various approaches has recently been published.[1] Figure 8.1 shows volume fraction profiles for two extreme cases. These correspond to adsorption (in the plateau of the adsorption isotherm) and depletion and were calculated using the Scheutjens–Fleer mean-field model.[1] For the case of adsorption, the volume fraction profile falls monotonically to the level of the bulk polymer concentration. The widths of both these profiles are of the order of the solution radius of gyration of the free chain. Among the many techniques that have been proposed to examine the adsorbed layer structure in colloidal systems, small-angle neutron scattering is the only method that has given detailed information on the form of the volume fraction profile.

8.4   SMALL-ANGLE NEUTRON SCATTERING

In small-angle neutron scattering (SANS) neutrons are elastically scattered by a sample, and the resulting pattern is analyzed to provide information about the sample size, shape, and orientation. Neutron radiation is available over a range of wavelengths typically from 0.1 to 30 Å, which defines the "size range" over which useful measurements can be made. This size range is given approximately as $2\pi/\lambda$, that is, 20 to 6000 Å. In polymer–colloid systems, we are generally interested in both particles and droplets whose sizes may vary between 100 and 20,000 Å as well as

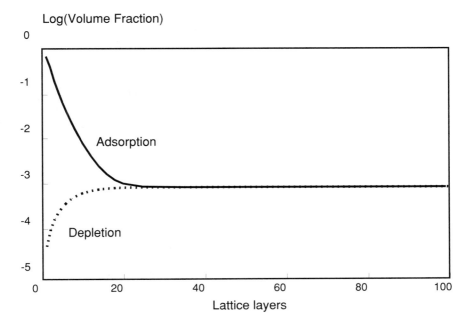

FIGURE 8.1 Schematic volume fraction profiles calculated using the Scheutjens–Fleer model for polymer adsorption.

interfacial layers whose widths vary between 10 and 1000 Å . Clearly, an adsorbed polymer layer falls in the range accessible by SANS, whereas particle size measurements are necessarily limited to these dimensions.

The neutron-scattering process can be coherent or incoherent and varies irregularly with elements and their isotopes. X-rays and light radiation are both scattered by electrons, but neutrons are scattered by the nucleus. This single phenomenon has several important consequences, which we develop below.

The efficiency of the scattering process is determined by the *neutron-scattering cross section*, $\sigma = 4\pi b^2$ where b is the scattering length. An important case arises for hydrogen, which has a coherent scattering cross section, σ_{coh}, of 1.75×10^{-8} Å^2 while deuterium has a value of 5.6×10^{-8} Å^2. This difference enables us to highlight the scattering from part of a complex structure, by selective deuteration, while suppressing that from the rest. Because atomic nuclei are some 10^4 to 10^6 times smaller than typical neutron wavelengths, the nuclei effectively act as point scatterers. The result of this is that the nuclear scattering remains constant as the scattering angle increases, allowing scattering patterns to be collected over the full range from forward to backward angles. For the systems we shall describe here, the scattering is spherically symmetric. The interaction of neutrons with matter is weak, and the absorption of neutrons by most materials is correspondingly small: Neutron radiation is therefore very penetrating, facilitating the use of complex sample environments.

In a SANS experiment, a beam of collimated, though not necessarily monochromatic, radiation is directed at a sample, illuminating a small volume, V ($= At_s$), where A is the cross-sectional area of the beam and t_s is the pathlength of the sample, (typically < 0.5 cm^3 for solvated systems). Some of the incident radiation is transmitted by the sample, some is absorbed and some is scattered. A detector, or detector element, of dimensions $dx \times dy$ positioned at a distance, L_{sd}, and scattering angle, θ, from the sample then records the flux of radiation scattered into a solid angle element, $\Delta\Omega$ ($= dx\, dy/L_{sd}^2$ This flux, $I(\lambda,\theta)$, may be expressed in general terms by[3]

$$I(\lambda,\theta) = I_0(\lambda)\Delta\Omega\eta(\lambda)TV\frac{\partial\sigma}{\partial\Omega}(Q) \qquad (8.1)$$

where I_0 is the incident flux, η is the detector efficiency (sometimes called the response), T is the sample transmission, and $(d\sigma/d\Omega)(Q)$ is a function known as the *(microscopic) differential cross section*. The first three terms of Eq. (8.1) are clearly instrument specific while the last three terms are sample dependent. The quantity Q, is the modulus of the resultant between the incident, $\overline{\mathbf{k}}_i$, and scattered, $\overline{\mathbf{k}}_s$, wave vectors, and is given by

$$Q = |\overline{\mathbf{Q}}| = |\overline{\mathbf{k}}_s - \overline{\mathbf{k}}_i| = \frac{4\pi\mathbf{n}}{\lambda}\sin(\theta/2) \qquad (8.2)$$

Typically n, the neutron *refractive index* is ~ 1; Q has dimensions of (length)$^{-1}$, normally quoted in reciprocal angstroms (Å^{-1}). The objective of a SANS experiment is to determine the differential cross section, since it is this that contains all the information on the shape, size, and interactions of the scattering bodies (assemblies of scattering centers) in the sample. The differential cross section is given by

$$\frac{\partial\sigma}{\partial\Omega}(Q) = N_p V_p^2(\Delta\rho)^2 P(Q)S(Q) + B_{inc} \qquad (8.3)$$

where N_p is the number concentration of scattering bodies (given the subscript p for particles), V_p is the volume of one scattering body, $(\Delta\rho)^2$ is the square of the difference in *neutron-scattering length densities* of the particle and the solvent (the *contrast*), $P(Q)$ is a function known as the *form* or *shape factor*, $S(Q)$ is the *interparticle structure factor*, Q is the modulus of the *scattering vector* and B_{inc} is the (isotropic) incoherent background signal; $(d\sigma/d\Omega)(Q)$ has dimensions of (length)$^{-1}$ and is normally expressed in units of reciprocal centimeters (cm^{-1}).

The neutron-scattering length density, ρ, of a molecule of i atoms may be readily calculated from the simple expression

$$\rho = \sum_i \mathbf{b}_i \bullet \frac{\mathbf{DN}_A}{\mathbf{M}} = \mathbf{N} \bullet \sum_i \mathbf{b}_i \qquad (8.4)$$

TABLE 8.1 Scattering Length Densities

Material	ρ (H form) $(\times 10^{-5} \text{ Å}^{-2})$	ρ (D form) $(\times 10^{-5} \text{ Å}^{-2})$
Water	-0.56	$+6.38$
Toluene	$+0.94$	$+5.66$
Poly(ethylene oxide)	$+0.64$	$+6.46$
Polystyrene	$+1.42$	$+6.42$

where D is the bulk density of the scattering body, M is its molecular weight, and N_A is Avogadro's number. For polymers, it is only necessary to calculate ρ for one repeat unit. The contrast is simply the difference in ρ values between that part of the sample of interest, ρ_p, and the surrounding medium or matrix, ρ_m, all squared; that is, $(\Delta\rho)^2 = (\rho_p - \rho_m)^2$. When $\Delta\rho$ is zero, the scattering bodies are said to be at *contrast match*. The technique of contrast matching can be used to highlight the scattering from an adsorbed layer while suppressing that from the substrate. Some typical values of ρ[4] are given in Table 8.1. As can be seen from the values listed in Table 8.1, D-polystyrene (e.g., a latex) can be *contrast matched* almost exactly to pure D_2O, and such a dispersion would only show minimal scattering.

The form factor $P(Q)$ is a function that describes how $(d\sigma/d\Omega)(Q)$ is modulated by interference effects between radiation scattered by different parts of the same scattering body. Consequently, it is very dependent on the shape of the scattering body, and forms exist for describing most common topologies. We shall ignore the interparticle structure factor $S(Q)$ in this chapter as the samples are presupposed to be dilute, that is, less than ~ 5% volume fraction when $S(Q)$ is equal to unity.

8.5 SANS FROM ADSORBED LAYERS

A properly designed and executed SANS experiment may provide various estimates of the thickness of the adsorbed layer, measures of how much polymer is adsorbed, and a description of how the polymer is arranged at the interface.

The scattering from a particle coated with a polymer layer is given by[5]

$$\frac{\partial\sigma}{\partial\Omega}(Q) = N_p P(Q)S(Q) + B_{\text{inc}} \tag{8.5}$$

where $P(Q) = [(\rho_p - \rho_m)F_p(Q) + (\rho_p - \rho_m)F_l]^2$ and $F_p(Q)$ represents the intraparticle structure factor for the core particle and $F_l(Q)$ the intralayer structure factor for the adsorbed polymer. These equations are only strictly valid if the curvature of the surface is small compared to Q, that is, if $QR_p \gg 1$.

Expanding Eq. (8.5) further gives

$$P(Q) = [(\rho_p - \rho_m)F_p(Q)^2] \tag{8.6}$$

$$+ [2(\rho_p - \rho_m)(\rho_l - \rho_m)F_p(Q)F_l(Q)] \tag{8.7}$$

$$+[(\rho_l - \rho_m)^2 F_l(Q)^2] \tag{8.8}$$

from which it can be seen that Eq. 8.6 describes the contribution to $P(Q)$ arising from the core particle (p), Eq. (8.8) describes the contribution from the adsorbed layer (l), and Eq. (8.7) is a particle–surface interference term. These three terms have been evaluated explicitly as:

$$I_p(Q) = (\rho_p - \rho_m)^2 \frac{2\pi A_p}{Q^4}\left[1 + \frac{1}{Q^2 R_p^2}\right] \tag{8.9}$$

$$I_{pl}(Q) = (\rho_p - \rho_m)(\rho_l - \rho_m)\frac{4\pi A_p}{Q^4}\left[\int_0^t \phi(z)\cos(Qz)\,dz - QR_p\int_0^t \phi(z)\sin(Qz)\,dz\right] \tag{8.10}$$

$$I_l(Q) = (\rho_l - \rho_m)^2 2\pi A_p\left[\frac{1}{Q^2}\left|\int_0^t \phi(z)\exp(iQz)\,dz\right|^2 + \tilde{I}_0\right] \tag{8.11}$$

where A_p is the surface area per unit volume of a particle (usually expressed as S_p/V_p), t is the maximum extent of the adsorbed layer, and $\tilde{I}_0$ is related to the density–density correlation function that describes fluctuations in the adsorbed layer. Auvray and Cotton[6] have shown that it is possible to draw analogies between the scaling descriptions for semidilute polymer solutions and densely grafted adsorbed layers. With this premise they show that any concentration fluctuations in a grafted adsorbed layer manifest themselves as a Lorentzian contribution to the scattering, that is

$$\tilde{I}_0 \propto \frac{1}{1 + (\xi Q)^2} \tag{8.12}$$

where ξ approximates to some average separation between adsorbed chains. Clearly, in the low-Q limit $\tilde{I}_0 \to 1$, and so this term can be safely ignored. The problem comes in the high-Q limit where $\tilde{I}_0 \propto Q^{-2}$. In this regime $(d\sigma/d\Omega)(Q)$ is normally of the same order as B_{inc}. The data are, therefore, frequently subject to large statistical errors that render any meaningful interpretation difficult except in a few select systems.

Using scaling arguments the predicted Q dependence of the fluctuation term is given by

$$\tilde{I}_0(Q) \propto A_p a (aQ)^{-4/3} \tag{8.13}$$

where a is the size of a polymer segment.

Equation (8.9) is simply a statement of the well-known Porod law.[7] When $\rho_l = \rho_m$, this term can be used to determine the surface area-to-volume ratio of the substrate.

Provided $QR_p \gg 1$, Eq. (8.10) reduces to the sine Fourier transform of $\phi(z)$. Thus, by subtracting the scattering obtained at contrast match for the particles ($\rho_p = \rho_m$) from the scattering obtained "off-contrast" ($\rho_l \neq \rho_p \neq \rho_m$), multiplying the result by Q^3, and taking the Fourier transform, it is, in principle at least, possible to obtain $\phi(z)$. The very first profile was obtained in this way.[8]

A more straightforward way to obtain $\phi(z)$ is from the scattering when the particles are at contrast match with solvent. Unfortunately, because it is the *modulus* of the integral that is measured experimentally, the transformation of the first bracketed term in Eq. (8.11) is complicated by the need to introduce a phase factor. A discussion of this is beyond the scope of this chapter.[5] Under conditions where the $\tilde{I}_0$ can be ignored, and if the data can be extrapolated successfully at both high and low Q, an inversion of the data to provide a model-free profile can be accomplished. Alternatively, Eq. (8.11) can be used to fit the experimental data with plausible forms of the volume fraction profile using nonlinear least-squares methods.[9]

Figure 8.2 shows scattering data obtained from polystyrene terminally grafted onto silica particles dispersed in carbon tetrachloride. The data has been analyzed both by direct inversion and by model fitting. The fit shown is the least-squares result, but a reinversion of the profile gives an identical fit.[10] The resultant volume fraction profiles are shown in Figure 8.3 and these are also identical within experimental error.

Provided $R_p \gg t$, the exponential term in Eq. (8.11) can be expanded to yield

$$I_0(Q) = (\rho_l - \rho_m)^2 2\pi A_p \left\{ \frac{1}{Q^2} \left| M \left[1 + iQz - \frac{(Qz)^2}{2} + \cdots \right] \right|^2 + \tilde{I}_0 \right\} \tag{8.14}$$

where the normalization constant, M, is given by

$$M = \int_0^t \phi(z)\, dz = \frac{\Gamma}{D} \tag{8.15}$$

where D is the polymer density.

Neglecting fluctuations and in the limit when $Q\sigma < 1$, Eq. (8.14) then approximates to

I(Q)/ cm⁻¹

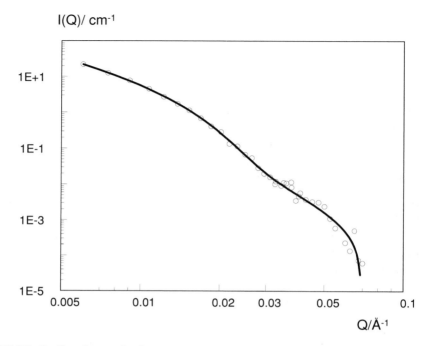

FIGURE 8.2   Small-angle scattering from (47,000 molecular weight) polystyrene grafted on silica from tetrahydrofuran. The solid line is a least-squares fit to the Gaussian profile shown in Figure 8.3.

φ(z)

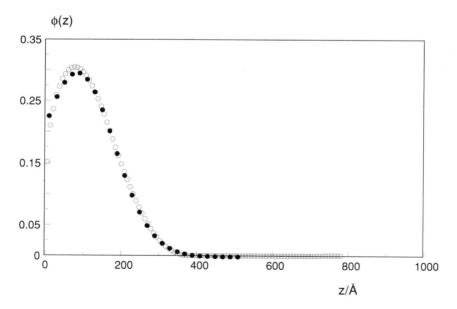

FIGURE 8.3   Volume fraction profiles for (47,000 molecular weight) polystyrene grafted on silica from tetrahydrofuran. (●) least-squares fit and (○) numerical inversion using the Crowley method.

$$I_0(Q) \approx (\rho_l - \rho_m)^2 \frac{2\pi A_p M^2}{Q^2} \exp(-Q^2\sigma^2) \tag{8.16}$$

where σ, *the second moment about the mean, or standard deviation, of the volume fraction profile* is defined as

$$\sigma^2 = \langle z^2 \rangle - \langle z \rangle^2 \tag{8.17}$$

and the root-mean-square thickness, δ_{rms}, is defined as

$$\sigma_{rms} = \langle z^2 \rangle^{1/2} \tag{8.18}$$

where

$$\langle z^n \rangle = \int_0^t \phi(z)z^n \, dz / \int_0^t \phi(z) \, dz$$

Physically, σ provides an estimate of the average distance of the center of mass of an adsorbed polymer from the interface. For a block profile of thickness t, $\sigma^2 = t^2/12$. Equation (8.16) can be used to obtain model-independent values of Γ and σ if the values of ρ are known.

An additional parameter that can be determined with knowledge of $\phi(z)$ is the average bound fraction, $\langle p \rangle$, given by

$$\langle p \rangle = \int_0^a \phi(z) \, dz / \int_0^t \phi(z) \, dz \tag{8.19}$$

where a is the thickness of the layer of segments bound at the interface. Although intuitively it may seem sensible to choose a to be the length of one segment, say, ≤ 0.5 nm, in practice it is better to choose a larger value in order to offset the effects of any interfacial inhomogeneities.

8.6   EXAMPLES OF VOLUME FRACTION PROFILES AND DERIVED PARAMETERS OBTAINED FOR ADSORBED POLYMERS

Many experimental systems have been used in scattering studies and some typical results are given here. The reader is referred to the original studies for more details.[1] The first set of systems studied systematically were homopolymers, and Figure 8.4 shows data for poly(ethyleneoxide) adsorbed onto D-polystyrene latex in D_2O.[11] The effect of increasing molecular weight can be seen as an extension of the profile into the bulk. The integral of the profiles also demonstrates that the adsorbed amount is

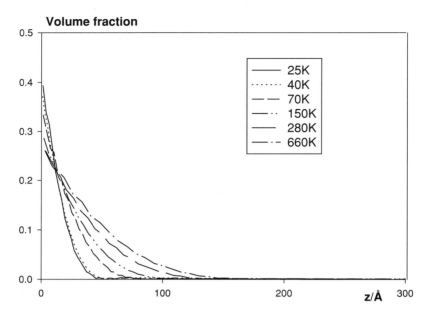

FIGURE 8.4 Volume fraction profiles obtained for poly(ethylene oxide) adsorbed on D-polystyrene latex in D-water. The legend shows the various molecular weights.

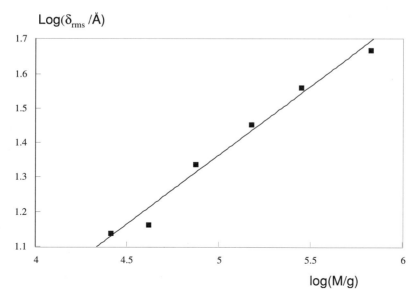

FIGURE 8.5 Root-mean-square thicknesses obtained by integrating the profiles given in Figure 8.4 plotted as a function of molecular weight. The fitted straight line has a slope of 0.4 ± 0.02.

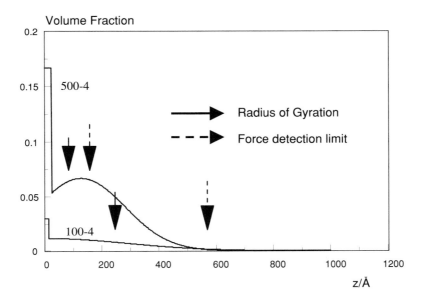

FIGURE 8.6 Volume fraction profiles for block copolymers of polystyrene (PS) and poly(ethylene oxide) (PEO) adsorbed in porous vycor glass from toluene. Two molecular weights are shown: PS, 100,000; PEO, 4000 and PS, 500,000; PEO, 4000. The solid arrows represent the free solution radius of gyration of the poly(ethylene oxide) block and the dotted arrows the distance at which a force is first observed under compression in the surface forces apparatus.

increasing with molecular weight. These results are in good qualitative agreement with theoretical predictions and other experimental measurements. Figure 8.5 shows the derived values of δ_{rms} from the data given in Figure 8.4 as a function of molecular weight. The linearity of the double logarithmic plot is in agreement with a scaling law of the form $\delta \sim M^{\alpha}$, where the value of α is found to be 0.4 ± 0.01. This value is also consistent with values found from ellipsometry. Figure 8.6 shows volume fraction profiles for block copolymers of polystyrene and poly(ethyleneoxide) adsorbed in Vycor porous glass from solution in toluene. The volume fraction profiles are highly extended even given the high molecular weight of the polystyrene. For comparison, the radius of gyration of the free polystyrene block is also shown. Data from the surface forces apparatus for these two polymer samples adsorbed on mica is also included and indicates that the first evidence of a repulsion occurs at the periphery of the adsorbed layer.

8.7 CONCLUSION

In this chapter we have shown the utility of SANS in determining the volume fraction profiles of adsorbed polymers and cited results from three different experimental systems. The technique works for particulates, porous solids, and for emulsions[12,13] and gives a unique insight into the structure of adsorbed layers.

REFERENCES

1. G.J. Fleer, M.A. Cohen Stuart, J.M.H.M. Scheutjens, T. Cosgrove, B. Vincent, *Polymers at Interfaces*, Chapman & Hall, London, 1993.

2. D. Napper, *Polymeric Stabilisation of Colloidal Dispersions*, Academic, London, 1989.

3. J.S. Higgins and H.C. Benoit, *Polymers and Neutron Scattering*, Oxford Series on Neutron Scattering in Condensed Matter, Vol. 8, Clarendon, Oxford, 1994.

4. V.F. Sears, *Neutron News*, **3**, 26 (1992).

5. T.L. Crowley, D.Phil. Thesis, University of Oxford, 1984.

6. L. Auvray and J.P. Cotton, *Macromol.* **20**, 202 (1987).

7. P. Linder and Th. Zemb, *Neutron, X-ray and Light Scattering*, North Holland, Amsterdam, 1991.

8. K.G. Barnett, T. Cosgrove, T.L. Crowley, B. Vincent, and Th.F. Tadros, in Th. F. Tadros, ed., *The Effect of Polymers on Dispersion Stability*, Academic Press, London, 1981, p. 183.

9. T. Cosgrove, T.G. Heath, and K. Ryan, *Langmuir.* **20**, 3500 (1994).

10. T.G. Heath, Ph.D. Thesis, Bristol, 1989.

11. T. Cosgrove *J. Chem. Soc Faraday Trans* **86**, 1323 (1990).

12. T. Cosgrove, L.M. Mallagh, and K. Ryan, in K.L. Mittal, ed., *Surfactants in Solution*, Vol. 9, Plenum, New York, (1989), p. 455.

13. C. Washington, S.M. King, and R.K. Heenan, *J. Phys. Chem.* **100**, 7603 (1996).

PART III
Methods for Investigating Polymer Adsorption

9   Nuclear Magnetic Resonance of Surface Polymers

FRANK D. BLUM

Department of Chemistry, University of Missouri-Rolla, Rolla, Missouri 65409-0010

9.1   INTRODUCTION

Nuclear magnetic resonance (NMR) has been shown to be a particularly useful, though underused, technique for studying adsorbed polymers. Early in its history, NMR techniques were primarily employed by chemists to elucidate both the structure and dynamics of molecules in solution. Later, the application of various solid-state NMR techniques to study the behavior of chemical species in bulk became more commonplace. In fact, the solid-state techniques are particularly important since the solid state is often the state in which polymers are used. The techniques used in many of the solution and bulk NMR experiments are applicable to the study of surface-bound polymers; however, the experimental application of these techniques and the interpretation of their results are more complicated than in homogeneous systems.

Nuclear magnetic resonance offers many advantages that can be exploited in surface studies. These include selection of various species via different nuclear spins, selective polarization of certain nuclei, insensitivity to the optical clarity of the material, and ability to probe "inside" the material. In addition, NMR can also be used to probe either structure or dynamics of the surface species.

Unfortunately, NMR techniques suffer from some serious limitations. The most difficult to contend with is limited sensitivity. This limitation comes about from the Boltzmann factor and, combined with the need for a support (surface), makes the experiments even more difficult. Nevertheless, these problems can often be overcome with high sensitivity, high-field spectrometers, specialized probes, and "fancy" pulse sequences. The use of high surface area substrates is also often required. This means that experiments are often done on experimental surfaces rather than say, a glass plate

Colloid–Polymer Interactions: From Fundamentals to Practice, Edited by Raymond S. Farinato and Paul L. Dubin
ISBN 0-471-24316-7   © 1999 John Wiley & Sons, Inc.

or fiber. Fortunately, a number of oxide surfaces are available as high surface area powders with surface properties similar to the bulk materials. In many cases, it is difficult to distinguish between bulk and surface-bound material. The contrast between these may be low because the surface and bulk species are chemically identical. Selective polarization of different nuclei and, in some cases, isotope labeling can make these different regions distinguishable.

There have been a few reviews of the use of NMR to study surface-bound polymers.[1-5] These report the study of polymers at solid surfaces in air, solvent, or other organic media. In this chapter, the author addresses how NMR can be and has been used to study the mobility of polymers adsorbed on solid surfaces. It is not the intent of this chapter to be a comprehensive review. The focus will be on: (i) how NMR can be used to probe surface behavior, (ii) how the nature of the polymer affects the dynamics, for example, homopolymer versus block copolymer, and (iii) how the surroundings, for example, presence of solvents, influence the behavior of the polymers. Examples will be taken from the laboratory of the author and others.

9.2   NUCLEAR MAGNETIC RESONANCE BACKGROUND

The use of NMR to study both structure and dynamics of chemical species may now be considered a mature field. A number of excellent books exist on the subject that are dedicated to both liquids and solids NMR[6-8] along with some focused strictly on polymers.[9-11] These serve as a good introduction to the field. For those readers who are unfamiliar with the basics of NMR, this section will serve as a brief introduction and should be sufficient for a cursory understanding of the material in this chapter.

Spin angular momentum is a fundamental property of a nucleus that can be characterized, to a large extent, by a spin quantum number, I, that can take on integer or half integer values. Atomic species with $I = 0$ are not NMR active, and these include many common nuclei such as ^{12}C, ^{16}O, ^{28}Si, and ^{32}S. They neither affect the spectrum nor directly affect the magnetic resonance of neighboring nuclei. Nuclei with $I = \frac{1}{2}$ such as ^{1}H, ^{13}C, ^{19}F, ^{15}N, and ^{31}P possess a magnetic dipole and interact with other NMR-active nuclei as well as external magnetic fields. Nuclei with $I > \frac{1}{2}$, such as ^{2}H, ^{14}N, and ^{27}Al, possess a nonspherical charge distribution resulting in an electric quadrupole moment in addition to the magnetic moment. This means that these nuclei are affected by electric field changes at the nucleus as well as by a magnetic field.

The magnitude of the nuclear magnetic moment is proportional to the magnetogyric ratio, γ (rad/T s), which is different for each type of nucleus. The basic frequency, ν (in Hertz), of the NMR experiment (often called the Larmor frequency) for a given nuclear species is determined by the magnetogyric ratio and applied magnetic field, H, by

$$\nu = |\gamma|H/2\pi \tag{9.1}$$

The basic proportionality to field strength [1 T (tesla) = 10,000 G (gauss)] is the primary difference between different NMR spectrometers that run at different magnetic fields. Current NMR spectrometers operate in the range of roughly 1 to 17 T, resulting in frequencies for nuclei in the range of 10 to 750 MHz.

A listing of a few NMR-active nuclei of relevance to surface species is given in Table 9.1 along with their isotope abundance, relative sensitivity, and Larmor frequencies. It is clear from Table 9.1 that a broad range of radio frequencies need to be used in multinuclear NMR. The corresponding energies associated with the radio-frequency transitions (ΔE) are very small. They are much smaller than typical thermal energies (kT, where k is the Boltzmann constant and T is the temperature) except at very low temperatures, on the order of a few degrees Kelvin. The resulting population distribution of nuclear spins into quantized energy levels is given by the Boltzmann factor, $e^{-\Delta E/kT}$. At room temperature, the ground and excited states are almost equally populated. For example, for protons at roughly 100 MHz, about 10 nuclei out of 10^6 would be in excess in the ground state. The NMR spectrum, with its intensity proportional to this excess, is then rather weak in comparison to the strength of the signals obtained from other techniques. The need for higher magnetic field strengths is primarily because of the increased signal intensities and improved spectral dispersion obtained at higher fields. The latter is crucial in the characterization of biomolecules in solution. The low energies of the NMR experiment also mean that the thermal motions of molecules are not directly perturbed by the adsorption of radio-frequency energy, although exposures to intense radio-frequency energy can lead to sample heating.

The sensitivities given in Table 9.1 reflect only those from the Boltzmann factor. The *overall* sensitivity (not shown) is the product of the natural abundance and the sensitivity listed in Table 9.1. Low natural abundance suggests the use of isotope enrichment. Low abundant nuclei naturally have small backgrounds. The best opportunities for enrichment at reasonable costs are found with low atomic mass isotopes, with ^2H, ^{13}C, and ^{15}N being the most common. It should also be noted that different isotopes of the same atomic species have different NMR properties. They

TABLE 9.1 Partial Listing of NMR Active Nuclei

Isotope	Abundance	I	Relative Sensitivity[a]	ν (MHz) at 1 T
^1H	99.98	1/2	1.000	42.58
^2H	0.0156	1	0.00964	6.536
^{13}C	1.108	1/2	0.0159	10.70
^{14}N	99.64	1	0.00101	3.08
^{15}N	0.365	$(-)1/2^b$	0.00104	4.32
^{19}F	100.00	1/2	0.834	40.06
^{27}Al	100.00	5/2	0.21	11.09
^{29}Si	4.70	$(-)1/2^b$	0.0785	8.46

[a]For equal number of nuclei at constant field.
[b]The negative sign indicates that these nuclei precess in the opposite sense.

resonate at different frequencies, which are almost always outside the range of the others.

The state of a material often determines how NMR experiments need to be performed. Simplistically, one can divide NMR experiments into those that require "liquids" and "solids" techniques. Liquids NMR techniques take advantage of the averaging of many interactions through molecular motion. This averaging results in the formation of very narrow resonances. Detailed structural information can be obtained from these studies. In contrast, in solids, non- or partially averaged effects from anisotropic chemical shifts, dipolar interactions, or quadrupolar interactions may dominate. These interactions can be reduced or exploited by a variety of different techniques such as magic angle spinning (MAS), cross polarization (CP), or wide-line (WL) NMR. Generally, different hardware is required for each of these techniques, although spectrometers can be configured to perform most of them.

The time scale of the NMR experiment varies, depending on the kind of experiment performed and the kind of interaction to which the experiment is sensitive.[12] For relaxation experiments, this time scale is usually fast with the so-called longitudinal relaxation time, T_1, being sensitive to motions on the order of the Larmor frequency (10 to 800 MHz). Other relaxation times, such as transverse, rotating frame, or quadrupolar relaxation times (T_2, $T_{1\rho}$, T_{1Q}, respectively), depend on the frequencies associated with those interactions as well, typically from low frequencies to those in the kilohertz range. Line shape analyses of solids usually depend on dipolar or quadrupolar couplings, or chemical shift anisotropies that are typically in the 1- to 100-kHz range. Lastly, two-dimensional exchange NMR can be used to detect much longer range motions that can be as long as seconds. Thus NMR can be used in different forms to measure motions of very different time scales. Unfortunately, as previously mentioned, these different experiments require different spectrometer modes. Mobile species at a gaseous or liquid interface may require solution techniques, while solid-state NMR techniques are usually required at solid interfaces or interphases.

9.3 DYNAMICS OF SURFACE POLYMERS

Polymers, unlike small molecules, may adsorb at interfaces in conformations that exclude most of the polymer segments from the surface. The adsorption can be driven by covalent or ionic bonds, polar, or van der Waals forces. These forces are quite short range, on the order of Ångstroms, and only directly affect segments that are very close to the surface. Segments not directly bound to the surface are indirectly affected because of their sequential attachment to the bound segments. The range of this effect is not entirely clear. However, the overall motions of the adsorbed polymer molecule and the smaller segmental motions must be affected by the adsorption of some of the groups.

For simplicity, we distinguish two different modes of adsorption for attached polymers, namely random or end-attached polymers. There are a variety of other ways to attach chains to interfaces. However, these two limits are instructive for distinguishing different kinds of behavior. Homopolymers or random copolymers generally attach themselves to interfaces in a random fashion. This means of attachment can be altered by the presence of certain functional groups. When high-affinity groups are at the ends of the polymer, such as in block copolymers, an end-adsorbed polymer results. In both cases, the packing of large molecules at the interface means that only a fraction of molecules will be in direct contact with the surface at any time. The rest of the segments will be indirectly attached.

For randomly attached polymers, it is convenient to refer to the adsorbed polymer segments as being part of *trains* (segments directly, or nearly directly, attached to the interface), *loops* (segments between two trains), or *tails* (the ends). The mobility of these different segments are expected to be different. When the interaction of the polymer segments with the solid (and rigid) surface is attractive, trains would be expected to have reduced mobility compared to loops and tails. One might expect these other surface species (loops and tails) to have reduced mobility as well, but this is not necessarily the case (vide infra). A schematic representation of a hypothetical adsorbed polymer is shown in Figure 9.1.

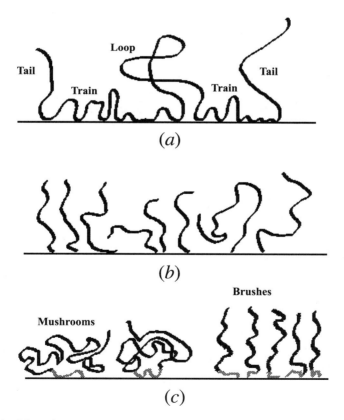

FIGURE 9.1 Schematic representation of the segments of (*a*) a randomly adsorbed polymer showing trains, loops, and tails; (*b*) terminally attached polymers; and (*c*) block copolymers in brush and mushroom conformations.

In the case of terminally attached polymers, the attachment may be through a single group (ionic or chemically selective) or a block of segments (as in a block copolymer) where the block may be considered to be adsorbed. The nature of the behavior of the conformation of the polymer chains will be a function of the nature, composition, and adsorbed amount of the polymer.[13] At the two extremes, the overall conformation of the molecule can be considered as described as a "brush" or "mushroom."[14] These are shown in Figure 9.1 and are affected by the interactions of both segments with the surface, the solvent, and each other.

There are many ways to characterize the behavior of adsorbed polymers.[3] Knowledge of the amount adsorbed, layer thickness, and density profile are difficult to measure in a consistent way. Simple ways to measure trains, loops, and tails, for example, may give different results depending on the technique used. Some techniques, such as infrared spectroscopy, count only directly bound moieties, while microcalorimetry and magnetic resonance count "close" groups, which may not be directly bound, but highly restricted. In principle, the results from these different techniques should be complementary.

It is appropriate to point out here that electron spin resonance (ESR) has also been shown to be a powerful technique in probing the dynamics of adsorbed polymers.[15] In principle, it yields information that is very similar to that from NMR. For ESR, a covalently attached spin label, usually a nitroxide, is required. ESR has the advantage that it is more sensitive than NMR and the experiments can be done on simpler equipment. Its disadvantage is that the attachment of the label must be done synthetically, and the interpretation of the dynamics of the polymer is complicated because of the label's internal motions and its perturbation of the motion of the polymer. Nevertheless, it has proven to be a powerful technique for studying surface polymers.

In the next two sections, we explore some of the NMR studies used to probe the mobility of these interfacial chains at solid–liquid and solid–air interfaces. In doing this, we have excluded the behavior of polymers at the solid–solid (as in a composite) and solid–elastomer interface (as in a filled rubber). More information on these systems can be found elsewhere.[4,5,10] The main focus is on how the dynamics of the polymer at the surface differs from that of the corresponding polymer in bulk or in solution.

9.3.1   Polymers at Solid–Liquid Interfaces

Surface-adsorbed polymers are used in a wide variety of applications such as coatings, dispersion stabilizers, and adhesives. The interactions of all of the components (polymer, solvent, and surface) with each other are critical for the correct performance of the system. Nuclear magnetic resonance provides a microscopic probe of the chains at the interface. Depending on the degree of solvent penetration and the inherent nature of the polymer, a polymer may behave as a solid or liquid. Thus, in some situations, it may be possible to observe liquidlike behavior of the polymer chains.

Perhaps the first NMR study of adsorbed polymers at the solid–liquid interface was that of Miyamoto and Cantow[16] who studied poly(methyl methacrylate) (PMMA) on silica in deuterochloroform and other solvents. With low adsorbed amounts of PMMA on silica, no protons from the polymer were observed, while at higher adsorbed amounts, liquidlike spectra were found. The polymer segments in the low-adsorbed-amount samples were primarily in trains, while with higher coverages, tails and loops were observed with narrow resonances. The narrowing of the resonances with increasing coverages was consistent with the trains and loops being more mobile (narrower resonances) at higher coverages. Interestingly the isotactic PMMA samples were more motionally restricted (broader) than actactic ones. The trains were more solidlike (broader), and their presence was not revealed in a high-resolution (liquid) spectrum.

The difference in the dynamics of trains versus loops and tails not only manifests itself in broad and narrow components, but it also results in different decays of the magnetization. Cosgrove and Barnet[17] used this behavior to develop the "driven equilibrium" method for measuring the intensity of solidlike and liquidlike spins. The technique is based on a solid echo to refocus all protons followed by liquid echos that refocused only mobile protons. In this way, a spin count of solid (trains) and liquid (loops and tails) spins could be made. The application of this technique to poly(vinyl pyrolidone) (PVP) in D_2O provided measurements of bound fractions in agreement with ESR studies.[18] In comparison, the results with end-attached polystyrene (PS) on carbon black yielded a much lower fraction of bound segments.

This technique was also extended to other systems where it was used with neutron scattering and light scattering to more fully characterize the systems. For random ethylene–vinyl acetate copolymers,[19] preferential adsorption of vinyl acetate groups was found for both methyl- and cyclohexyl-treated silica surfaces. For poly(vinyl alcohol) on polystyrene latex in water, a small amount of bound polymer was found.[20] In contrast, for end-labeled poly(ethyleneoxide) (PEO) on silica, no bound fraction was found, consistent with a terminally attached polymer, probably in an extended conformation.[21,22]

Relaxation times have also been shown to be sensitive to the dynamics of surface polymers. Carbon-13 and proton relaxation behavior for PEO on silica in benzene identified two different polymer components.[23] Both components were relatively narrow, (i.e., not solid-like) but suggested anisotropic motion. The proton behavior also suggested sensitivity to the local concentration of segments. Other measurements have shown that a concentration profile could be made for adsorbed PEO based on the T_2 behavior of the PEO in solution and the neutron scattering profile.[24–26] The comparison was based on the assumption that the T_2 values would be the same for the solution and surface-bound polymer of the same local segment concentration. The agreement between the predicted and experimentally measured T_2 profile suggested the appropriateness of the assumption. This approach was also extended to poly(styrene sulfonate) (PSS) adsorbed on silica.[27] Since PSS was a nonabsorbing polymer, the results were found to be useful in determining the thickness of the depletion layer.

In the aforementioned studies, little attempt was made to accurately measure the rates of reorientation of the surface polymer. In our lab, several recent studies have focused on this. Much of this work has involved the use of deuterium NMR on specifically labeled polymers. Deuterium measurements have the advantage of selectivity and sensitivity (due to the presence of the label) along with speed (due to the shortness of the deuterium relaxation times). We have found that the motional model of Hall and Helfand (HH)[28] and the log-normal (LN) distribution[29] were most useful in developing a more quantitative measurement of the motions of the polymer segments based on relaxation data. The HH model yields two correlation times related to faster and slower motions of the polymer chains. The LN model fits the relaxation times to a mean correlation time (τ_0) and a width (σ) parameter. Both of these models work well over a wide range of polymer conditions.

We observed the behavior of two different types of deuterium-labeled systems. The two systems studied were the block copolymer system of poly(styrene-*b*-vinylpyridine) (SVP) and the homopolymer poly(methyl acrylate). Each polymer was labeled with deuterium on the backbone so that information on backbone (C–D bond vector) reorientation was obtained directly. For the SVP (sometimes referred to as VPS), we found that when the polymer was adsorbed on silica and then swollen with styrene, the styrene segments gave very narrow ^{13}C resonances, while the vinylpyridine segments gave very broad ones. This is shown in Figure 9.2 where all of the VP and styrene (S) segments can be distinguished in the solution spectra. In contrast, the VP segments in the surface spectra are so broad that they are absent.[30] This is consistent with the view that the VP groups are attached to the surface and the styrene segments are able to rapidly reorient away from the surface.

To find out how fast the S segments reoriented, we also measured the relaxation times for the S-backbone deuterons of a few SVP copolymers.[30] The results are shown in Table 9.2 for two SVP polymers with the S-units deuterated in the middle (VPDSS) and at the end (VPSDS) of the styrene segments. Both polymers were about 20,000 g/mol and about 75% S. The results from the two different labels were only slightly different, and were consistent with a small motional (and concentration) gradient away from the VP segments. It was also observed that, if a comparison was made between the surface-bound polymers in the presence of toluene and the same polymer in solution, the T_1/T_2 ratios for the surface-bound polymers were much smaller than those for the solution polymers. In fact, the behavior of the surface-bound polymer was more similar to that of a small molecule (typically $T_1/T_2 \approx 1$) than that of a polymer (typically $T_1/T_2 > 1$). We attributed this "enhanced" mobility to the highly extended brush structure of the polymer. This structure allowed the C–D bond vector to be more completely averaged than in solution. Subsequent studies verified that this enhanced mobility was lost in other solvent systems where the styrene was not in an extended brush conformation.[31] In addition, this behavior did not occur for a lower molecular weight SVP polymer,[32] highlighting the role of packing on the surface. This packing is a function of the molecular weight and the relative amounts (asymmetry ratio) of the two components in the block copolymer.[13]

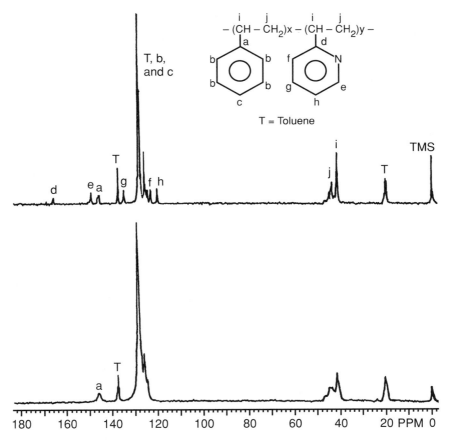

FIGURE 9.2   Proton decoupled ^{13}C spectra of poly(2-vinylpyridine-*co*-styrene) in toluene solution (top) and on silica swollen with toluene (bottom). Reprinted with permission from Ref. 30; copyright 1990, American Chemical Society.

The comparison of two polymers at similar T_1's suggested that the polymers had similar fast, short-range motions. However, the T_2 relaxation times had contributions from slower, long-range motions. Thus, we believed that the short-range motions were similar for the surface and solution polymer, but the longer range motions were different, with *the solution polymer being in a more restrictive environment*. In addition to the ratio of relaxation times, the mean correlation times, τ_0, and width parameter, σ, for the LN are also listed in the table. These are consistent with the qualitative predictions from the relaxation time ratios. This behavior may not be what one might naively expect. The difference is most likely due to the structure of the polymers in the different environments. In solution, the more coiled nature of styrene segments result in limited mobility due to segment-segment interactions. These appear to be lessened in the styrene segments in the more highly stretched brushes and a higher mobility results.

TABLE 9.2   Relaxation Behavior of Different Solvent-Swollen Systems

System	State	T_1	T_1/T_2	$\tau_0{}^a$	σ
VPDSS^b/toluene					
In solution	3.63	2.75	1.32	1.21	2.35
On silica	3.63	3.38	1.07	0.78	1.45
VPSDS^b/toluene					
In solution	3.94	3.25	1.21	0.93	2.14
On silica	3.94	3.60	1.09	0.74	1.64
PMA-d/toluene					
In solution	5.45	4.69	1.16	0.62	2.20
On silica	5.40	2.20	2.45	2.84	3.75

^{a}In 10^{-9} s from the log-normal distribution.
^{b}The difference between the two polymers is the positions of the labels on the styrene segments.

For comparison with the SVP polymers, we listed the relaxation behavior of backbone-deuterated poly(methylacrylate) (PMA) in Table 9.2 as well.[33] This polymer was randomly adsorbed and upon adsorption had the opposite effect from that shown by SVP. The adsorbed polymer showed a much higher ratio of T_1/T_2 for the surface polymer than the solution polymer. This was consistent with the more restricted nature of the surface polymer, which had many segments bound to the surface. This effect was also apparent in the mean correlation from the LN model.

There are also direct practical applications of this type of work. Relaxation times in the poly(vinyl pyrrolidone)-co-(vinyl acetate) (PVP-co-VAc) alumina system, as measured by Parker et al.,[34] correlated with the dispersion stability of the system. It was suggested that the relaxation time measurements were sensitive to selective displacement for certain functional groups on the surface. It has also been shown that diffusion of solvents in swollen polymer systems could be used to determine the size and structure of adsorbed polymers.[35] The technique was based on the additional retardation of solvent diffusion due to association with stationary adsorbed polymer. With appropriate modeling the adsorbed layer thickness can be obtained.

9.3.2   Polymers at Solid–Air Interfaces

The problems of polymers at the solid–air interface were a bit different from those where solvent was present. Without the swelling of the solvent, these polymers were likely to be more bulklike than when swollen with solution. Consequently, solids NMR techniques were more appropriate for these systems. In addition, the broader resonances and potentially long relaxation times made the experiments more difficult as well. In this section, a few examples will be cited of how NMR was used to probe the interaction with the surface, the effect of the surface on the conformation of the polymer, and the changes in the dynamics of the polymer due to the surface.

The interaction between substrate and polymer was probed with polarization transfer from the protons in adsorbed poly(vinyl alcohol) (PVA) to ^{29}Si nuclei on silica.[36] The experiment consisted of preparing the proton spins (on the polymer) in a known state and then transferring that spin polarization (via a process called cross polarization) to the ^{29}Si spins on the silica. The rates and amounts of polarization transferred gave information about distances between the exchanging species. Upon deuteration of the exchangeable PVA and silanol protons, the polarization transfer was much less efficient. The authors concluded from this that the hydroxyl groups in the PVA were the "driving mechanism" for the interaction of PVA with the silica.

For the adhesion of polyimides (PI) to alumina,[37] ^{15}N-labeled PI was used. The solid-state chemical shift was followed as a function of time and temperature with MAS NMR. For thin layers of 30 to 150 Å (2 to 10 monolayers), evidence was found for polymer backbone cleavage at 150 to 255 °C along with amine production. The adhesion of the PI on alumina appeared to be enhanced with higher temperature treatments, consistent with bonding between the alumina and the polymer. The treatment of the alumina with silane coupling agents "passivated" the alumina to this reaction.

Carbon-13 solid-state chemical shifts were used to probe the conformational changes of adsorbed poly-L-lysine and poly-L-glutamic acid on silica and on hydroxyapatite.[38] These shifts were consistent with more extended structures compared to the bulk material (as opposed to α-helicies). Deuterium NMR experiments marked the reduction of the mobility of the side chains on the surface while spin-locking experiments identified reduced mobilities of both the backbone and side chain upon adsorption.

Relaxation times for dry adsorbed polymers are also different from those in bulk. It was shown that the T_1's of poly(dimethyl siloxane) (PDMS) chains adsorbed on modified silica gel increased with surface coverage.[39] At a surface coverage of about 1.1 g PDMS/g silica, the T_1 was comparable to that in the bulk polymer, thus giving a distance scale to the range of this effect.

The mobility of surface polymers can be deduced from line shape changes. It is well known that many NMR experiments are sensitive to higher frequencies than many mechanical techniques. For example, the glass transition temperatures (T_g's) from NMR experiments are typically 30 to 50 °C higher than those obtained from differential scanning calorimetry (DSC) experiments.[40] This increase is because the onset of faster motions (in the kilohertz range) where the solids line shapes change occur at higher temperatures than the slower motions (in the hertz range) to which the DSC is sensitive.

A good example of the so-called T_g effect can be found for poly(isopropyl acrylate) (PIPA). This polymer has a T_g(DSC) just below room temperature. At room temperature, where the polymer is rubbery, the cross polarization-magic angle spinning (CP-MAS) solid-state ^{13}C spectrum with dipolar decoupling (DD) of the protons of the bulk polymer shows quite broad and sometimes missing resonances from the polymer backbone.[1,41] Adsorption of this polymer onto silica at saturation coverage from toluene and subsequent drying results in a material whose ^{13}C spectra shows clear distinct resonances for each different carbon type. In this case, the backbone molecular motion in the bulk polymer was sufficient to either render the effects of magic angle spinning or dipolar decoupling useless. When the polymer was adsorbed on the surface, this motion was reduced and the spectrum of a "glassy" polymer was revealed. For this polymer, mobility of the majority of the segments became reduced upon adsorption.

As suggested, the intensities of surface–polymer spectra are sometimes problematic. Typically, samples of surface species are not limited by amount, but rather by how many nuclei one can get in the sample coil. In our laboratory, we have tried to overcome this problem for solid-state deuterium NMR by using a specially built probe with an 8-mm sample coil. From geometric considerations, this results in an additional factor of 2.5 in sensitivity. The probe must also be capable of accepting the higher pulse power required to excite the spins.

To illustrate the effect of dilution of nuclei in surface samples, we show in Figure 9.3, the spectra of poly(methyl acrylate)-d_3 in bulk and on silica. The line shape obtained is called a Pake pattern and it is obtained with a quadrupole-echo pulse sequence.[6,11] The spectra are representative of a methyl group rapidly rotating about its symmetry axis. In this case the backbone motions of the polymer make no contribution to the line shape of the methyl deuteron because the backbone motions are slow compared to the reciprocal of the splittings. The line shape is comprised of two overlapping powder patterns for the spin $I = 1$ deuterons. The spectrum for the bulk sample shows excellent sensitivity and the experiment can be performed in a few minutes.[42] At the same number of scans, the surface spectrum shows reasonable, but significantly less signal to noise. Approximately 750 scans are needed for the signal to noise in the surface spectrum to be as good as the bulk spectrum with 16 scans.

The behavior of the quadrupole-echo spectra for poly(vinyl acetate)-d_3 are shown in Figure 9.4 for the bulk polymer as a function of temperature. At low temperatures, the spectra are essentially the same as those in Figure 9.3 where the backbone motion of the polymer has little effect on the spectra. As the temperature increases, the spectra gradually sharpen as the backbone motions of the polymer start to further average the quadrupole tensor of the deuterons. One can consider the region of the transition from the Pake pattern to a single resonance as the T_g(NMR). This is about 40 °C above the T_g (DSC).[43] It is important to note that the spectra are such that they could be classified as due to a single component, that is, all of the polymers undergo more or less the same motions. Even though the motions of the polymer are complex, all of the polymers behave similarly.

In contrast to the behavior of polymers in bulk, adsorbed polymers show multicomponent behavior. The spectra of PVAc-d_3 adsorbed on silica from toluene, and then dried, are shown in Figure 9.5. At the lowest temperature, the surface and bulk spectra are similar. However, instead of a gradual change and broadening of the spectra with temperature, the broad component remains roughly the same, while a second, narrow component builds in. This narrow component was interpreted as due to material at the air–polymer interface. Having few constraints, the polymer segments in this region have enhanced mobility compared to bulk. In reality, the

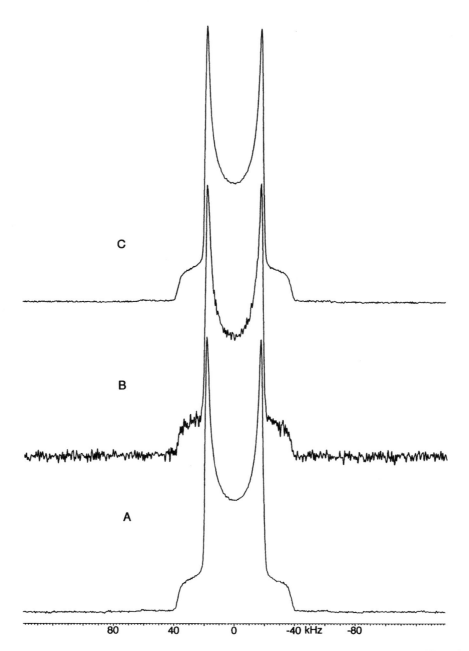

FIGURE 9.3 Deuterium NMR spectra of poly(vinyl acetate)-d_3 (A) in bulk, 16 scans, (B) on silica, 16 scans, and (C) on silica, 750 scans.

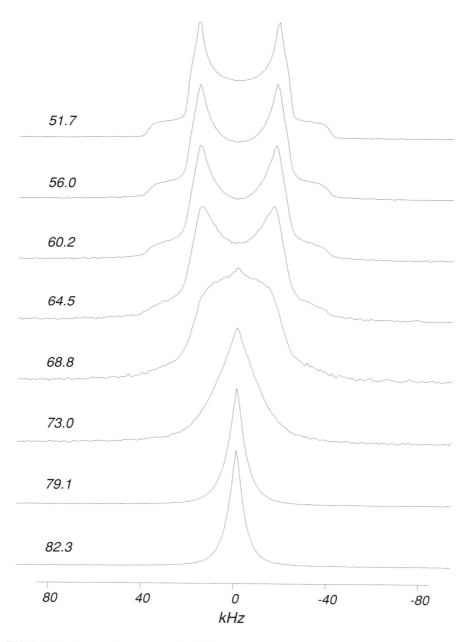

FIGURE 9.4   Quadrupole-echo deuterium NMR spectra of poly(vinyl acetate)-d_3 in bulk as a function of temperature. Reprinted with permission from Ref. 43; copyright 1996, American Chemical Society.

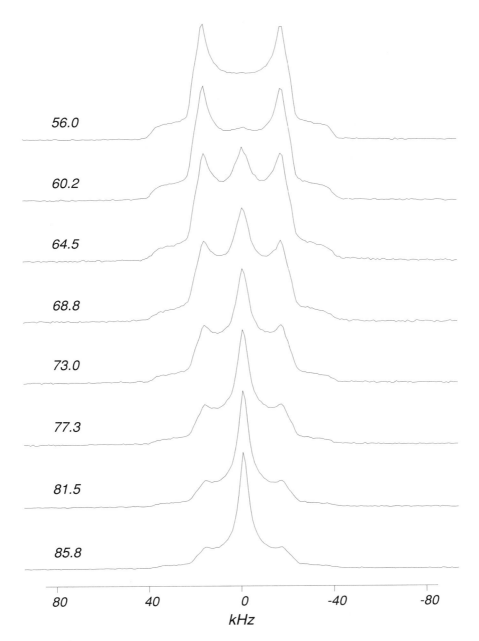

FIGURE 9.5   Quadrupole-echo deuterium NMR spectra of poly(vinyl acetate)-d_3 at saturation coverage on silica as a function of temperature. Reprinted with permission from Ref. 43; copyright 1996, American Chemical Society.

problem is more complicated than just two components, and there exists a gradient in mobility. The mobility varies from relatively rigid at the silica–polymer interface to more mobile at the air–polymer interface. This knowledge helps us better understand the adhesion process where mobility plays an important role.

9.4 CONCLUSIONS

Nuclear magnetic resonance can be an important source of information on the behavior of surface polymers. It can help elucidate the mechanisms of interactions with the surfaces and to probe the dynamics of the polymer at the interface. Selectivity of the interfacial layer can be achieved with the transfer of magnetization or selective isotope labeling. The segmental motions of polymers at interfaces may be enhanced or retarded compared to bulk. The changes are related to a variety of factors including the nature of the polymer adsorbed, the interaction with the surface, the molecular weight, or packing at the interface. Interfaces graded in terms of the mobility of the segments in a single molecule may be observed.

Acknowledgments   The author wishes to thank the National Science Foundation (DMR-900926) for support of some of the work reported here. He also acknowledges the work of his co-workers listed in the references.

REFERENCES

1. F.D. Blum, *Colloids Surf.* **45**, 361 (1990).
2. T. Cosgrove and P.C. Griffiths, *Adv. Colloid. Interface Sci.* **42**, 175 (1992).
3. G.J. Fleer, M.A. Cohen-Stuart, J.M.H.M. Scheutjens, T. Cosgrove, and B. Vincent *Polymers at Interfaces*, Chapman & Hall, London, 1993.
4. F.D. Blum, *Ann. Rep. NMR Spectros.* **28**, 277 (1994).
5. F.D. Blum, in H. Ishida ed., *Characterization of Composite Materials*, Vol. 12 in *Practical Guides for the Surface, Interfacial and Micro Analysis of Materials*, R. Brundle and C. Evans eds., Manning Publications, Greenwich, CT, 1994, pp. 64–79.
6. C.A. Fyfe, *Solid State NMR for Chemists*, CFC Press, Guelph, Ontario, 1983.
7. R.K. Harris, *Nuclear Magnetic Resonance Spectroscopy*, Wiley, New York, 1985.
8. J.K.M. Sanders and B.K. Hunter, *Modern NMR Spectroscopy*, Oxford University Press, Oxford, 1987.
9. L.J. Mathias, ed., *Solid State NMR of Polymers*, Plenum, New York, 1991.
10. V.J. McBrierty and K.J. Packer, *Nuclear Magnetic Resonance in Solid Polymers*, Cambridge University Press, Cambridge, UK, 1993.
11. K. Schmidt-Rohr and H.W. Spiess, *Multidimentional Solid-State NMR of Polymers*, Academic, New York, 1994.
12. R.G. Byrant, *J. Chem. Edu.* **60**, 933 (1983).
13. M. Tirrell, in S.E. Webber, P. Munk and Z. Tuzar, eds., *Solvents and Self-Organization of Polymers*, Kluwer Academic Publishers, Boston, 1996, pp. 281.

14. P.G. de Gennes, *Adv. Colloid Interface Sci.* **27**, 189 (1987).

15. H. Hommel, *Adv. Colloid Interface Sci.* **54**, 209 (1995).

16. T. Miyamoto and H.J. Cantow, *Makromol. Chem.* **162**, 43 (1972)

17. T. Cosgrove and K.G. Barnett, *J. Magn. Reson.* **43**,15 (1981).

18. K.G. Barnett, T. Cosgrove, B. Vincent, D.S. Sissons, and M. Cohen-Stuart, *Macromolecules* **14**, 1018 (1981).

19. T. Cosgrove, N. Finch, B. Vincent, and J. Webster, *Colloids Surf.* **31**, 33 (1988).

20. T. Cosgrove, N. Finch, and J. Webster, *Colloids Surf.* **45**, 377, (1990).

21. K.G. Barnett, T. Cosgrove, B. Vincent, A.N. Burgess, T.L. Crowley, T. King, J.D. Turner, and T.F. Tadros, *Polymer Comm.* **22**, 283 (1981).

22. T. Cosgrove, T.L. Crowley, B. Vincent, K.G. Barnett, and T.F. Tadros, *J. Chem. Soc., Faraday Symp.* **16**, 101 (1981).

23. L. Facchini and A.P. Legrand, *Macromolecules* **17**, 2405 (1984).

24. T. Cosgrove and K. Ryan, *J. Chem. Soc., Chem. Comm.* **21**, 1424 (1988).

25. T. Cosgrove and K. Ryan, *Langmuir* **6**, 136 (1990).

26. T. Cosgrove, T.G. Heath, K. Ryan, and B. van Lent, *Polym. Comm.* **28**, 64 (1987).

27. T. Cosgrove, T.M. Obey, and K. Ryan, *Colloids Surf.* **65**, 1 (1992).

28. C. K. Hall and E. Helfand, *J. Chem. Phys.* **77**, 3275 (1982).

29. R.D. O'Connor and F.D. Blum, *Macromolecules* **27**, 1654 (1994).

30. F.D. Blum, B.R. Sinha, and F.C. Schawb, *Macromolecules* **23**, 3592 (1990).

31. B.R. Sinha, F.D. Blum, and F.C. Schawb, *Macromolecules* **26**, 7053 (1993).

32. M. Xie and F.D. Blum, *Langmuir* **23**, 5669 (1996).

33. M. Liang and F.D. Blum, *Macromolecules* **29**, 7374 (1996).

34. A.A. Parker, G.H. Armstrong, and D.P. Hedrick, *J. Appl. Polym. Sci.* **47**, 1999 (1993).

35. T. Cosgrove and P.C Griffiths, *Colloids Surf. A* **84**, 249 (1994).

36. N. Zumbulyadis and J.M. O'Reilly, *Macromolecules* **24**, 5298 (1991).

37. C.W. Chu and P.D. Murphy, *J. Adhesion Sci. Technol.* **6**, 1119 (1992).

38. V.L. Fernandez, J.A. Reimer, and M.M. Denn, *J. Am. Chem. Soc.* **114**, 9634 (1992).

39. J. Van Alsten, *Macromolecules* **24**, 5320 (1991).

40. D.W. McCall, Nat. Bur. Stand. (US), Spec. Publ., 1969, 301, 475. *Acc. Chem. Res.* **4**, 223 (1971).

41. F.D. Blum, R.B. Funchess, and W. Meesiri, in L. Mathias ed., *Solid State NMR of Polymers* Plenum, New York, 1991, p. 271.

42. For a comparison, see Figure 2 in Ref. 1, which shows spectra for a similar polymer taken at 30.7 MHz in a 5-mm sample coil.

43. F.D. Blum, G. Xu, M. Liang, and C.G. Wade, *Macromolecules* **29**, 8740 (1996).

10 Radiochemical Methods for Polymer Adsorption

JOSEPH B. SCHLENOFF

Department of Chemistry and Center for Materials Research and Technology (MARTECH), Florida State University, Tallahassee, Florida 32306

10.1 INTRODUCTION

There are a wide variety of techniques for characterizing polymer adsorption.[1] The kind of information obtained depends strongly on the method employed. Indirect measurements involve determining the solution concentration of polymer before and after addition of sorbant, which yields the surface coverage (milligrams per square meter or milligrams per gram). High specific surface area particles, such as silica,[2] alumina, activated carbon,[3] glass beads,[4] metal powders,[5] and clays, facilitate such difference measurements. Direct methods are able to distinguish between polymer on the surface and dissolved polymer. Optical techniques such as ellipsometry[6,7] (fixed or variable wavelength) and reflectometry (with X-rays, neutrons, and visible light) are popular for in situ measurements on planar surfaces:[8–10] Both the film thickness and adsorbed amount may be deduced. Neutron scattering[11] provides information on adsorbed layer thickness and amount as well as the segment density profile. Internal reflectance infrared spectroscopy[12] and the quartz crystal microbalance[13] provide for direct determination of surface excess with good time resolution. Hydrodynamic methods[14] give the hydrodynamic layer thickness. Magnetic resonance, such as electron spin resonance (ESR)[15] and nuclear magnetic resonance (NMR),[16,17] have also been employed in the study of polymer adsorption. These, the latter described elsewhere in this monograph, are capable of probing the molecular dynamics of adsorbed molecules. Also detailed in other chapters are recent advances in neutron scattering, reflectometry, and fluorescence methods as applied to the study of polymer adsorption.

Colloid–Polymer Interactions: From Fundamentals to Practice, Edited by Raymond S. Farinato and Paul L. Dubin
ISBN 0-471-24316-7   © 1999 John Wiley & Sons, Inc.

Radiochemical methods present many desirable features for the measurement of small quantities of dissolved or surface-localized materials. Since naturally occurring levels of background radiation are low, radioanalytical detection is potentially very sensitive: It is possible to measure percents of a monolayer. Surfaces may be smooth and planar or rough and particulate. Reduction of the raw data (counts per second) to reliable surface coverages is straightforward. The measurement apparatus is stable and precise over long periods of time. The disadvantages of radiochemical techniques are that a labeling step must be included in polymer synthesis and that there are significant safety and regulatory concerns associated with the use of radioisotopes.

A survey of the literature reveals that most prior radioanalytical work for macromolecule adsorption has focused on biomolecules such as proteins. This is presumably due to the availability of radiolabeled biomolecules, or facile labeling schemes, and the interest of the medical and biochemical community in developing assays for enzymes, proteins, and the like coupled with the traditional emphasis on scintillation counting as a sensitive means of detection.[18] Thus, avidin,[19] globulin,[20] insulin,[21] glucogen,[22] IgG,[23] albumin,[24] protein A,[25] labeled with ^{125}I have been adsorbed at the solid–liquid interface. The commercial availability of small and large biomolecules with ^3H, ^{125}I, ^{14}C, ^{35}S, and ^{32}P labels is extensive. Particularly relevant to our discussions is a small body of work on the adsorption of proteins at liquid–liquid and air–liquid interfaces.[26–29]

Prior work on the use of radiolabels for studying the adsorption of *synthetic* polymers is limited to indirect or ex situ methods: Counting is performed in a separate operation subsequent to adsorption. Hensley and Inks[30] followed the adsorption of ^{14}C-methoxycellulose on textile fibers including wool, cotton, rayon, cellulose acetate, and nylon. Howard and Wood[31] used tritiated polystyrene to demonstrate that higher-molecular-weight polymer is preferentially adsorbed onto silica from cyclohexane at 35°C (θ solvent). The theoretical basis for this preference was provided by Roe[32] and Cohen Stuart et al.[33] Also using ^3H-polystyrene, Grant et al.[34] examined kinetics and isotherms for adsorption onto chrome and mercury from a θ solvent. Adsorption on chromium took at least a day to reach equilibrium whereas adsorption onto mercury was complete in an hour. More recent studies on adsorption of polystyrene (PS) to gold from cyclohexane have confirmed that the time for adsorption of the PS–cyclohexane is closer to 1 h. Grant et al.[34] also measured the desorption of PS on chrome or mercury into pure solvent and found very slow apparent desorption rates. As Fleer et al. point out,[35] the lack of polymer desorption on exposure to fresh solvent does not necessarily imply irreversibility since the adsorption isotherm is such high affinity that even the smallest solution concentrations yield pseudo-constant adsorbed amounts.

Varoqui, Pefferkorn, and coworkers have made extensive use of ^3H-labeled polyacrylamides[36–39] to study their adsorption to the solid–liquid interface. A flow-through stirred tank containing glass beads as sorbant was mainly employed, with off-line scintillation counting of effluent. Information on adsorption kinetics and equilibria was obtained. Exchange of adsorbed solution polymers was found to be slow for polystyrene,[40–42] polyacrylamide, and a polystyrene–polyvinylpyridine (PVP) diblock copolymer.[43]

Sulfur-functionalized tritiated polystyrenes were prepared by Stouffer and McCarthy.[44] Both thiol-terminal groups and propylene sulfide blocks caused polystyrene to adsorb to gold from a good solvent [tetrahydrofuran (THF)]. Thiol-terminated polymers with a molecular weight beyond ca. 2 to 5×10^5 did not adsorb, since the sulfur–gold interaction was not sufficient to overcome the enthalpy and entropy loss of polymer sticking to the surface.

Radiolabeled diblock copolymers of styrene and 2-vinylpyridine were described by Pefferkorn et al.[43] and by Parsonage et al.[45] The latter group performed a systematic study of the dependence of adsorbed amount on PS and PVP block length. The substrates employed were silicon and mica: The PVP block adsorbs to these from toluene, whereas the PS block does not. For copolymers of moderate asymmetry the surface density, σ, was found to be influenced more strongly by the length of the PVP block, whereas for copolymers with large PS blocks σ was controlled by the molecular weight of the PS block.

The balance of this chapter deals with in situ methods. For the in situ study of small molecules and ions at the solid–liquid interface the reader is referred to the extensive work by the groups of Varga, Horanyi, and Wieckowski.[46–52] These researchers have paid particular attention to the adsorption of sulfate ion on polycrystalline and single-crystal metals. The principles enumerated within these works are fully applicable to the study of macromolecules, since the ranges of the decay particles are much longer than the dimensions of a typical polymer at a surface. These ideas have recently been extended in our laboratory to the in situ study of adsorption of synthetic polyelectrolytes at the liquid–liquid and solid–liquid interfaces. In situ analysis offers better time resolution than ex situ techniques, as well as better quality (less scatter of data, more data points). One finds that the study of polymer adsorption at the *liquid–liquid* interface, of much practical significance,[53,54] has been eclipsed by extensive research on adsorption at the *solid–liquid* interface, primarily due to the limited experimental access to the "buried" interface in the former system. Experimental techniques for liquid–liquid interfaces that have met, nevertheless, with some success include surface pressure measurements[55] and neutron[56,57] or x-ray reflection.[58] In the work that is described here, particular attention has been paid to the liquid–liquid interface.

10.2 EXPERIMENTAL METHODS

10.2.1 Apparatus

Our experiments are performed in a light-tight enclosure equipped with a photomultiplier tube (PMT) powered by a high-voltage supply and connected to a counter interfaced to a computer. Provision is made for remote injection of reagents within the dark box. For our current setup we use a 2-inch diameter RCA 8850 end-on

PMT with a silica window biased to 2200 V with a Bertan 313B high-voltage supply and a Philips PM 6654C frequency meter. The pulse height threshold on the counter is set to reject noise from the PMT. The background count rate, mostly noise from the PMT, is ca. 1 cps. Coincidence counting could, in principle, lower this background, but most of our experimental count rates are well above this level. One could also use the ^{14}C channel of a commercial liquid scintillation counter, although it would have to be adapted to provide for stirring if desired. Magnetic fields affect PMTs so mu-metal shielding is advisable if stirring with a magnetic stir bar.

10.2.2 Scintillators

Crookes and Regener[59] introduced the concept of scintillation counting in 1908, using phosphorescent zinc sulfide screens. Solid naphthalene was the first organic scintillator (discovered by Kallmann[60] in 1947), and Ageno et al.[61] reported scintillation pulses from certain organic liquid solutions. For in-depth information on scintillators and scintillation counting the reader should consult a comprehensive monograph by Birks on the topic.[62]

Liquid Scintillator   Any fluorescent compound ("fluor") is a potential scintillator or component of a scintillator system. Berlman[63] provides an extensive list of candidate molecules. Thus, a common feature of scintillators is that they have unsaturated or conjugated bonds. A small fraction of the impinging ionizing radiation excites the solvent or main constituent of scintillator and places π-electrons in the singlet excited state. The bulk of the energy is dissipated nonradiatively. Energy is transferred to a primary fluor, present in ca. 1% concentration, chosen such that its absorption overlaps the emission spectrum of the solvent. In order to decrease self-absorption within the scintillator (to provide a longer pathlength for emitted light) and to bring the wavelength to a region optimum for PMT performance, a secondary scintillator is added in much lower concentration (e.g., < 0.1%). Toluene is widely used as a solvent for liquid scintillators. Dioxane, often loaded with naphthalene, is used as a water-compatible scintillator. The duo of 2,5-diphenyloxazole (PPO), and 1,4,-*bis*[5-phenyloxazole-2-yl]benzene (POPOP), as primary and secondary fluor is popular for liquid scintillation "cocktails." We use *p*-terphenyl, PTP (1 wt %) and 1,4,-*bis*[2-methylstyryl]benzene, *bis*-MSB (0.1 wt %), both scintillation grade, and high-pressure liquid chromatography (HPLC)-grade toluene. Since impurities tend to collect at interfaces, this mixture is shaken with water and the organic layer is separated for further use.

Plastic Scintillator[64]   Solid scintillators include naphthalene crystals and inorganic materials (glass or doped salt crystals). Tl-doped NaI, for example, has a peak emission at 410 nm, decay time of 250 nS, and a light output of about 4 photons per 100 eV deposited.[65] Following the success of liquid scintillators, Schorr and Torney[66] introduced a solid polymeric ("plastic") scintillator that was composed of polystyrene containing *p*-terphenyl. Swank and Buck[67] then showed that scintillator based on poly(vinyltoluene) was a little more efficient. The base plastic in which the dye is dispersed needs aromatic rings for energy transfer. Poly(methyl methacrylate) (PMMA) can be used, but must be loaded with aromatic molecules such as naphthalene. One obtains ca. 1 photon per 100 eV of energy deposited in plastic scintillator.[65] The material is used extensively by the particle physics community to detect high-energy particles. The base polymer absorbs the energy of a particle traversing it and transfers a small fraction of the energy to a primary dye (via Förster energy transfer) then to a secondary dye that is chosen to fluoresce in the blue to match the wavelength of maximum efficiency of a photomultiplier. Scintillating plastic is made by mixing the fluors and monomer together, then heating to produce polymer. If polymerization is done between two sheets of glass, the polymer surface retains the smoothness of the glass: Measurements using atomic force microscopy (tapping mode) have indicated a surface roughness of about 5 Å.

The counting efficiency, when used with a low-noise photomultiplier, such as an RCA 8850, is good. Plastic scintillator is comparable in speed and efficiency to liquid scintillators. The inherent limitation, when using plastic scintillator to count films, is that half the decay particles go away from the scintillator. Thus the maximum counting efficiency is 50%. The lifetime of the pulse is < 10 ns. (Fluorescence lifetimes tend to be a little longer in the solid state, but the quantum efficiency is also enhanced somewhat.) A plastic scintillator is also reasonably radiation hard, making it a good candidate for detectors in particle accelerators, but the radiation fluxes that are typical for labeling experiments are orders of magnitude lower than the fluxes found in accelerators. Sometimes plastic scintillators are crosslinked (e.g., polystyrene crosslinked with divinyl benzene) to make them stiffer (especially at higher temperatures). For our purposes, crosslinked plastic is required to make a charged gel. Clearly, the use of plastic or organic liquid scintillator limits the choice of solvent from which adsorption occurs. It would not be feasible to use toluene, dimethyl formamide (DMF), or THF, for example, whereas water is ideal.

10.2.3 Measurement Principles

A review by Muramatsu[68] on radiotracers in surface and colloid science, focusing mostly on small-molecule surfactants, provides several examples of adsorption at the air–water interface. Krauskopf and Wieckowski[52] describe more recent work on radiochemical methods to measure adsorption at polycrystalline and single-crystal metal surfaces.

In the approach used extensively here, scintillators are employed as proximity detectors. Since β particles have a finite range in water, most labeled polymer in the bulk of the aqueous solution is not detected, whereas decay events from adsorbed polymer are observed (Scheme 10.1). However, the observed count rate contains a contribution from labeled material in a thin layer of liquid adjacent to the interface. This contribution, which is here termed "solution background," is present in all samples and is directly proportional to the solution concentration of labeled polymer. Solution background is distinct from "instrument background," which is the count rate

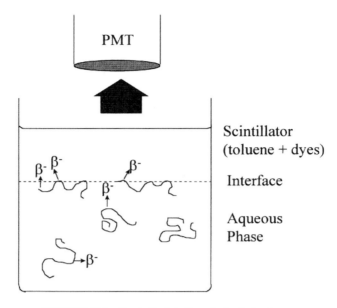

SCHEME 10.1 Proximity detection of adsorbed polymer.

in the absence of added radiolabel. For an experiment where concentration is a variable (i.e., for an adsorption isotherm) one may project the following scenarios, summarized in Figure 10.1: No adsorption to the interface occurs; the count rate comes only from the solution background, which is proportional to the solution concentration of polymer (curve *A*). (2) High-affinity adsorption with a constant surface excess over all measured concentration; count rate has a constant contribution from adsorbed polymer on a rising background; *y* intercept yields adsorbed amount (curve *B*). (3) Increasing adsorption with concentration, count rate rises and eventually the surface excess reaches a (pseudo) constant value (curve *C*). In all cases, background (curve *A*) contributes to the observed count rate and must be removed to yield the true surface excess.

Experimental error is greater at high concentrations since solution background dwarfs contributions from surface-localized species. Thus, it is seen that polymers, having high-affinity isotherms, are *particularly suited to this type of detection.* Small molecules,[68] with saturation coverages at higher concentration, would have considerable solution backgrounds in the plateau region of the isotherm. It is clear that the selection of isotope becomes important in the balance of surface versus solution signals: Different isotopes emit particles with different ranges. For example, the maximum range in water for ^{3}H, ^{14}C, ^{35}S, and ^{45}Ca β's is 6, 300, 340, and 650 μm, respectively.[68] The higher-energy emissions, though easier to detect, yield correspondingly higher solution backgrounds. Note that the energy of ^{3}H β's is sufficiently weak (17.9 keV) that the range is significantly smaller than the others listed below.

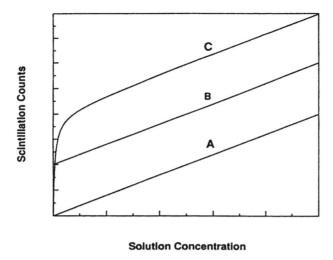

Solution Concentration

FIGURE 10.1 Generalized response (counts per second as a function of solution concentration) at an aqueous–organic interface for the adsorption of radiolabeled, water-soluble polymer. Curve *A* represents no adsorption of polymer; signal from solution only ("solution background"). Curve *B* is expected for high-affinity adsorption with constant surface excess. Curve *C* is for increasing adsorbance with concentration, with a pseudo-constant adsorbance at higher solution concentration.

Since the only restriction in the proximity detection scheme is that the adsorbed polymer must be within a few micrometers of the scintillator, it is possible to coat the scintillator with thin films. For example, we have used plastic scintillator coated with a thin film of gold (vida infra). Adsorption studies for other, thick, substrates are possible by arranging to bring the electrode in contact with the scintillator during measurement. Weickowski[50–52] describes an apparatus for squeezing single crystals of metals against a scintillator in a prescribed geometry during counting. This method reduces the solution background somewhat, since only a thin film of liquid remains trapped between surface and scintillator.

Determining Solution Background   In principle, if the response function of detector, counting efficiency, geometry, the specific activity of the radiolabeled polymer, and the attenuation of the β particle are known, it should be possible to calculate the contribution to the signal from solution background. Since geometries differ between experiments and other parameters are not known with precision, this background is determined operationally. For example, Aniansson,[69] for water–air adsorption, describes a method whereby a species that is known *not* to adsorb is added to the aqueous phase in a separate experiment. This species, such as acetate ion, contains the same label as the surface-active species. The disadvantages with this approach are that one needs access to another radiolabeled molecule, and the exact relative specific activities of the surface-active and non-surface-active molecules are required.

In our experiments solution background is determined on an individual basis by one of the following methods:

1. The background slope (curve A in Fig. 10.1) is measured by continuing the experiment to sufficiently high solution concentration.
2. A small aliquot of solution containing a large excess of unlabeled material is added and allowed to self-exchange. The remaining counts are mainly from solution background.
3. If self-exchange does not occur to completion, the experiment is repeated with the same amount of activity per unit volume of solution, but much lower *specific* activity (Ci mol^{-1}).

The ramp measurement, method [1], is useful here because polymer adsorption is high affinity and the adsorption pseudo-plateau is quickly reached. Also, for some cases, we are able to define conditions where adsorption of polymer is minimal (i.e., no salt). For small ions, such as $^{45}Ca^{2+}$, self-exchange is rapid and method [2] works well.

Standards A critical step in establishing accurate surface excesses is the use of a reliable and reproducible standardization technique for converting raw counts per second (cps) into milligrams per square meter (mg m^{-2}). For standards a known mass of labeled material must be distributed uniformly over the scintillator. Corrections must be made to account for backscattering of β particles at the scintillator interface.[52] These corrections are significant for scatter from materials of high atomic number, Z (such as platinum), but for the interface between water and plastic, the average Z's are similar such that backscattering (which we have determined experimentally) is minimal. For standards at the liquid–liquid interface a small aliquot of polymer is allowed to dry on a glass cover slip that is then suspended near the interface within the toluene layer. A correction of about 10% is required to account for additional electrons scattered off the glass and back into the scintillator. For this reason, the accuracy of the adsorbance values at the liquid–liquid interface is quoted to ±15% and values for the plastic–liquid interface are ±10%.

Which Isotopes to Use A plastic scintillator can be used to detect α, β, or γ radiation. However, the distance these particles travel in plastic (or water) is very different: α particles have a very short range and the β's of useful isotopes penetrate up to a few hundred micrometers in plastic; γ-rays have significantly longer attenuation lengths. Thus, the first two particles will deposit all their energy in the several millimeters of scintillator normally used, but γ's will not. Furthermore, as discussed above, the proximity detection scheme relies on a very limited path length for decay particles in water, restricting the utility of γ emitters (and the more energetic β emitters). For α particles, we have found that counting efficiency is rather low with our apparatus, and organic macromolecules are not likely to contain atoms that can be replaced with an α-emitting isotope. The following lists common β emitters:

Label	Energy (MeV)	Half-life
^{14}C	0.156	5730 years
^{35}S	0.167	87.4 days
^{32}P	1.709	14.3 days
^{45}Ca	0.257	163 days
^{3}H	0.0186	12.4 years
^{90}Y	2.28	64.1 h
^{36}Cl	0.710	3×10^5 years

For our purposes, the most useful of these are ^{45}Ca, ^{35}S, and ^{14}C. For ^{14}C and ^{35}S the "average" thickness of the solution layer from which decay is observed (solution background) is about 35 μm.

Making Labeled Polymers   Labeling schemes should be a one-step process and should require as inexpensive a label as possible. To this end we use ^{35}S-H_2SO_4 to

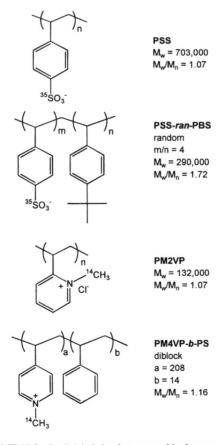

PSS
$M_w = 703,000$
$M_w/M_n = 1.07$

PSS-*ran*-PBS
random
$m/n = 4$
$M_w = 290,000$
$M_w/M_n = 1.72$

PM2VP
$M_w = 132,000$
$M_w/M_n = 1.07$

PM4VP-*b*-PS
diblock
$a = 208$
$b = 14$
$M_w/M_n = 1.16$

SCHEME 10.2   Radiolabeled polymers used in the present work.

sulfonate polystyrene to obtain negative polyelectrolytes, and we methylate 2- or 4-vinylpyridines with $^{14}CH_3I$ for positive polyelectrolytes (Scheme 10.2). These polymers are available as narrow molecular-weight-distribution standards, making interpretation of polyelectrolyte data easier. Scheme 10.3 outlines some of the approaches that have been taken to the synthesis of labeled polymer.

Dual Scheme for Labeling Sites and Polymers   Polyelectrolyte   adsorption   at oppositely charged surfaces involves ion exchange of a small counterion at the adsorption site with a charged polymer segment.[70] It is possible to label the counterion and follow the *decrease* in count rate as the polymer displaces it from the surface, as shown in Figure 10.2. Coupled with experiments employing labeled polymer, one obtains a comprehensive picture of site occupancy and adsorbed amount. For example, $^{45}Ca^{2+}$ can be used as a countercation for negatively charged surfaces. When a positive polyelectrolyte adsorbs, it displaces the $^{45}Ca^{2+}$ from the site and the count rate decreases. The great advantage of this technique is that no synthesis is required for labeling, the $^{45}Ca^{2+}$ isotope is one of the least expensive, and one can

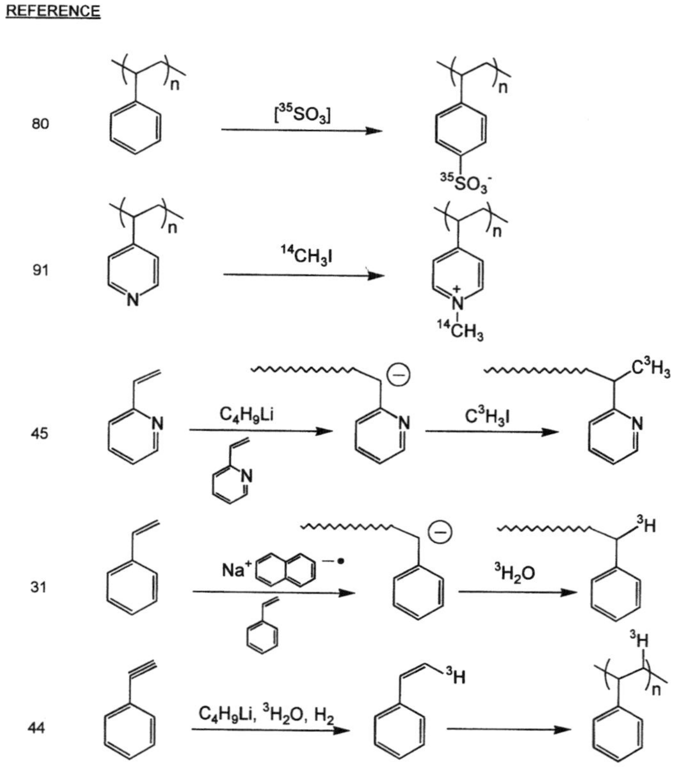

SCHEME 10.3   Select labeling approaches for radiolabeled polymers.

make the specific activity (and thus the count rate) as high as required. The minimum detectable limit depends on whether the measurement is in situ or ex situ: for in situ measurements one has the contribution from bulk species that are close to the surface. We can measure in situ surface coverage of $^{45}Ca^{2+}$ to less than 1% of a monolayer.

Counting Statistics; Activity vs. Time Resolution Tradeoffs   The apparatus we use has a background of about 4 cps. This background comes from γ-rays and cosmic rays inducing scintillation, and (mostly) from PMT noise. The precision of the measurement depends on the total number of counts collected and is governed by Poisson statistics: The error is (number of counts)$^{1/2}$. Thus, for 100 counts the error is 10 counts, or 10%. For 1000 counts the error is 32 counts, or 3%. For a count rate of 1 cps one can reduce the error to a few percent by counting for several minutes. In kinetics experiments one has to weigh the need to acquire data at set time intervals against the need for precision. For a 100-cps count rate the precision will be 10% if a data point every second is required, but if it is possible to wait for a minute between points, at this count rate, the error goes down to 1%. Unfortunately, the count rate for a labeled polymer adsorbing will be lowest at the beginning of the experiment, where the fastest measurements are needed. Error bars will be larger at the beginning of adsorption. An advantage to the "probe displacement" method (see Fig. 10.2) is that the count rate is highest at the beginning of adsorption.

10.3   RESULTS AND DISCUSSION

10.3.1   Adsorption of Small Ions and Polyelectrolytes on Charged Scintillator

We provide here the first in a series of examples of the kind of information that can be gleaned using the radiochemical techniques described above. Exposure of a crosslinked polystyrene-based plastic scintillator to sulfonating agents, such as H_2SO_4/SO_3, yields a thin polysulfonate layer.[71] This is the composition of a classical resin-based cation exchanger, and it is thus possible to perform ion-exchange-type measurements with radiolabeled cations. For example, $^{45}Ca^{2+}$ rapidly self-exchanges with unlabeled Ca^{2+} as shown in Figure 10.3. Other cations, such as H^+, will compete for the sulfonated sites and displace Ca^{2+}. In general, ions with a higher charge are preferred by an exchanger,[72] thus Y^{3+} is particularly effective at displacing Ca^{2+} (Fig. 10.4). Since the exchanging layer is so thin, mass transport was found to be limited by diffusion through a thin film of stagnant liquid adjacent to the substrate.[71] We have determined, using reflectance Fourier transform infrared (FTIR) spectroscopy, that sulfones, a common by-product of sulfonations, are not formed to a measurable extent on the scintillator surfaces.

If concentrated H_2SO_4 is employed for the sulfonation at room temperature, the sulfonation is limited to ca. 50% of a monolayer. The resulting surface has a root-mean-sequence (RMS) roughness of ca. 0.7 nm (determined by "tapping mode"

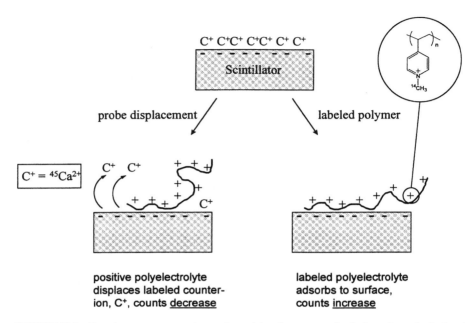

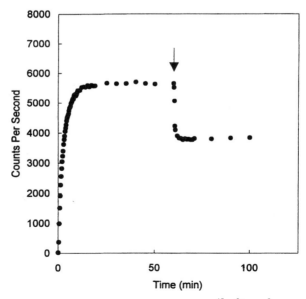

FIGURE 10.2   Complementary approaches to determining the amount of polyelectrolyte adsorbed and the extent of counterion exchange at a charged interface.

FIGURE 10.3   Count rate vs. time during the self-exchange of $^{45}Ca^{2+}$ (10^{-3} M, 2 Ci mol^{-1}) at room temperature in a sulfonated layer of polystyrene 0.4 μm thick on the surface of a 2% crosslinked polystyrene scintillator. At the point indicated, an aliquot of concentrated HCl was added to yield a concentration of 0.03 M HCl.

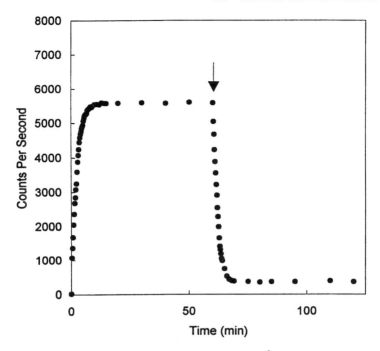

FIGURE 10.4   Count rate vs. time for self-exchange of calcium (10^{-3} M) as in Figure 10.3, followed by displacement of $^{45}Ca^{2+}$ by the addition of an equal concentration (10^{-3} M) of yttrium (as YCl_3).

atomic force microscopy) and is thus effectively flat on the scale of a polymer molecule. The high negative charge density, which is not pH dependent, leads to strong coulombic interactions with oppositely charged polyelectrolytes. Figure 10.5 depicts the adsorption to surface-sulfonated scintillator of ^{14}C-poly(N-methyl 2-vinylpyridinium), PM2VP, from aqueous solutions containing various concentrations of salt. The adsorption follows "screening-enhanced" behavior as defined by Steeg et al:[73] Salt screens polymer–polymer repulsions and polymer–surface attractions to allow the macromolecule to adsorb with more loops into solutions and fewer trains on the surface. Adsorption kinetics show a rapid regime at short time followed by a more gradual increase in surface excess, in accordance with prior work on polyelectrolyte adsorption, which postulates slow rearrangement of polymer on the surface.[74–77] Self-exchange was probed by the addition of an excess of unlabeled PM2VP at the point indicated in Figure 10.5. Partial exchange was observed only for polymer adsorbed from solutions with high salt concentration. This implies that the bulk of the polymer is adsorbed in a thermodynamically irreversible state.

Since polyelectrolyte adsorption at an oppositely charged surface is essentially an ion-exchange process, as depicted in Figure 10.2, one would expect that at sufficiently high salt concentration the small ions would compete with the polyelectrolyte segments for the surface and polymer adsorption should be suppressed. Clearly, from

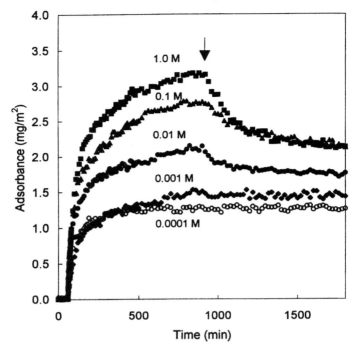

FIGURE 10.5 Adsorption of ^{14}C-labeled poly(N-methyl-2-vinylpyridinium) to surface-sulfonated scintillator from 2.5×10^{-5} M aqueous solutions containing different concentrations of NaCl, as shown. At the point indicated, a 10-fold excess of unlabeled polymer was added.

Figure 10.5, this limit is not achieved for our system. When radiolabeled salt ions, such as ^{45}Ca^{2+}, are employed, a direct measure of the degree of site occupancy by polymer segment is provided. For example, Figure 10.6 shows that all sulfonated sites are occupied by polymer segments. There is no evidence for isolated sites due to limited or excluded polymer conformations.

10.3.2   Polyelectrolyte Multilayers

Radiolabeled polyelectrolytes are useful for characterizing other systems, such as those made from alternating layers of positively and negatively charged polymers[78,79] (including proteins). For example, Figure 10.7 depicts the buildup of an alternating multilayer of ^{35}S-labeled poly(styrene sulfonate) (PSS) and ^{14}C PM2VP on scintillator from solutions containing either 0.1 M NaCl or no salt.[80] The deposition rapidly becomes linear with the number of times the substrate is cycled between positive and negative polyelectrolytes, and coverage is strongly salt dependent, as observed previously by other methods.[79] Summation of total negative and positive charges on

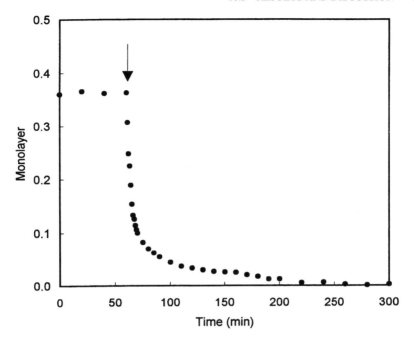

FIGURE 10.6 Displacement of $^{45}Ca^{2+}$ countercations adsorbed from 1×10^{-5} M CaCl$_{2(aq)}$ by 5×10^{-5} M (unlabeled) PM2VP added at the point indicated. The coverage of $^{45}Ca^{2+}$ is given in terms of monolayers of Ca^{2+} where one monolayer is 2×10^{-10} mol Ca^{2+} cm^{-2} or 2.5×10^{14} charges cm^{-2}. The background has been established by self-exchanging $^{45}Ca^{2+}$ with unlabeled Ca^{2+}. Within the detection limit, all surface sites are eventually occupied by polymer segments.

polymers reveals that they are equal, which implies that charge within the multilayer is not balanced by salt ions occluded within the structure.

10.3.3 Adsorption of Copolymers

Random Copolymer The adsorption of a random copolymer of PSS and poly(*t*-butyl styrene) (PBS) at the toluene–aqueous interface[81] shown in Figure 10.8 can be converted to an adsorption isotherm by subtracting the ramp due to solution background, as described above. The resulting Figure 10.9 shows a high-affinity isotherm, rounded as expected for polydisperse polymer.[33] Adsorption, driven by the strong partitioning of the *t*-butyl group into the organic phase, is a strong function of salt concentration. Equilibration to a steady-state surface excess following the addition of each aliquot of polymer required less than an hour.

The use of radiolabeled polymer provides an opportunity to evaluate the thermodynamic reversibility of the system. Since equilibria are dynamic, unlabeled polymer added to the system should (self-) exchange with labeled polymer already at

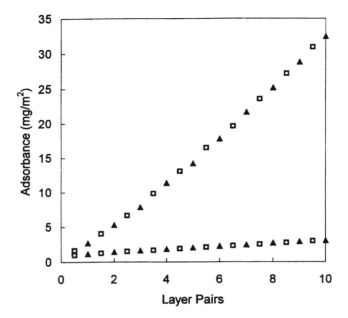

FIGURE 10.7   Buildup of an alternating multilayer of ^{35}S-PSS and ^{14}C- PM2VP on surface-sulfonated scintillator from 1×10^{-4} M polymer solutions containing salt (upper) and no salt (lower).

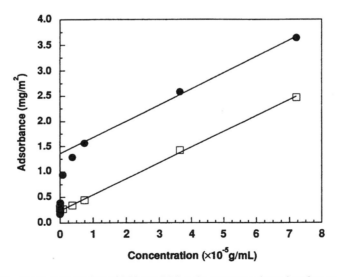

FIGURE 10.8   Apparent adsorption of PSS-*ran*-PBS at the aqueous–toluene interface as a function of concentration in water (squares) and 1 M NaCl (circles). Uncorrected for solution background.

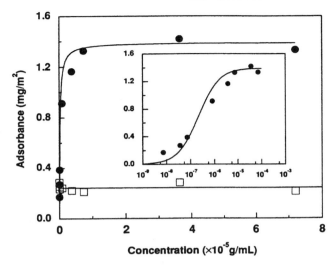

FIGURE 10.9 Room temperature adsorption isotherms for PSS-*ran*-PBS at the aqueous–toluene interface from water (squares) and 1 M NaCl. Absorbance (from Fig. 10.8) has been corrected for solution background. Inset shows a semilog plot of the same.

the interface. (Examples of this behavior for small ions is shown in Figure 10.3.) Figure 10.10 shows that an excess of unlabeled PSS-*ran*-PBS (identical to labeled material in other respects) will displace the interfacial polymer within a few hours. Most of the polymer exchanges rapidly. The remaining polymer may be exchanged

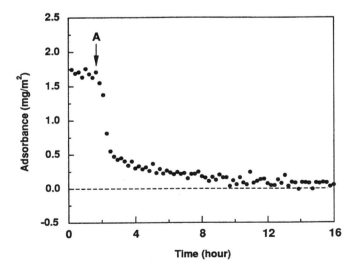

FIGURE 10.10 Self-exchange of ^{35}S labeled PSS-*ran*-PBS (solution concentration 7.2×10^{-5} g mL^{-1}) at the toluene–1 M NaCl$_{aq}$ interface with 10-fold excess of unlabeled polymer added at point A. The equilibrium coverage of labeled polymer is expected to be 9% of the original amount.

over the course of several hours. Detailed studies on the adsorption and exchange of polystyrene and deuteriostyrene at silica from cyclohexane revealed displacement time constants of hours and an exponential dependence on molecular weight.[82] The slow part of the exchange may thus reflect desorption of the longest chains.

The adsorption of PSS-*ran*-PBS copolymer at the polymer–water interface was also studied. In this case, the substrate is simply unmodified (uncharged) polystyrene scintillator. Since polystyrene is chemically very similar to toluene, similar hydrophilic/hydrophobic driving forces and adsorption energies are expected. The major difference is in the viscosity of the substrate. The adsorption isotherm of PSS-*ran*-PBS on polystyrene was similar to that for the toluene–water interface. Self-exchange was also demonstrated, although the rate was somewhat slower.

Diblock Copolymer   Diblock copolymers, where one of the blocks has a strong preference for the surface, have been used extensively in polymer adsorption.[83] For example, poly(ethylene oxide)/polystyrene (PEO/PS) copolymers adsorb to the solid–liquid[84–86] or air–liquid interface.[87] In the former example, PEO adsorbs to silica or mica from toluene, whereas PS does not. A similar preference for the surface is

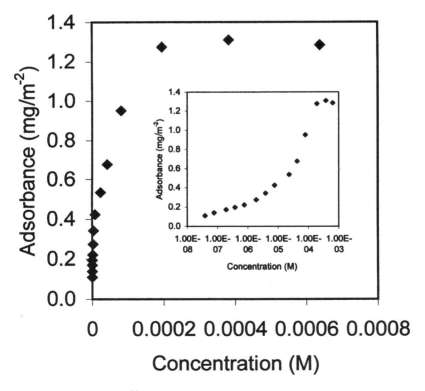

FIGURE 10.11   Adsorption of ^{14}C-PM4VP-*b*-PS diblock copolymer at the toluene–aqueous interface from 1 M NaCl.

exhibited by the vinylpyridine blocks in a poly(vinylpyridine)/PS copolymer.[45] Water-soluble copolymers having a preponderance of styrene sulfonate units with a short styrene block have similar surface active properties.[88,89] High surface densities of such diblock polymer yields brush architectures.[90]

In our work, a diblock polymer having 14 styrene and 208 4-vinylpyridine units was methylated with $^{14}CH_3I$.[91] The polystyrene block provides a hydrophobic anchor for immobilizing the polymer at interfaces. Toluene (with scintillator dyes) was used again in conjunction with aqueous salt solutions to produce a liquid–liquid interface. Surface excess as a function of solution concentration for PM4VP-*b*-PS in 1M $NaCl_{(aq)}$ is presented in Figure 10.11. The polymer *appears* to give an isotherm with a well-defined pseudo-plateau of ca. 1.3 mg m^{-2}. This may be compared with a surface coverage of 2.0 ± 0.5 mg m^{-2} for polysiloxane-*block*-poly(vinylpyridinoxide) adsorbing at the hexane water, as determined by neutron reflection.[57]

The semilog inset in Figure 10.11 clearly shows that polymer appears at the interface even at the lowest solution concentrations. Salt concentration is also an important variable for this system. In Figure 10.12 the salt concentration covers the range 10^{-5} to 1 M. Minimal adsorption is detected in the absence of salt, and there appears to be a critical concentration of ca. 0.01 M beyond which coverage rises

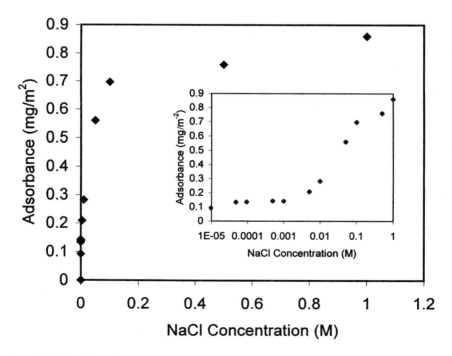

FIGURE 10.12 Effect of NaCl concentration on the adsorption of PM4VP-*b*-PS at the toluene–aqueous interface. Polymer concentration, 3.84×10^{-5} M.

rapidly. A similar phenomenon was observed by Zhang et al.[89] for a block copolymer of styrene sulfonate and *t*-butyl styrene adsorbing to mica from aqueous salt solutions.

There are some inconsistencies that are observed in comparing surface coverages of PM4VP-*b*-PS among different experiments. For example, the coverage from adding increasing salt to a fixed concentration of polymer was consistently higher than that from adding more polymer to a fixed concentration of salt (compare the adsorbance at 1.0 M in Fig. 10.12 with the corresponding point on Fig. 10.11). These are, in fact, clues to kinetic, rather than thermodynamic, control over the adsorption of polymer. The lack of self-exchange confirmed the irreversible nature of this system, as seen in Figure 10.13. Only a small fraction of labeled polymer exchanges, even after several days. When unlabeled polymer is added first to the interface, followed by labeled polymer, the latter does not displace the former. The microscopic picture of brush adsorption and exchange kinetics have been discussed by Milner,[92] and Ligoure and Leibler:[93] Adsorption is initially diffusion controlled, then, when overlap and crowding come into play, there is a substantial energy barrier to the approach of the adsorbing end of a chain to a surface. Presumably, this energy penalty, coupled with long-range electrostatic repulsion between surface and incoming polyelectrolyte, suppresses self-exchange.

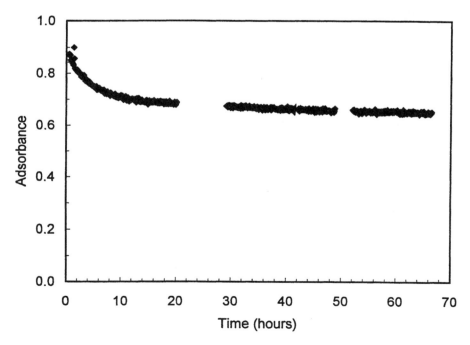

FIGURE 10.13   Attempted self-exchange of labeled PM4VP-*b*-PS at the toluene–aqueous (1 M NaCl) interface with 100-fold excess of solution unlabeled PM4VP-*b*-PS.

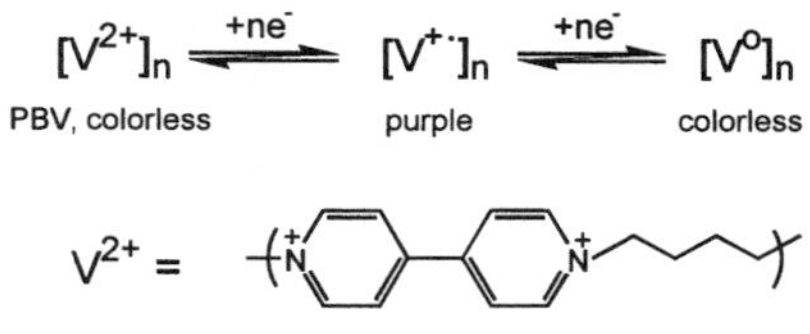

$$[V^{2+}]_n \underset{}{\overset{+ne^-}{\rightleftarrows}} [V^{+\cdot}]_n \underset{}{\overset{+ne^-}{\rightleftarrows}} [V^0]_n$$

PBV, colorless   purple   colorless

$$V^{2+} =$$

SCHEME 10.4 Polyviologen reduction.

10.3.4 Small Ion Transport

As a final example of the use of radiochemical techniques, we have made polyelectrolyte multilayers from an electrochemically active polycation, poly(butanyl viologen), PBV, and PSS. These multilayers[94] are made by exposing a surface in an alternating fashion to positive and negative polyelectrolytes, as described by Decher et al.[78,79] Since PBV is redox active,[95–97] the charge and ion balance within the multilayer can be controlled via electrochemistry. The PBV can be reduced with one electron per repeat unit to yield a highly colored radical cation.[98] A further electron reduces the PBV to a colorless state (Scheme 10.4). The introduction of electrons must be compensated by the migration of anions out of the multilayer and/or the insertion of cations from solution into the multilayer. If the counterions are radiolabeled,

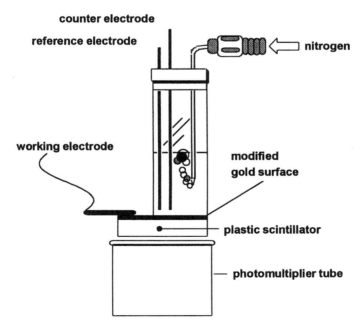

FIGURE 10.14 E-Rad cell for measurement of simultaneous electrochemistry and radiochemistry. A thin (ca. 100-nm) film of gold, evaporated onto plastic scintillator, serves as the working electrode. A rectangular polystyrene cuvette defines the area exposed to electrolyte.

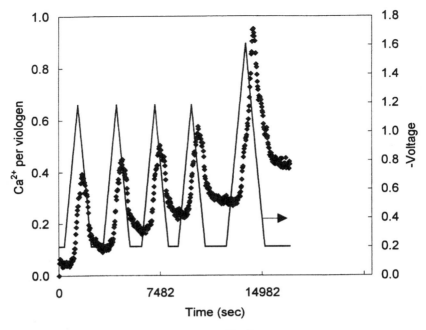

FIGURE 10.15   In situ measurement of ion transport ($^{45}Ca^{2+}$) in an eight-layer pair PBV/PSS multilayer on gold substrate immersed in 1 mM CaCl$_2$. The number of calciums per repeat unit is depicted vs. time as the polymer is switched four times between the $[V^{2+}]_n$ and the $[V^{+•}]_n$ oxidation states, then once to the $[V^0]_n$ state.

proximity detection methods can be employed using a combined electrochemical-radiochemical (E-Rad) cell as depicted in Figure 10.14. In this device, plastic scintillator is vapor coated with a thin layer of gold on which polymer thin films, such as polyelectrolyte multilayers, are deposited. Electrical contact is made to the gold film, which also serves as the working electrode in an electrochemical cell.

In Figure 10.15, $^{45}CaCl_2$ has been used as the electrolyte while the potential is scanned between -0.2 V and a value sufficient to reduce the viologen by one electron (-1.2 V) or two electrons (-1.6 V). It is clear that the $^{45}Ca^{2+}$ cations are responsible for charge compensation on the initial reduction. With further cycling, part of the ion flow includes chloride anions, and neutral CaCl$_2$ gradually builds up within the multilayer. Thus, one can control the concentration of salt ions within the multilayer by the potential and the number of redox cycles.

10.4   CONCLUSIONS

In the study of polymer adsorption, radiochemical methods offer good sensitivity and straightforward interpretation of raw data. In this chapter, we have emphasized the use of labeled polyelectrolytes coupled with in situ methods, as well as the possibilities of studying the liquid–liquid interface. It is feasible, in addition, to use self-exchange as a diagnostic tool for the reversibility of the adsorbed system. A random copolymer, interacting with the interface via hydrophobic/hydrophilic forces, was found to be reversibly adsorbed, whereas a polyelectrolyte at an oppositely charged interface and a diblock hydrophobic/hydrophilic copolymer adsorbed irreversibly.

Acknowledgments   This work was supported by the National Science Foundation. The author acknowledges the contributions of Ming Li, Hiep Ly, Delphine Laurent, and Ruiming Wang.

APPENDIX

Sources for Plastic Scintillator

Bicron NE
12345 Kinsman Road
Newbury, OH 44065-9677
Tel: (216) 564-2251
http://www.bicron.com

National Diagnostics
305 Patton Drive
Atlanta, GA 30336
(800) 526-3867
Fax: (404) 699-2077

Kuraray Inc. Japan
200 Park Avenue
New York, NY 10166-3098
Tel: (212) 986-2230

Crismatec
104 Rue de Larchant
BP 521
77794 Nemours CEDEX
France

National Diagnostics
Fleet Business Park
Itlings Lane
Hessle, Hull
England

Kuraray Co. Ltd
Maruzen Building
3-10, 2-Chome, Nihonbashi, Chuo-Ku
Tokyo 103
Japan

REFERENCES

1. G.J. Fleer, M.A. Cohen Stuart, J.M.H.M. Scheutjens, T. Cosgrove, and B. Vincent, *Polymers at Interfaces*, Chapman and Hall, London, 1993.

2. K. Furusawa and K. Yamamoto, *J. Coll. Interfac. Sci.* **96**, 268 (1983).

3. H.L. Frisch, M.Y. Hellman, and J.L. Lundberg, *J. Polym. Sci.* **38**, 441 (1959).

4. S. Ellerstein and R. Ullman, *J. Polym. Sci.* **55**, 123 (1961).

5. J. Koral, R. Ullman, and F.R. Eirich, *J. Phys. Chem.* **62**, 541 (1958).

6. F.L. McCrackin and J.P. Colson in E. Passaglia, R.R. Stromberg, and J. Kruger, eds., *Ellipsometry in the Measurement of Surfaces and Thin Films,* Nat. Bureau Stand. Misc. Publ. Vol. **256**, 1964, p. 61.

7. R.B. Davis, R.M.A. Azzam, and G. Holtz, *Surf. Sci.* **96**, 539 (1980).

8. J. Penfold and R.K. Thomas, *J. Phys. Condens. Mat.* **2**, 1369 (1990).

9. T. Cosgrove, T.G. Heath, J.S. Phipps, and R.M. Richardson, *Macromolecules* **24**, 94 (1991).

10. A.R. Rennie, E.M. Lee, E.A. Simister, and R.K. Thomas, *Langmuir* **6,** 1031 (1990).

11. T. Cosgrove, T.G. Heath, K. Ryan, and T.L. Crowley; *Macromolecules* **20**, 2879 (1987).

12. D.J. Kuzmenka and S. Granick, *Coll. Surf.* **31**, 105 (1988).

13. H. Xu and J.B. Schlenoff, *Langmuir* **10**, 241 (1994).

14. M.A. Cohen Stuart, F.H.W.H. Waajen, T. Cosgrove, T.L. Crowley, and B. Vincent, *Macromolecules* **17**, 1825 (1984).

15. K.K. Fox, I.D. Robb, and R. Smith, *J. Chem. Soc. Farad. Trans. 1* **70**, 1186 (1974).

16. F. Blum, *Coll. Surf.* **45**, 361 (1990).

17. T. Cosgrove and P. Griffiths, *Adv. Coll Interfac. Sci.* **42**, 175 (1992).

18. M. Horisberger and M. Vauthey, *Histochemistry* **80**, 13 (1984)

19. N.D. Tolson, R. Boothroyd, and C.R. Hopkins, *J. Microsc.* **123**, 215 (1981).

20. R.B. Dickson, M.C. Willingham, and I. Pastan, *J. Cell. Biol.* **89**, 29, (1981).

21. G.A. Ackerman and K.W. Wolken, *J. Histochem. Cytochem.* **29**, 1137, (1981).

22. G.A. Ackerman, J. Yang, and K.W. Wolken, *J. Histochem. Cytochem.* **31**, 433 (1983).

23. S.P. Kent, T.C. Caldwell, and A.L. Siegel, *Ala. J. Med. Sci.* **16,** 22 (1979).

24. J.B. Warchol, R. Brelinska, and D.C. Herbert, *Histochemistry* **76**, 567 (1982).

25. S.L. Goodman, G.M. Hodges, L.K. Trejdosiewicz, and D. Livingston, *J. Microsc.* **123**, 201 (1981).

26. D.E. Graham, L. Chatergoon, and M.C. Philips, *J. Phys. E* **8**, 696 (1975).

27. D.E. Graham and M.C. Philips, *J. Coll. Interfac. Sci.* **70**, 403 (1979).

28. D.E. Graham and M.C. Philips, *J. Coll. Interfac. Sci.* **70**, 415 (1979).

29. J.R. Hunter, P.K. Kilpatrick, and R.G. Carbonell, *J. Coll. Interfac. Sci.* **142**, 429 (1991).

30. J.W. Hensley and G.G. Inks, *Textile Res. J.* **29**, 505 (1959).

31. G.J. Howard and S.J. Woods, *J. Polym. Sci.* **A2**, 1023 (1972).

32. R.J. Roe, *Polym. Sci. Technol.* **12B**, 629 (1980).

33. M.A. Cohen Stuart, J.M.H.M. Scheutjens, and G.J. Fleer, *J. Polym. Sci. Polym. Phys. Ed.* **18**, 559 (1980).

34. W.H. Grant, L.E. Smith, and R.R. Stromberg, *J. Chem. Soc. Farad. Disc.* **59**, 209 (1975).

35. G.J. Fleer, M.A. Cohen Stuart, J.M.H.M. Scheutjens, T. Cosgrove, and B. Vincent, *Polymers at Interfaces*, Chapman and Hall, London, 1993, p. 279.

36. E. Pefferkorn, A. Carroy, and R. Varoqui, *Macromolecules* **18**, 2252 (1985).

37. E. Pefferkorn, A. Carroy, and R. Varoqui, *J. Polym. Sci. Pt. B., Polym. Phys. Ed.* **23**, 1997 (1985).

38. E. Pefferkorn, A.C. Jean-Chronberg, G. Chauveteau, and R. Varoqui, *J. Coll. Interfac. Sci.* **137**, 66 (1990).

39. E. Pefferkorn, A.C. Jean-Chronberg, and R. Varoqui, *Macromolecules* **23**, 1735 (1990).

40. E. Pefferkorn, A. Haouam, and R. Varoqui, *Macromolecules* **22**, 2677 (1989).

41. E. Pefferkorn, A. Haouam, and R. Varoqui, *Macromolecules* **21**, 2111 (1988).

42. A. Elaissari, A. Haouam, C. Huguenard, and E. Pefferkorn, *J. Coll. Interfac. Sci.* **149**, 68 (1992).

43. E. Pefferkorn, A. Elaissari, and C. Huguenard, *Macromol. Rep.* **A29**, 147 (1992).

44. J.M. Stouffer and T.J. McCarthy, *Macromolecules* **21**, 1204 (1988).

45. E. Parsonage, M. Tirrell, H. Watanabe, and R.G. Nuzzo, *Macromolecules* **24**, 1987 (1991).

46. K. Varga, E. Maleczki, and G. Horanyi, *Electrochim. Acta* **31,** 1667 (1986).

47. G. Horanyi, *Electrochim. Acta* **25**, 43 (1980).

48. K. Varga, P. Baradlai, and A. Vertes, *Electrochim. Acta* **42**, 1143 (1997).

49. A. Kolics, E. Maleczki, K. Varga, and G. Horanyi, *J. Radioanal. Nucl. Chem.* **158**, 121 (1992).

50. K. Varga, P. Zelenay, and A. Wieckowski, *J. Electroanal. Chem.* **330**, 453 (1992).

51. P. Zelenay and A. Wieckowski, *J. Electrochem. Soc.* **139**, 2552 (1992).

52. E.K. Krauskopf and A.Wieckowski, in J. Lipowski and P.N. Ross, eds., *Adsorption of Molecules at Metal Electrodes*, VCH, New York, 1992, Chap. 3, p. 119.

53. I. Piirma, *Polymeric Surfactants*, Dekker, New York, 1992.

54. B. Idson, in M.M. Rieger, ed., *Surfactants in Cosmetics,* Dekker, New York, 1985.

55. J.H. Brooks and B.A. Pethica, *Trans. Farad. Soc.* **60**, 208 (1964).

56. T. Cosgrove, J.S. Phipps, and R.M. Richardson, *Coll. Surf.* **62**, 199 (1992).

57. J.S. Phipps, R.M. Richardson, T. Cosgrove, and A. Eaglesham, *Langmuir* **9**, 3530 (1993).

58. S.J. Roser, R. Felici, and A. Eaglesham, *Langmuir* **10**, 3853 (1994).

59. E. Rutherford, J. Chadwick, and C.D. Ellis, *Radiations from Radioactive Substances,* Cambridge University Press, Cambridge, 1930.

60. H. Kallmann, *Natur Technik*, July 1947.

61. M. Ageno, M. Chiozzotto, and R. Querzoli, *Anal. naz. Lincei* **6**, 626 (1949).

62. J.B. Birks, *The Theory and Practice of Scintillation Counting*, Macmillan, New York, 1964.

63. I.B. Berlman, *Handbook of Fluorescent Spectra of Aromatic Molecules*, Academic, New York, 1971.

64. J.B. Birks, *The Theory and Practice of Scintillation Counting*, Macmillan, New York, 1964, Chap. 9.

65. Review of Particle Physics, *Phys. Rev. D*, **54**(1), (1996).

66. M.G. Schorr and F.L. Torney, *Phys. Rev.* **80**, 474 (1950).

67. R.K. Swank and W.L. Buck, *Phys. Rev.* **91**, 927 (1953).

68. M. Muramatsu in E. Matijevic, ed., *Surface and Colloid Science*, Wiley, New York, 1973.

69. G. Aniansson, *J. Phys. Coll. Chem.* **55**, 1286 (1951).

70. M.A. Cohen Stuart in J. Daillant, P. Guenoun, C. Marques, P. Muller, J. Tran Tranh Van, eds., *Short and Long Chains at Interfaces*, Editions Frontieres, Gif-sur-Yvette, 1995.

71. M. Li and J.B. Schlenoff, *Anal. Chem.* **66**, 824 (1994).

72. F. Helfferich, *Ion Exchange*, McGraw-Hill, New York, 1962.

73. H.G.M. van de Steeg, M.A. Cohen Stuart, A. de Keizer, and B.H. Bijsterbosch, *Langmuir* **8**, 2538 (1992).

74. E. Pefferkorn and A. Elaissari, *J. Coll. Interfac. Sci.* **138**, 187 (1990).

75. A.W.M de Laat, G.L.T. van den Heuvel, and M.R. Bohmer, *Coll. Surf.* **98**, 61 (1995).

76. N.G. Hoogeveen, M.A. Cohn Stuart, and G.J. Fleer, *J. Coll. Interfac. Sci.* **182**, 133 (1996).

77. N.G. Hoogeveen, M.A. Cohen Stuart, and G.J. Fleer, *J. Coll. Interfac. Sci.* **182**, 146 (1996).

78. G. Decher and J.D. Hong, *Ber. Bunsenges. Phys. Chem.* **95**, 1430 (1991).

79. G. Decher in J.P. Sauvage, M.W. Hosseini, eds., *Templating Self-Assembly and Self-Organization,* Pergamon, Oxford, 1996.

80. J.B. Schlenoff and M. Li, *Ber. Bunsenges. Phys. Chem.* **100**, 943 (1996).

81. R. Wang and J.B. Schlenoff, *Macromolecules* **31**(2), 494 (1998).

82. P. Frantz and S. Granick, *Macromolecules* **27**, 2553 (1994)

83. A. Halperin, M. Tirrell, and T.P. Lodge, *Adv. Polym. Sci.* **100**, 31 (1992).

84. H. Motschmann, M. Stamm, and Ch. Toprakcioglu, *Macromolecules* **24**, 3681 (1991).

85. J.R. Dorgan, M. Stamm, C. Toprakcioglu, R. Jerome, and L.J. Fetters, *Macromolecules* **26**, 5321 (1993).

86. D.A. Guzonas, D. Boils, C.P. Tripp, and M.L. Hair, *Macromolecules* **25**, 2434 (1992).

87. H.D. Bijsterbosch, V.O. de Haan, A.W. de Graaf, M. Mellema, F.A.M. Leermakers, M.A. Cohen Stuart, and A.A. van Well, *Langmuir* **11**, 4467 (1995).

88. C. Amiel, M. Sikka, J.W. Schneider, Jr., Y Tsao, M. Tirrell, and J.W. Mays, *Macromolecules* **28**, 3125 (1995)

89. Y. Zhang, M. Tirrell, and J.W. Mays, *Macromolecules* **29**, 7299 (1996).

90. S.T. Milner, *Science* **251**, 905 (1991).

91. J.B. Schlenoff and R. Wang, *PMSE Preprints, ACS Proc.* **77**, 654 (1997).

92. S.T. Milner, *Macromolecules* **25**, 5487 (1992).

93. C. Ligoure and L. Leibler, *J. Phys. (Paris)* **51**, 1313 (1990).

94. D. Laurent and J.B. Schlenoff, *Langmuir* **13**, 1552 (1997).

95. A. Factor and G.E. Heinsohn, *Polym. Lett.* **9**, 289 (1971).

96. C.L. Bird and A.T. Kuhn, *Chem. Soc. Rev.* **10**, 818 (1981).

97. H. Ohno, N. Hosoda, and E. Tsuchida, *Makromol. Chem.* **184**, 1061 (1983).

98. M.S. Simon and P.T. Moore, *J. Polym. Sci. Polym. Chem. Ed.* **13**, 1 (1975).

11 Measurement of Colloidal Interactions Using the Atomic Force Microscope

PATRICK G. HARTLEY

Advanced Mineral Products Special Research Centre, School of Chemistry,
University of Melbourne, Victoria 3052, Australia

11.1 INTRODUCTION

Over the last 30 years, the ability to directly measure surface physical and chemical properties has played a prominent role in developing both academic and technological understanding of colloidal phenomena. As in many other fields, this ability has frequently been facilitated by the availability of innovative computer-controlled instrumentation. The appearance and subsequent widespread use of the atomic force microscope (AFM) in surface and colloid science laboratories worldwide provides a good example of this process.

Evolving from scanning tunnelling microscope (STM),which won the Nobel prize for its creators,[1] the AFM first appeared in 1986,[2] and is one of a group of devices now generically referred to as scanned probe microscopes (SPMs). It was designed as a method for topological mapping of nonconductive surfaces at subnanometer resolution, over areas up to hundreds of microns in X-Y dimension. The AFM functions in all its imaging modes by controlling to some extent the forces between the tip and the sample. It is easy to see, therefore, that the control systems used in the device may easily be adapted to measure those same forces directly. This ability forms the basis of the studies discussed in this chapter.

The measurement of surface forces has for many years been the focus of much interest in the field of colloid and surface science.[3] This is principally because long-range interactions between the surfaces of particles frequently dictate the stability and rheology of colloidal dispersions.

Colloid–Polymer Interactions: From Fundamentals to Practice, Edited by Raymond S. Farinato and Paul L. Dubin
ISBN 0-471-24316-7 © 1999 John Wiley & Sons, Inc.

To a large extent, AFM measurements of surface forces overlap with research carried out using the surface force apparatus, and the reader is referred to the chapter in this volume which discusses the use of this device in many applications similar to those described here.

The one significant advantage of the AFM over the surface force apparatus is that it is not confined to studying optically transparent surfaces that are molecularly smooth over a large surface area. Its major disadvantage is that it does not provide an absolute measure of separation between the surfaces of interest, a problem discussed in detail later. However, it is the flexibility in sample type, the ability to image the topology of samples under study, and the relative ease of use of the AFM that have rapidly established it as a powerful tool in this for surface force measurement.

11.2 OPERATION OF THE ATOMIC FORCE MICROSCOPE IN IMAGING SURFACE TOPOLOGY

The term microscope generally implies the utilization of some form of focused light or electron beam, passing through, or reflecting off a sample of interest, and casting an image on a suitable viewer. While the AFM does employ laser light, the light itself does not interact with the sample directly, and hence the AFM is not a microscope in its truest sense. In this case the laser is used in the measurement of the deflection of a microfabricated cantilever due to its interaction with surface topological features. This contrast in imaging mechanisms serves to highlight two advantages of the AFM over more conventional microscopy. The first of these is that nonconducting, nontransparent surfaces can be imaged. Second, imaging can be performed under ambient and even aqueous conditions, since no staining or vacuum is required. These features impart particular benefits to the colloid scientist. There are, however, disadvantages of AFM as well. Since it is truly a topographic method, it is unable to visualize the internal features of a sample. Also, since it relies on intimate contact between a tip and the sample, damage of one or other is common, particularly with softer samples. This problem may be reduced using newer imaging techniques such as tapping mode.[4]

Figure 11.1 shows a schematic of the basic features of a widely used AFM. Some differences exist between the designs of different manufacturers, but the central features are conserved, and these are:

1. A piezoelectric drive, enabling the displacement of the sample surface to be controlled to Angstrom resolution over several micrometers of X, Y, and Z motion

2. A microfabricated cantilever assembly (usually silicon nitride)

3. A diode laser and lens/prism/mirror optical path allowing focusing of laser light onto the cantilever, and subsequent gathering of the reflected light

4. A position-sensitive photodiode capable of detecting changes in position of the laser light due to deflection of the cantilever.

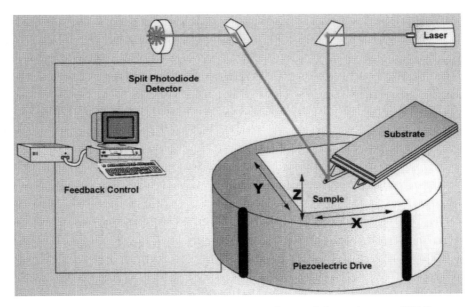

FIGURE 11.1 Schematic of the basic components of an atomic force microscope (AFM).

5. An electronic feedback loop that (optionally) responds to changes in cantilever deflection by adjusting the tip–sample separation, such that the cantilever returns to its original deflection. This is known as "constant force" imaging. In an alternative mode of operation, the efficiency of this feedback loop is reduced, and the deflection of the tip itself is used to map the surface. This is the so-called error signal mode.

In a typical imaging run, the AFM tip is moved toward the surface (in the Z dimension in Fig. 11.1) using a coarse drive motor. Contact with the surface is signaled by a tip deflection being measured at the position-sensitive photodiode. The feedback loop immediately engages, and maintains the tip at a preset force with respect to the surface. The piezoelectric tube is then used to raster the sample rapidly back and forth in the $x/{-}x$ direction. After each complete x scan, the sample is advanced by one increment in the y direction. In this way, a square of user-defined dimensions is mapped out. As the tip encounters a feature on the surface, it begins to deflect. Once again, the feedback control rapidly adjusts the position of the sample relative to the tip, such that the deflection is minimized. The feedback loop thereby ensures that the tip exerts a constant force on the sample. For assembly of a three-dimensional image of the surface, Z-dimension data at each X-Y position is derived from the feedback loop, which corresponds to the displacement of the surface required to keep the force between tip and sample constant. Figure 11.2 shows a three-dimensional surface topological image generated in this way.

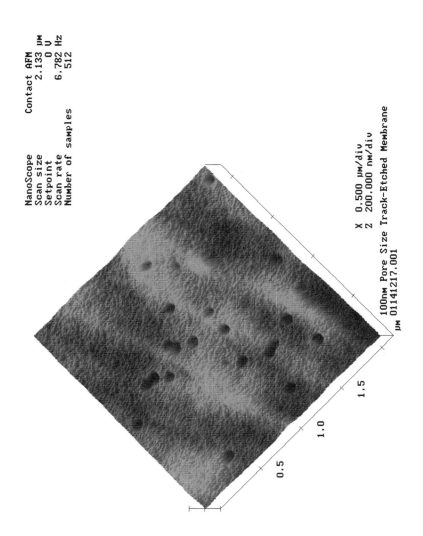

FIGURE 11.2   Typical topological image produced using the AFM. The image shows a 2×2 µm scan of the surface of a track-etched ultrafiltration membrane (Osmonics Inc.). The darker colors correspond to lower heights. The black holes represent the individual pores in the membrane, the nominal diameter of which is 0.1 µm.

11.3   OPERATION OF THE ATOMIC FORCE MICROSCOPE IN SURFACE FORCE MEASUREMENT

In its simplest form, surface force measurement in the AFM simply consists of using the piezoelectric drive tube to hold the sample and tip at one X-Y position relative to each other, and then to ramp them together and apart in the Z-dimension only. Any deflection of the tip is monitored using the split photodiode, and a plot of piezo displacement z versus photodiode voltage V may be generated.

The three regions of a displacement–photodiode voltage plot are summarized in Figure 11.3. At large tip–sample separations, there exists no interaction between tip and sample, and thus no deflection of the cantilever. This region of the plot (region I in Fig. 11.3) gives the baseline photodiode voltage (V_0). At a certain tip–sample separation, surface forces may act, and the tip deflects due to them. This deflection generates a change in voltage at the photodiode (V_z, region II in Fig. 11.3). At a certain travel position, the tip contacts and becomes coupled to the movement of the surface. Assuming a rigid, smooth sample, any further motion of the tip toward the sample results in a cantilever deflection equal to that motion. That is, the tip and sample are in "constant compliance." This is signaled by a straight line of nonzero gradient in the deflection (V_z) versus piezo displacement (z) plot (region III in Fig. 11.3). Acquisition of this portion of the plot is a critical requirement for the calculation of a force versus separation profile. The gradient of this straight-line portion (dV/dz) in units of volts deflection per nanometer piezo travel, allows the photodiode voltage (V_z) to be converted to a tip deflection (D_z) at any position since:

$$D_z = (V_z - V_0)/(dV/dz) \quad \text{(units of distance)} \quad (11.1)$$

The force acting on the cantilever at any position (F_z) may now be readily obtained providing its spring constant, K_s, is known (see below), using Hooke's law:

$$F_z = K_s D_z \quad \text{(units of force)} \quad (11.2)$$

Finally, the real separation between tip and sample (h_z) is calculable for each piezo travel position, z, away from the constant compliance line (z_0):

$$h_z = (z - z_0) + D_z \quad \text{(units of distance)} \quad (11.3)$$

11.3.1   Measurement of Cantilever Spring Constants

The size of standard AFM cantilevers (~100 μm in length) means that measuring their spring constant (K_s) presents certain problems. The spring constant of a macroscopic cantilever such as that used in the surface forces apparatus is readily obtained by using Hooke's law [Eq. (11.2)]. A series of different weights may be attached to the cantilever, and the deflection of the spring from its equilibrium position can be

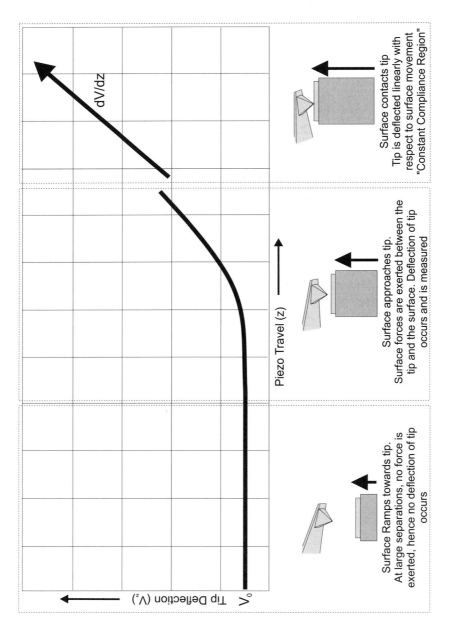

FIGURE 11.3   Schematic of the raw output data required to construct a force *vs.* distance profile in the AFM (see text for details).

measured microscopically. While this method is in principle suitable for AFM cantilevers,[5,6] the requirement of mass loading and measuring deflections on such a small object presents something of a challenge. Several methods have been developed to circumvent this problem. Some have used theoretical predictions of the spring constant of simple rectangular cantilevers based on bulk measurements of the Young's modulus of the cantilever material (silicon or silicon nitride),[6] while other work has focused on combining this approach with measurements of the unloaded resonance frequency of the more widely used V-shaped silicon nitride cantilevers.[7] Such methods are, however, complicated by the requirement for accurate values of the dimensions of the cantilevers and coatings, as well as the less than reliable assumption that the stoichiometry of the bulk material and thus its elastic modulus is preserved during cantilever production.[7,8]

Another approach to the direct measurement of spring constant involves measuring the deflection of the cantilever against a second beam of known spring constant. The beam (a glass filament or another cantilever) is mounted in the AFM beneath the cantilever of interest as shown in Figure 11.4.[6] When a displacement (Δh) is applied to the piezo drive, it is absorbed by deflection of both the beam (Δb) and the cantilever (Δc), that is,

$$\Delta h = \Delta c + \Delta b \tag{11.4}$$

Hooke's law relates the spring constants of the cantilever (K_s) and beam (K_b) to the force (F_{cb}) between them, which must be balanced, thus:

$$F_{cb} = K_s \, \Delta c = K_b \, \Delta b \tag{11.5}$$

Thus, by monitoring the deflection of the cantilever Δc, under the displacement Δh, the spring constant K_s is accessible, provided the spring constant of the beam (K_b) is known, since by substitution Eq. (11.4) gives

$$K_s = K_b(\Delta h/\Delta c - 1) \tag{11.6}$$

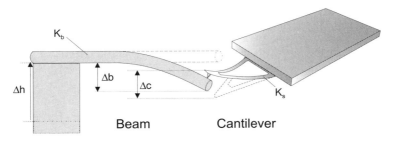

FIGURE 11.4   Method for measuring the spring constant of an AFM Cantilever (see text for details).

The Cleveland Method The most common form of direct measurement developed by Cleveland et al.[8] relies on measurement of the shift in resonance frequency of the cantilever in response to its loading with known masses of a dense material, such as tungsten microspheres. Its principal advantage is that several masses can be applied to the same cantilever, thereby reducing the errors associated with single measurement techniques. It also has the advantage that it does not require the use of the piezo drive of the AFM, since a calibration of cantilever deflection versus piezo displacement is not required.

The resonance frequency shift is related to the mass applied to the cantilever, thus:

$$M = K_s(2\pi\nu)^{-2} - m^* \tag{11.7}$$

where M is the applied mass, ν the resonance frequency, and m^* the effective mass of the cantilever. A plot of $(2\pi\nu)^{-2}$ versus M therefore gives a straight line of slope K_s, the cantilever spring constant.

For older AFM equipment, measurement of resonant frequencies of the cantilever required the split photodiode output to be intercepted and fed into an external spectrum analyzer. However, since the advent of resonance imaging techniques such as tapping mode, more recent devices have this function built into the standard AFM software.

11.3.2 Geometrical Considerations

Force-measuring devices are critically dependent on knowledge and/or control of the area of interaction of interest. This is particularly true if data are to be compared between different techniques, as well as with theoretical work that employs potential energy calculations. The standard mathematical method for interpreting force–distance data is the Derjaguin approximation,[9] which enables the expression of force–distance data in a highly "portable" form for a variety of interaction geometries. The most common of these is the "sphere approaching flat plate" geometry employed by both the surface forces apparatus and AFM:

$$F(D)_{\text{sphere-plate}} = 2\pi R W(D)_{\text{plate-plate}} \tag{11.8}$$

Normalizing the forces between a spherical and flat surface at each separation $[F(D)]$ to the radius of curvature of the spherical surface (R), therefore, gives a direct measurement of the interaction energy between flat plates of the same materials $[W(D)]$, provided the surface separation is smaller than the radius of the sphere. Assuming a flat sample surface, which can readily be confirmed using AFM in imaging mode, interpreting AFM force–distance data in this way only requires a knowledge of the radius of curvature of the force probe. The availability and ease of use of standard AFM tips for these measurements has driven the development of both theoretical and experimental methods for determination of a notional radius of curvature at the apex of the tip pyramid, which acts as one of the interacting surfaces.[10]

Drummond and Senden[11] have proposed a method that involves measuring the forces between a silicon nitride tip of unknown curvature and a "standard" surface bearing the adsorbed surfactant, hexadecyltrimethylammoniumbromide (CTAB). The measured force profile was fit to a known sphere-flat geometry interaction profile by adjusting the effective radius of the tip (R_{eff}). Arai and Fujihara[12] have suggested that modeling the tip apex as a spherical surface is only appropriate for tip cone angles greater than 30°. Such indirect approaches introduce added degrees of error into the experiment, particularly since the Derjaguin approximation does not hold for small radii surfaces. More careful control of this critical parameter was achieved by Ducker et al.[13] who attached individual colloidal spheres at the end of AFM cantilevers using a minute quantity of inert glue. Accurate determination of the radius of these particles was then made using electron microscopy. A scanning electron micrograph of one such "colloid probe" from our laboratory is shown in Figure 11.5.

Another possible error arises with the larger interaction area employed in the colloid probe technique, that of surface roughness. Should hard (constant compliance) contact between the two surfaces occur on an asperity that is extended away from the plane of origin of the force on either surface, the apparent maximum magnitude of that force will be lower than that recorded if true contact was achieved, that is, if the planes of origin of the force on the surfaces actually touch. Since the force data is scaled to the contact position, then the entire force–distance profile will be reduced in magnitude. Asperities close to the interaction area can thus cause significant attenuation of the measured forces, particularly when the decay length of the force is comparable to the asperity height. This problem has been discussed with reference to long-range electrostatic interactions between different surfaces.[14–16]

11.3.3 Compressible Surfaces: Zero Distance Problem

As described above, the constant compliance region of the force plot is a critical ingredient in the calculation of a force–distance profile from the raw deflection versus piezo travel plot, since it defines zero separation as well as internally calibrating the tip displacement–photodiode output relationship. A problem, however, arises when either of the surfaces under investigation is not rigid, as is the case for polymer, surfactant, or biomolecule bearing surfaces. In such a case, the first contact between the colloid probe and the surface does not correspond to zero separation as it would for a hard surface. Instead, this contact may be followed by deformation and/or desorption of surface-associated molecules, acting to dissipate the energy of compression. Three possible mechanisms and their associated deflection–piezo travel plots are shown in Figure 11.6. In the first of these (Fig. 11.6a), the surface may be considered analogous to a nonlinear spring. Under lower compressional forces, the surface exerts an increasing force on the cantilever. At a certain point, this increase in force with respect to separation (i.e., the spring constant of the surface) exceeds the spring constant of the AFM cantilever, and causes it to deflect as if in contact with a solid surface. In this case the constant compliance line is linear, accurately reflecting the displacement versus photodiode relationship, but not zero separation. Instead it

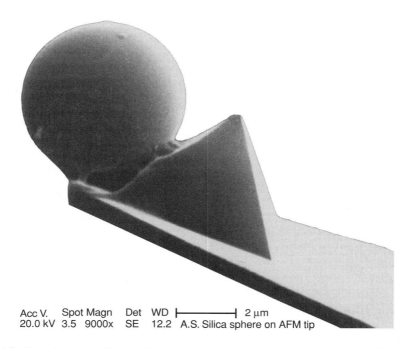

Acc V.   Spot Magn   Det   WD |————————| 2 μm
20.0 kV   3.5   9000x   SE   12.2   A.S. Silica sphere on AFM tip

FIGURE 11.5   Typical spherical silica colloid probe attached to the imaging pyramid of a standard silicon nitride imaging cantilever. The probe has a radius of 2.5 μm.

depends on the final spring constant of the surface.[17] This problem of "pseudo-compliance" also arises with any interaction-force law that has a rapidly increasing gradient, and has been observed in the measurement of repulsive van der Waals interactions (see Hilling et al.[18] and later). By employing multiple cantilevers with different spring constants in the same experiment, this artefact may be readily identified.[18]

Other possibilities exist for surfaces bearing adsorbed species. Here, a compressible or noncompressible material may be displaced from the interaction area under a certain threshold force, giving rise to discontinuities, or "jumps," in the data. This is shown schematically in Figure 11.6b. The separation change associated with each jump may in fact be of interest, since it reflects the actual thickness of a compressed layer as it is "squeezed out" of the gap between the tip and surface. This has been used to some effect in studies of adsorbed surfactant layers, with agreement between such discontinuities and expected dimensions of different aggregate morphologies obtained; see also Ducker and Wanless[19] and Figure 11.11.

Provided a hard "constant compliance" contact is finally reached, the force data in this case may be interpreted meaningfully, but caution is advised particularly for systems where multilayering exists, since several of these discontinuities may be recorded before a "true" constant compliance corresponding to zero separation is measured.

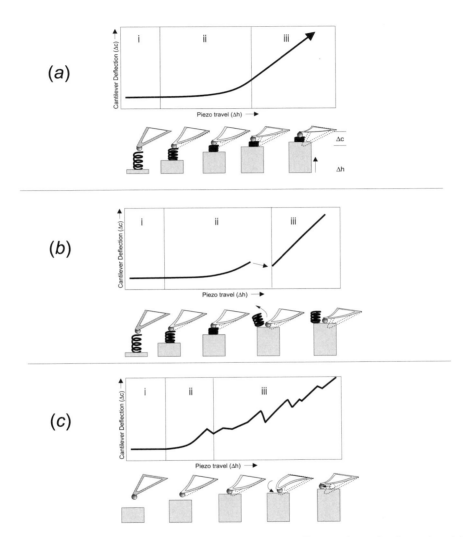

FIGURE 11.6 Schematic of some possible raw data outputs for different probe–surface interactions. (*a*) Contact with an elastic surface or adsorbed species allows measurement of compressional forces (region ii), followed by pseudo-compliance (region iii), when the effective spring constant of the surface exceeds that of the AFM cantilever. (*b*) Compression of an adsorbed species on the surface (region ii) leads to its displacement from the tip–sample contact at a certain pressure. This is reflected by a jump in the data (regions ii–iii), which corresponds to the thickness of the compressed layer. (*c*) Surface forces between probe and surface are measured in region ii, but increasing the pressure between the probe and the sample after contact leads to rolling and/or sliding of the probe on the sample. This may appear as jumps or bends in the constant compliance line (region iii).

Another dissipative mechanism is the rolling and/or sliding of the tip on the surface. This is shown schematically in Figure 11.6c and can occur in the absence of adsorbed species. Such behavior may lead to the appearance of jumps in the data reminiscent of Figure 11.6b due to stick-slip motion of the probe along the surface.

In many cases therefore, a value for zero separation is not accessible directly from the force data and may only be achieved by recourse to another technique. This is of particular importance in studies of surfaces bearing adsorbed species and presents problems for the deconvolution of steric interactions between, for example, polymer-coated surfaces.

11.4   ATOMIC FORCE MICROSCOPE FORCE MEASUREMENT: CONTROL OF TIP–SAMPLE INTERACTIONS DURING IMAGING

The first uses of force measurement in the AFM were concerned with improving the quality of imaging and primarily looked at the adhesion between tip and sample. Such adhesions were known to cause damage to both tip and sample during imaging, particularly in the case of softer substrates. The first advance in this area came with the realization that capillary forces between tip and sample (due to condensation of moisture/contaminant in the tip–sample junction) were significant, and that by immersing the tip and sample in fluid, these attractions could be minimized.[20]

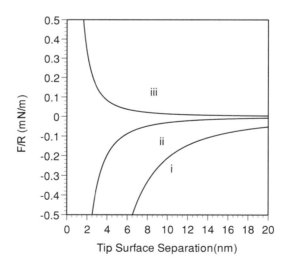

FIGURE 11.7   Theoretical force–distance profiles calculated using Eq. (11.10) for nonretarded van der Waals interactions between a silicon nitride tip and silica surfaces in different media. The forces (F) are normalized to the radius of curvature of the apex of the tip (R), generally thought to be approximately 10 nm. (i) Silicon nitride–air–silica interaction. Nonretarded Hamaker constant 10.38×10^{-20} J taken from Ref. 27. (ii) Silicon nitride–water–silica interaction. Nonretarded Hamaker constant 1.9×10^{-20} J taken from Ref. 27. (iii) Silicon nitride–diiodomethane–silica interaction. Nonretarded Hamaker constant -0.8×10^{-20} J taken from Ref. 39.

Working in vacuo or in reduced relative humidity conditions has also been shown to be effective in minimizing these forces.[20] The role of attractive van der Waals interactions was also found to be important. Here the governing factor for the interaction is the Hamaker constant, which is discussed in more detail later. By altering the dielectric properties of the medium between the tip and the sample, reductions in the Hamaker constant have been shown to improve imaging properties due to lowered adhesions.[21,22] Some typical theoretical tip–sample van der Waals interactions are shown in Figure 11.7. In more complex systems, chemical modification of AFM tips to alter their wettability, surface charge, or to enhance specific interactions with the sample have also been employed to improve imaging contrast.[23–25]

11.5   ATOMIC FORCE MICROSCOPE FORCE MEASUREMENT BETWEEN SOLID SURFACES: VAN DER WAALS INTERACTIONS

At a fundamental level, van der Waals interactions are of great importance in colloid and polymer science. In many cases, it is these forces alone that determine the stability of colloidal suspensions. Frequently polymer additives are used to impart stability to a suspension solely by presenting a steric barrier around the particles, which reduces the effective magnitude of van der Waals attractions. The dispersion energy contribution dominates van der Waals forces, so much so that the two terms are frequently used interchangeably. It arises due to the formation of instantaneous polarization within atoms, molecules, and surfaces, and since polarizability is an inherent property of all of these, it is ubiquitous. A simple relation describes how the van der Waals interaction energy [$W(D)$] scales with the separation (D) between two surfaces:

$$W(D) = -A_H R / 12\pi D^2 \tag{11.9}$$

By application of the Derjaguin approximation [Eq. 11.8], this becomes

$$F(D) = -A_H R / 6D^2 \tag{11.10}$$

The decay in force is thus as a function of $1/D^2$. This gives the effective range of the interaction, and signifies that provided its magnitude is large enough, it is well within the nanometer regime in which the AFM operates. The magnitude of the force is controlled by A_H, the Hamaker constant. This property is unique to the two interacting surfaces and the medium between them. It is determined by the polarizability of these three components and can be calculated using additivity or, more accurately, using Lifshitz theory from bulk optical data.[26,27]

Several theoretical studies have been devoted to predicting tip substrate van der Waals interactions, both in vacuo and in gaseous or liquid environments, with a particular emphasis on improvements in imaging performance as described above.[21,28,29] Using simulations, Argento and French[30] have suggested a method for accurate Hamaker constant and tip curvature determination by parametric fitting of force curves. Measurements of van der Waals interactions have been conducted using both conventional deflection AFM and tip resonance techniques.[31] These have also largely focused on AFM tip–substrate interactions.

Burnham and others[32–34] have investigated long-range tip–sample attractions in dry gaseous environments between a variety of surfaces such as tungsten, diamond or gold tips and alumina, diamond, graphite, poly(tetrafluoroethylene) (PTFE), or nickel surfaces.[32,33] Initially these studies attempted to describe the long-range attractive force–distance data in terms of purely van der Waals interactions,[32] but later analyses included patch charge, dipole–dipole, and capillary interactions.[33,34] Correlation between attractive and adhesive forces and the surface energy of Langmuir–Blodgett coatings of fluorinated and nonfluorinated stearic acids was also observed.[33] In all cases the attractions were longer range than could be expected from van der Waals interactions assuming reasonable tip radii. As the surfaces closely approached, contact mechanical models were also invoked to explain short-range repulsions and adhesive forces. The influence of maximum loading force and contact asperities on the "true" contact area and thus the measured loading–unloading curves was also assessed. Blackman et al.[35] meanwhile investigated thin films of chemisorbed perfluoro-polyether surfaces and Langmuir–Blodgett films of calcium arachidate on clean silicon substrates with a view to the study of friction and lubrication. The authors found substantially weaker attractive forces than the previous study, which they ascribed to capillary, van der Waals, and other short-range forces.

Gauthier-Manuel[36] has measured van der Waals interactions between a tungsten tip held above a mica surface in air using a magnetic coil. In this way, instabilities due to the weak spring constants associated with AFM tips were avoided, and the full range of the attractive force curve was accessible. The measured value of the nonretarded Hamaker constant, derived from an approximate measurement of the radius of curvature of the tip was found to be 1.5×10^{-19} J, as compared to a calculated value of 2×10^{-19} J. Ducker and Clarke[37] measured an attractive van der Waals interaction between silicon nitride (Si_3N_4) tips and silicon nitride surfaces, which could be substantially eliminated by addition of a zwitterionic surfactant to the intervening medium.

The van der Waals force between tip and sample is usually attractive, as observed in the above studies. Theory and experiment show that in fact it should always be attractive between identical surfaces. A unique case, however, arises when the Hamaker coefficient of the interstitial medium lies between that of the two interacting surfaces. This results in a negative Hamaker constant for the interaction, and thus from Eq. (11.10), a repulsive force between the surfaces. Hartmann[21] has demonstrated this theoretically for silica probe–metal surface interactions occurring across glycerol and formamide (see also Fig. 11.7).

Tip–sample measurements of van der Waals interactions have also been performed between silicon nitride AFM tips and muscovite mica, silica (SiO_2), and silicon nitride (Si_3N_4) surfaces. By replacing media and surfaces such that the above unique Hamaker coefficient condition was met, attractive forces could be switched to repulsive and net zero van der Waals interactions, which qualitatively agree with the predictions of Lifshitz theory.[38,39]

Strict control of the geometry of the interaction zone is required for more accurate determinations of the Hamaker constant, as well as a realistic comparison of experiment with Lifshitz theory. Measurements of this type have been made using the colloid probe technique.

Gady et al.[40] measured interactions between a polystyrene microsphere attached to the end of an AFM cantilever and a graphite surface. At separations of less than 30 nm, a van der Waals attraction was observed that agreed with theory. Similar agreement with theory was obtained for measurements of purely attractive forces between a cantilever bearing a PTFE (Teflon) colloid and a PTFE surface interacting in air,[41] and between a gold-coated silica sphere and a gold surface in water.[42]

Milling et al.[18] investigated repulsive and attractive van der Waals interactions between a gold-coated colloidal particle attached to the end of an AFM cantilever and a PTFE surface in solvents of different polarity. For the repulsive interaction, the authors were able to calculate both a Hamaker constant and a true separation by fitting the data using Eq. (11.10). Weaker spring constants showed larger discrepancies between true and apparent final separations, demonstrating the presence of pseudo-compliance in the data (see earlier). In most cases, the values of Hamaker constants derived from experiment agreed reasonably with those calculated independently using Lifshitz theory. However, several of the polar solvents generated an attractive interaction between the tip and surface, which was thought to be due to a non–van der Waals force in these systems.

Van der Waals interactions in aqueous metal, metal oxide, and metal sulfide systems have wide-reaching consequences in the control of many industrially important processes such as ceramic engineering, mineral floatation, and coating technology. The determination of Hamaker constants in these systems is thus of great relevance. The AFM force measurements have been performed in several such systems. Here, the most successful technique has been to remove electrostatic repulsive interactions, thereby "unmasking" the van der Waals forces. Hillier et al.[43] managed to achieve this by minimizing the potential of a gold interface in an electrochemical cell and measuring the forces between it and a silica colloidal sphere. For insulating systems, surface potential minimization can be achieved by pH adjustment to neutralize surface-charged groups, such that the measurements are made at the isoelectric point (IEP) of the surfaces, or by screening the repulsions using increased electrolyte concentrations.

Van der Waals forces between silica colloids and silica surfaces have been studied both at the silica IEP[16] and by electrostatic screening using a high concentration (0.01 M) of calcium chloride.[44] Larson et al.[45] performed measurements between rutile (TiO_2) surfaces at their IEP (pH 5.6). Atkins and Pashley[46] also used the screening approach to study van der Waals interactions between a zinc sulfide (ZnS) particle and mica surface immersed in 0.1 M KCl solution. Good agreement between calculated and measured Hamaker constants has been found in most cases. Toikka et al.[47] investigated ZnS–ZnS interactions close to the IEP of the surfaces and at high electrolyte concentration, but were unable to measure a van der Waals attraction possibly due to small asperities on the surfaces preventing intimate contact between them. A similar mechanism was suggested to explain the disparity between theoretical Hamaker constants with those derived from measurements of works of adhesion between an oxidized iron colloid and a silica surface.[48]

11.6 ATOMIC FORCE MICROSCOPE FORCE MEASUREMENT BETWEEN SOLID SURFACES: SURFACE CHEMISTRY AND DLVO THEORY

The frequently good agreements between Lifshitz theory and force measurements in a range of systems demonstrates that van der Waals (VDW) interactions are essentially mediated by the bulk properties of interacting materials. A second classification of interactions are associated with a finite region close to the interface between solid surfaces and any surrounding liquid. As will be discussed, it is perhaps striking to consider that such interactions can be radically altered by the addition or removal of single layers of molecules, and as such are critically dependent on the chemical composition of interfacial molecules that often constitute a minute proportion of the mass of a material.

When a surface is immersed in a polar liquid, it usually acquires a net charge either by ionization of surface groups or by adsorption of ions from the bulk solution. In both cases, the charge becomes balanced by an equal but opposite atmosphere of counterions close to the surface. The concentration of counterions in this electrical double layer decay exponentially from the surface, with the decay described by the Poisson–Boltzmann distribution:

$$\frac{d^2\psi}{dx^2} = -\left(\frac{ze\rho_\infty}{\varepsilon\varepsilon_0}\right)\exp\left(-\frac{ze\psi_x}{kT}\right)$$ (11.11)

This equation shows that the electrostatic potential at any point x away from a surface, ψ_x, is a function of the number concentration of counterions (ρ_∞) of valency z and charge e in the bulk medium, as well as the permittivity of the medium ($\varepsilon\varepsilon_0$) and the thermal energy, kT.

When two identical surfaces, each surrounded by an electrical double layer are moved into close proximity, the overlap of these double layers gives rise to a repulsive pressure due to the increased concentration of counterions in the gap compared to the bulk medium. This double-layer force is the controlling parameter in a whole range of interfacial physicochemical and biological processes, as well as being the critical ingredient in colloidal stability in many systems. As a result of this importance, it is now probably the best understood of surface phenomena, and the AFM has proved a useful method for elucidating its nature, following on from the great success the surface force apparatus has had in this area.[49]

The DLVO (Derjaguin, Landau, Verwey, and Overbeek) theory[50,51] developed in the early 1940s provides the theoretical framework for interpreting the forces between surfaces in the presence of electrical double layers and van der Waals interactions. The overall force (V_T) was expressed as a summation of repulsive electrostatic (V_R) and attractive van der Waals (V_A) terms, thus:

$$V_T = V_R + V_A \qquad (11.12)$$

With the assumption of low surface potentials (less than 25 mV), the Poisson–Boltzmann distribution may be simplified, and the electrostatic repulsion between two surfaces may be defined thus:

$$V_R = \{[64kT\rho_\infty \tanh^2(e\psi_0/kT)]/\kappa\}\exp(-\kappa D) \qquad (11.13)$$

Eq. (11.13) shows that two factors govern the repulsive term in DLVO theory. The first is the Debye parameter (κ), which is a function of the counterion atmosphere surrounding the surface and is defined:

$$\kappa = \left(\frac{2z^2e^2\rho_\infty}{\varepsilon\varepsilon_0 kT}\right)^{1/2} \qquad (11.14)$$

The second governing factor is the surface potential (ψ_0) of the surfaces, which is determined by their surface chemistry. Simply expressed, $1/\kappa$ (units of distance) controls the decay length of the interaction, while ψ_0 (units of electrical potential) its magnitude.

In force measurements, $1/\kappa$ is readily controlled by adjusting the bulk electrolyte concentration and valency. It and the surface potential (ψ_0) may be extracted from force measurements by fitting the data with computed force–distance data plots.[52,53]

As stated above, it is surface chemical reactions that control the surface potential. The dominant mechanism is via dissociation equilibria of surface groups, with the simplest surface consisting of a balance between protonated and unprotonated groups on a negatively charged surface. By reducing the pH of the medium surrounding a negatively charged surface of this type, protonation of surface groups increases, and the net surface charge drops. This causes a concomitant reduction in the volume of the electrical double layer, and hence the repulsion between two surfaces carrying double layers also drops (see Fig. 11.8). At the isoelectric point, the net surface charge and thus surface potential are zero, and no repulsion is measurable. As Eq. 11.12 indicates, in many systems this allows van der Waals interactions to dominate. Knowledge of the conditions under which this occurs is therefore important in understanding coagulation in colloidal suspensions. The AFM force measurements have been used to determine electrical neutrality conditions between silicon nitride tips and substrates such as silicon nitride (IEP = pH 5 to 6)[27,54] and alumina (IEP = pH 8.1).[55] The fine positioning ability of the AFM also allows the measurement of differences in surface charge at discrete positions on a surface. This is of particular interest in the study of chemically heterogeneous patchy surfaces[54,56] and surfaces bearing locally ordered adsorbed species.[56]

Controlled geometry experiments have allowed direct comparison between DLVO theory and experiment. Several colloidal materials of known dimensions have been attached to AFM cantilevers in order to study their electrostatic interactions. The original work of Ducker et al.[13,57] concerned the electrostatic interactions between a silica colloid and an oxidized silicon wafer as a function of electrolyte concentration and pH. Agreement between the measured Debye length ($1/\kappa$) and the actual electrolyte concentration was found. In addition, the surface potential derived from the force measurements was found to decrease with pH as the silica surfaces approached their isoelectric point. An example of this phenomenon is shown in Figure 11.8 for the interaction of a silica sphere with a silica flat surface.

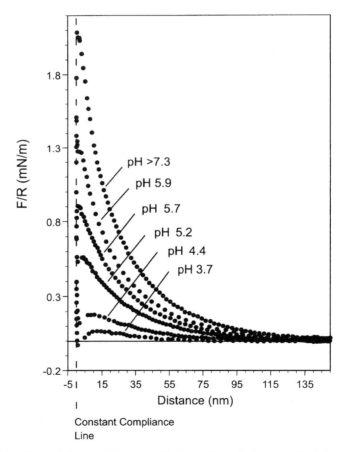

FIGURE 11.8   Measured force vs. distance profiles for a silica colloid sphere ($R = 2.5\ \mu m$) interacting with a flat silica surface in 10^{-4} M 1:1 electrolyte (NaNO₃) as a function of solution pH. The data have been converted to this form by assuming the data in the constant compliance line (see Fig. 11.3) corresponds to zero separation. Adapted with permission from Ref. 16.

Measurements below the isoelectric point of rutile titania (pH 5.6)[45] and zirconia (pH 7.7)[58] showed an increasingly repulsive force as the pH was reduced. This is consistent with an increased protonation of surface groups such that the surface reverses charge becoming positive, and an electrical double-layer repulsion between the surface reappears. Kekicheff et al.[59] have also observed charge reversal by adsorption of calcium on a silicon nitride tip and mica surface in a direct comparison with surface force apparatus measurements.

An alternative method for measuring the electrostatic potential of surfaces relies on measuring either colloid mobility in an electric field or the electro-osmotic effect close to a planar interface. These measurements provide the potential at the plane of shear (termed the zeta potential), which is thought to be located some finite distance in the solution away from the actual solid–solution interface.

The relationship between electrokinetically derived zeta potentials and the surface potentials derived from force measurements of several colloid–flat surface pairs has been investigated in our laboratory and by others. Silica, mica, titania, zirconia, zinc sulfide, and polystyrene latex surfaces have all been studied.[16,45,47,58,60] Generally good agreement between the potentials derived from the two types of measurement has been found, strongly suggesting a close relationship between zeta and surface potentials.

11.6.1 Dissimilar Surface Electrostatic Interactions

Theoretical embellishments[52,61] have allowed the full solution of the Poisson–Boltzmann distribution, and extended the predictive capabilities of DLVO theory to higher surface potentials and different geometries. Implicit in these models has been the necessary definition of boundary conditions relating to the surface chemical changes that occur during the interaction between the surfaces. The constant surface charge, constant surface potential, and surface charge regulation models describe the mechanism by which charged surfaces react to the presence of an impinging double layer from another surface.[61]

Other refinements have allowed computations of the interactions between dissimilarly charged surfaces that dominate electrostatic flocculation and other processes that rely on the removal of suspended matter, such as water treatment.[62] Surface force apparatus experiments in this area have been restricted by the limited number of surfaces available for study.[63] The lack of such limitations in the AFM colloid probe technique meanwhile has opened up research in this area considerably. Good agreement between theory and experiment has been found for silica-titania, silica-alumina, and silica-mica force measurements fit using dissimilar double-layer theory and independent zeta potential measurements of both the surfaces and colloids.[16,64,65] Meagher and Pashley[66,67] measured interactions between a silica colloid probe and a polypropylene surface before and after plasma treatment and inferred a significant negative surface potential (-55 mV in 10^{-4} M electrolyte) on the untreated polymer surface. The surface potential was also found to increase after plasma treatment. The mechanism for charging of the surfaces was thought to be through the specific adsorption of solution bicarbonate ions.

Meanwhile Hillier et al.[43] were able to externally control the surface potential of a gold interface in an electrochemical cell and measure the forces between it and a negatively charged silica colloidal sphere. An attraction between the surfaces was found when a positive potential was applied to the surface, while repulsion increased between the surfaces with the applied potential negative. Agreement with DLVO theory was once again observed.

One interesting prediction from theory that is unique to dissimilar double-layer overlap is the constant surface potential turnover, which occurs between surfaces of charges of the same sign but different magnitude.[52,62] Here, the presence of the greater magnitude double layer surrounding the more highly charged surface causes charge reversal of the weaker potential surface, resulting in a net attraction between the surfaces, despite their similar charge at large separations.

Atkins and Pashley[46] attempted to investigate this phenomenon using zinc sulfide (ZnS) particles and mica surfaces in electrolyte solutions. The authors were unable to observe constant surface potential turnover in this case. Recently, however, this phenomenon has been observed in our laboratory between a silica sphere and a polyelectrolyte coated mica surface. These results are discussed later, and a sample of the data is shown in Figure 11.10.

11.7 HYDRATION AND HYDROPHOBIC INTERACTIONS

There are many systems where the simple interplay between electrical double-layer repulsion and van der Waals attraction is neither a complete nor accurate description of interfacial interactions. This is because DLVO theory simplistically describes the solvent and solutes surrounding a surface as a dielectric continuum containing point charges. In reality, the majority of real technological and biological interactions occur in systems where the relationship between the surface and solvent–solute mixtures is more complex.

The simplest deviation from DLVO theory occurs when the continuum breaks down, that is, at separations between surfaces that approach molecular dimensions. The presence of a surface may induce structuring of solvent molecules that results in their ordered arrangement such that the free energy of the surface–solvent interaction is reduced. Layering of molecules close to an interface may then occur, which is reflected in oscillatory or structural forces as two such surfaces approach. Solvent structural forces were first directly measured between mica surfaces immersed in octamethylcylcotetrasiloxane (OMCTS) using the surface force apparatus.[68] Later, they were also observed in the AFM between a tip and a graphite surface in OMCTS and dodecane. In both cases, oscillatory forces of a period close to the solvent molecular diameters were observed.[69]

In other situations, specific attractions between a surface and surrounding solvent or solute may result in the immobilization of species at the interface in a process akin to adsorption. This process has been termed secondary hydration when used with reference to the adsorption of hydrated ions. The process of removal of these ions and/or bound water is thought to give rise to short-range repulsions of approximately 3 nm decay length between surfaces in concentrated electrolyte solutions.[70] Although problems with the designation of zero surface separation become apparent with short-range forces such as these, they have been observed between silicon nitride AFM tips and mica surfaces in 3 M magnesium chloride solutions.[55] Cleveland et al.[71] were also able to observe at high resolution ordering of structured water/ions between a silicon tip and calcite surface using a thermal fluctuation technique.

Other hydration forces have also been measured. These short-range repulsive forces have been observed in a limited number of cases, most notably between silica surfaces. They are either thought to be the result of the high affinity of water for surface moieties requiring application of a force for dehydration or to dissolution of surface species giving rise to a gel layer.[13,44]

Hydrophobic forces, meanwhile, occur between nonpolar surfaces in aqueous media, resulting in long-range attractions of indeterminate range between them. Development of a coherent understanding of the mechanism giving rise to hydrophobic surface forces continues to be a preoccupation of force measuring experiments,[72] yet bulk thermodynamic measurements such as liquid contact angles still provide the most reliable assessments of surface hydrophobicity. Surface force apparatus measurements have focused on hydrocarbon surfactants physisorbed on mica.[72,73] An electrostatic model for interpreting AFM measurements of long-range attractions between glass surfaces neutralized by adsorption of cationic surfactants has also been proposed. It was suggested that solute ion pairs could adsorb in discrete patches leading to attractions between opposing surfaces.[74] Other studies using uncharged surfactants in the AFM have also showed only shorter range attractions (<10 nm).[75]

The greater sample flexibility of the AFM has also facilitated measurements between chemically attached and bulk hydrophobic materials. Rabinovich and Yoon[6,76,77] have studied the forces in aqueous media between silica surfaces bearing chemically grafted alkyl silanes. These compounds were attached to the surfaces as aggregate domains, the size of which appeared to be a controlling factor in both the contact angle and a long-range attractive force between them. This force was found to be sensitive to electrolyte concentration, implying an electrostatic origin. However, increased concentration of dissolved gas in the medium caused increases in the range of the attraction. This lent support to a cavitation mechanism for the hydrophobic force in which bubbles or dissolved gas molecules may form bridges between surfaces.[72,78]

Meagher and Craig[79] have also investigated the interactions between a polypropylene surface and a polypropylene bead attached to an AFM cantilever. The advancing water contact angles on these surfaces were high (>100°) yet the attractive forces in dilute electrolyte were only of short range (<30 nm). Reducing the dissolved gas in the intervening solutions also reduced the effective range of the interaction.

11.8 SURFACES BEARING ADSORBED SPECIES

Solutes with an enhanced affinity for surfaces such as small organic species, uncharged and charged polymers, surfactants, and proteins, can adsorb on surfaces. If this affinity is high, the adsorption may continue until the surface is completely covered, and a new surface with the physicochemical properties of the adsorbed species alone is formed. In situations where surface coverage of the adsorbed species is lower, an intermediate condition exists. Here the properties of the adsorbed species as well as the underlying surface control the interactions with other surfaces. Such surfaces may be relatively homogeneous or patchy. The latter patchy surfaces have received recent attention both theoretically and experimentally, particularly with respect to the mysterious hydrophobic force discussed in the previous section.[80,81] The driving forces for adsorption are many and varied. They depend exquisitely on the relationships between solvent, solute(s), and surface and thus an all-embracing theoretical model for the process continues to be elusive.

Experimental studies of simple model adsorbates provide insights into adsorption mechanisms. Biggs et al.[82] have studied the adsorption of trisodium citrate, which has a strong affinity for the gold surface and is frequently used to stabilize colloidal gold materials by conferring a significant negative charge and thus an electrostatic repulsion between particles. In addition to long-range electrostatic repulsions, short-range repulsive forces were found between a gold colloid and gold surface with periodic steps that appeared to be of the approximate dimensions of the citrate molecule. This suggested a secondary barrier to coagulation due to multiple layers of citrate ions. The kinetic displacement of this adsorbed layer has also recently been observed using the colloid probe technique. The negatively charged citrate was displaced by uncharged pyridine or 4-dimethylaminopyridine, resulting in an observable decrease in the electrostatic potential of gold surfaces derived from force curves collected over a period of time.[83]

Uncharged polymers, which, at high surface density may have monomer units extending from the surface into solution, have also been studied. Overlap or compression of these segments between approaching surfaces gives rise to steric repulsions, which are a result of the exclusion of solvent from the overlap region (an osmotic component) as well as a reduction in the overall configurational entropy of the polymers (a structural component). The length scale for these interactions has been shown to depend on the radius of gyration of the polymers in free solution. Despite problems associated with the assignment of a zero separation position for compressible surfaces in AFM measurements, steric forces have been observed between uncharged polyethylene glycol polymers (molecular weight 2000) chemically grafted onto a silicon nitride surface and a clean silicon nitride tip. By reducing the quality of the aqueous solvent for the polymer using magnesium sulfate, collapse of the steric layer was observed as a reduction of the effective range of repulsive forces.[84] Braithwaite et al.[85] studied the time dependence of the adsorption onto glass surfaces of a similar polymer (polyethyleneoxide) of molecular weight 56,000 and found the buildup of a steric layer on each surface, which was up to six times thicker than the radius of gyration of the polymer in solution (~7.5 nm).

Charged polymers (polyelectrolytes) are additives in a wide range of industrial processes that involve dispersion or flocculation. Electrosteric repulsions between zirconia surfaces bearing low-molecular-weight polyacrylic acid (M_w = 2000) have been measured using the AFM. These anionic polymers were found to adsorb on a zirconia surface above its IEP giving rise to 8-nm-thick layers. Reduction in the pH to below the IEP of the zirconia was found to increase the polymer–surface attraction and reduce intrapolymer repulsion (by protonating polymer charge groups). This caused the collapse of the polymer closer to the surface, which was inferred from the reduced range of the intersurface repulsive forces.[86] A sample force profile is shown in Figure 11.9.

At lower surface coverage, bridging forces occur when two surfaces approach and adsorption of higher-molecular-weight polymer simultaneously occurs on both surfaces. This results in an attractive and adhesive force between them. Measurement of these forces with the AFM have been reported for the zirconia–PAA [poly(acrylic acid)] system by Biggs[87] and in the glass–poly(ethyleneoxide) (PEO) system by Braithwaite et al.,[88] who found a time-dependent transition between bridging interactions at shorter adsorption times (25 to 30 min) to purely steric repulsions at longer incubation times (24 h). At extended time scales (>24 h), adsorption of polymer–impurity aggregates on the surfaces was thought to result in extremely long range (>1.5 μm) repulsions between them. The presence of bridging forces was confirmed by incubation of only one surface in the polymer solution.

The electrostatic interactions between surfaces bearing low concentrations of oppositely charged polyelectrolytes have recently been studied in our laboratory. The affinity between the anionic surface (silica or mica in these studies) and a cationic polyelectroyte [poly(2-vinylpyridine)] is high and results in extremely thin (<2 nm) layers of the polymer on the surface. In this situation, the steric repulsion, and thus the zero-distance error in the measurements, is small. We investigated the effect of pH and electrolyte changes on the purely electrostatic interactions between these composite surfaces. The presence of the polyelectrolyte radically altered the IEP, increasing it from pH 2 to 3 for the uncoated to pH 6 for the coated surfaces. The charging process as a function of pH, however, was found to be controlled by the underlying surface. The constant surface potential turnover (see earlier) was also observed between a silica sphere of high negative surface potential and polyelectrolyte-coated mica surface of low negative surface potential. This phenomenon is shown in Figure 11.10.[89]

The role of adsorbed surfactants in modifying the electrostatic forces between a silicon nitride tip and a hydrophobic surface has been examined by Ishino et al.[90] in studies of the forces between silicon nitride tips and monolayers of differently charged surfactants, which were Langmuir–Blodgett deposited on silicon wafers rendered hydrophobic by hexamethyldisilazane.

Similar measurements have been part of detailed studies of surfactant aggregate structures. Aggregate thicknesses on solid surfaces have been measured by pushing through surface layers using the AFM tip. The aggregate thickness was determined from the tip displacement from the start of the push through at a force maximum to the constant compliance region.[37,91] A typical force–distance curve exhibiting this behavior is shown in Figure 11.11.[19] Longer-range tip–surface force measurements in these systems tended to show the properties of the surfactant coatings.[19] This suggested high surface coverage, which was again confirmed using AFM imaging.[92]

The buildup mechanism of surface aggregates as the concentration of surfactant is increased has also been studied by direct force measurement. Rutland and Senden[75] used the colloid probe technique to measure the forces between silica surfaces in different concentrations of a nonionic surfactant ($C_{12}E_5$). Low bulk/surface concentrations of the surfactant were found to induce hydrophobic attractive forces and adhesion between the surfaces. As the concentration was further increased, this adhesion became less, yet short-range repulsive forces were not measurable, suggesting incomplete coverage of the surfaces by the surfactant. Further concentration increase removed the adhesion, and steric forces resulting from mono- and bilayers of surfactant between the surfaces were observed. The long-range

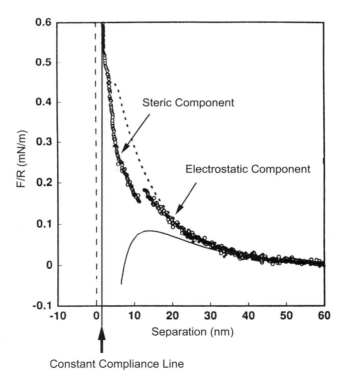

FIGURE 11.9 Electrosteric interactions between zirconia surfaces bearing an adsorbed layer of poly(acrylic acid) (molecular weight = 2000) at pH 7.15 in KNO_3. The solid and dotted lines represent theoretical DLVO fits to the data assuming identical surfaces of $\psi_0 = -26$ mV and (1:1electrolyte = 5×10^{-4} M) and a nonretarded Hamaker constant of 6×10^{-20} J. The offset between the measured constant compliance line and true zero separation of the uncoated surfaces was thought to be due to a compressed polymer layer (see also Fig. 11.6a). Adapted with permission from Ref. 86.

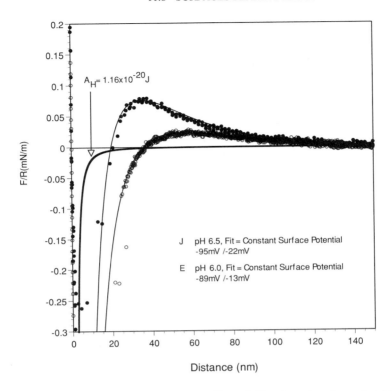

FIGURE 11.10   Measured force profiles between a silica colloid probe and a mica surface bearing an adsorbed layer of cationic polyelectrolyte (poly(2-vinyl pyridine)). The constant surface potential turnover is observed at two pHs leading to a long-range attraction between two surfaces that have the same sign but different magnitude of surface potential (see text). The predicted purely van der Waals interaction for the surfaces is shown for comparison. Adapted with permission from Ref. 89.

electrostatic forces under these final conditions were significantly reduced, suggesting masking of the silica surface charge by the uncharged surfactant layers.

Sharma et al.[93] used force measurements between a silicon nitride tip and a silica surface in conjunction with an imaging study of the buildup of surface aggregates of the cationic surfactant CTAB. Here also long-range attractions and adhesions observed at low concentrations were replaced at higher concentrations by repulsive electrostatic interactions and "squeeze outs" as in the earlier work. Johnson et al.[14] studied the electrostatic properties of silica surfaces at different CTAB concentrations using the colloid probe technique, as well as with independent electrokinetic measurements. The same approaches were used to investigate the role of surfactant counterion in a comparison between CTAB (bromide counterion) and CTAC (chloride counterion). At concentrations of electrolyte greater than 10^{-3} M, significant deviations between the AFM-derived and electrokinetic potentials were observed. This was rationalized as greater sensitivity of the former measurements to surface roughness.

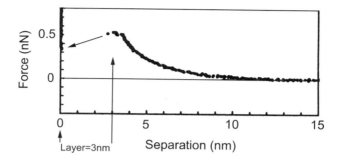

FIGURE 11.11   Steric squeezing out of a surfactant aggregate as a silicon AFM tip approaches a mica surface bearing adsorbed aggregates of the surfactant dodecyltrimethyl ammonium bromide (DTAB). The length of the jump in this data corresponds to the thickness of the aggregate (~3 nm). Adapted with permission from Ref. 19.

11.9   DEPLETION INTERACTIONS

The interactions between colloidal particles immersed in solutions of nonadsorbing polymer species are of technological importance to mineral processing, paint, and ink formulation. When the net interaction energy between a polymer and surface is repulsive, negative adsorption occurs, and the polymer is excluded from a region close the interface. The thickness of this region, known as the depletion layer thickness, is related to the radius of the polymer coil.[94] The overlap of depletion layers between two surfaces, at a surface separation of less than twice the depletion layer thickness, causes the volume of liquid between the surfaces to be at a higher chemical potential than the bulk liquid outside the gap. There is thus a driving force for the removal of fluid from the gap, which is equivalent to an attractive force between the surfaces. For colloidal systems in concentrated solutions of nonadsorbing polymer, this may result in depletion flocculation. Depletion flocculation is discussed in more detail elsewhere in this volume.

Depletion interactions have been explored using the colloid probe technique. The forces in a nonaqueous system, consisting of hydrophobic (stearylated silica) surfaces immersed in a solution of poly(dimethylsiloxane) in a good solvent (cyclohexane) showed an attractive depletion force comparable in range to twice the radius of gyration of the polymer, as predicted from theory.[95] A more complex aqueous system, involving electrostatic repulsion between the polymer [poly(styrenesulfonate)] and surface (silica) showed oscillatory forces of variable periodicity and magnitude. The periodicity could be finely tuned by changing the polymer molecular weight and concentration, as well as pH and background electrolyte conditions.[96] Oscillations beyond the final force maximum closest to the surface, which were not implicit in depletion theory were thought to be a result of electrostatically induced ordering of the polymers close to the interface. Reducing the molecular weight of the polymer decreased the range of the depletion attraction and increased the frequency of oscillations, presumably due to a reduction in the polymer radius of gyration and hence depletion layer thickness. Increases in polymer concentration increased the oscillation frequency and magnitude, with concomitant decreases in the range, and increases in the magnitude of the depletion force. This was thought to be due both to increased osmotic pressure in the bulk and a slight reduction in the depletion thickness due to confinement of the polymer coils. Reduction in pH weakened the repulsion between the surface and the polymer, allowing the polymer to approach the surface more closely, thereby reducing the depletion layer thickness. This was manifested as a reduction in both the range and magnitude of the measured attractive force. An increase in electrolyte concentration reduced the magnitude, but not the periodicity, of the structural forces, with reduction in the depletion interaction thought to be due to a reduction in the osmotic pressure of the solution.

Recently, we have also been able to observe a similar phenomenon in nonadsorbing surfactant solutions well above the critical micelle concentration (cmc). In these conditions, anionic sodium dodecyl sulfate (SDS) micelles have been found to produce periodic interactions between a silica colloid probe and mica surface (see Fig. 11.12). The frequency of these oscillatory forces is considerably higher than those

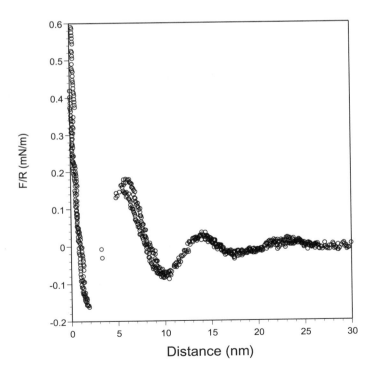

FIGURE 11.12 Oscillatory forces between a silica colloid probe and mica surface immersed in a 0.25 M micellar solution of sodium dodecyl sulfate (SDS).

observed in the previous polymer systems and is a reflection of the relative sizes of the micellar structures compared to the polymer radii of gyration in the previous studies.

11.10   ATOMIC FORCE MICROSCOPE FORCE MEASUREMENT IN BIOLOGICAL SYSTEMS

The interest in measuring surface forces in biological systems stems from their presumed importance in determining specific lock-and-key type of reactions, which the simple interaction force laws outlined above may not adequately describe. Added to this is the desire to understand the mechanisms of cell attachment to surfaces that play an important role in determining the biocompatibility of prosthetic implants, as well as "biofouling" in many technological processes. Surface forces may also be an early determinant in microbial pathogenesis and cellular differentiation through the control of membrane adhesion.[97]

Nonspecific protein–surface interactions were the focus of early AFM force measurements. The ability of the imaging tip to home-in on a localized biomolecular aggregate was utilized by Butt[56] in a study of the interactions between a silicon nitride tip and a two-dimensional crystalline protein surface formed from purple membranes adsorbed on alumina.

The most frequently employed system for the investigation of specific biomolecular interactions is that involving the high-affinity binding of the simple ligand biotin (vitamin H) to the proteins avidin or streptavidin. Both the receptor protein and the ligand have been immobilized on surfaces in a variety of ways via different chemical functionalities. The energies of adhesion between functionalized tips and surfaces have been calculated from the forces required to separate them. Reduction of the adhesive forces has also been achieved by blocking the protein binding sites with free ligands, as well as by the use of mutant proteins with lower binding affinities.[98–101] The specific interaction between antibodies and antigens[102,103] and specific cell adhesion molecules have been studied in a similar manner.[104]

11.11   ATOMIC FORCE MICROSCOPE FORCE MEASUREMENTS WITH DEFORMABLE SURFACES

Recently, the versatility of the AFM technique has led to the extension of force measurements to experiments involving the interactions of solid surfaces with deformable interfaces, such as those between a gas bubble and a liquid or an oil surface and aqueous medium. These studies are directly aimed at understanding the complex interactions that occur in froths and emulsions, which have implications for the food, mining, and polymer industries.

Ducker et al.[17] measured the forces between a silica sphere and a confined air bubble in the AFM. The sphere was attracted to the interface but not engulfed by it. A significant adhesive force was found following retraction. The authors were able to calculate an effective spring constant for the air–water interface of 0.065 N/m. Other workers found a purely repulsive interaction as a silica sphere approached an air–water interface, although an adhesive force was once again found between the colloidal sphere and the interface following retraction.[105]

Butt[106] also measured repulsive forces between a glass particle and air bubble in a similar experiment. These repulsions were overcome at a certain threshold force, leading to either adhesion to, or engulfment by, the bubble.

In all three cases, the experiments were compared to measurements between hydrophobic spheres and the air–water interface. Complete or partial engulfment of the hydrophobic sphere by the bubble was found in all cases, with gross deformations of the bubble on approach also observed in one study.[106]

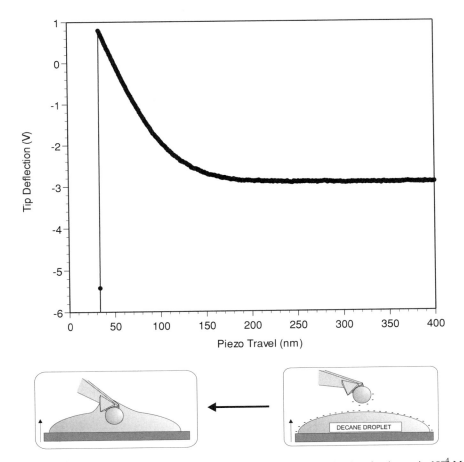

FIGURE 11.13   Interaction forces between a silica colloid probe and a droplet of *n*-decane in 10^{-4} M background electrolyte. The force curve shows an exponentially increasing repulsive force of a decay length concurrent with the theoretical Debye length, indicating an electrostatic origin. At a certain threshold force, the colloid probe instantaneously snaps into the droplet, and the laser signal is lost.

The long-range component of the forces between the surfaces was found to be either purely attractive[17,106] or electrostatically repulsive,[105] presumably reflecting differences in particle preparation. In one case, addition of the surfactant SDS was found to remove the attraction and induce a DLVO-type repulsive interaction at large separations. This was thought to be a result of surfactant adsorption rendering the bubble both hydrophilic and strongly charged.[17]

The forces between a silica surface and an oil droplet in water have been investigated in our laboratory.[107] A preliminary study involving an immobilized *n*-decane droplet containing an oil-soluble dye showed a short-range attraction, followed by a well-defined constant compliance region. Adhesion between the surfaces was also observed. Addition of SDS to this system resulted in a monotonically repulsive force that was not adequately described by DLVO theory. We have subsequently found a DLVO-type electrostatic repulsion between a silica sphere and pure *n*-decane droplet in electrolyte solutions, which is overcome at a certain threshold force, leading to particle engulfment. A sample of this data is shown in Figure 11.13. Addition of surfactant is also found to prohibit the engulfment process.[108]

Interactions between a hydrophobic polymer–inorganic laser printer toner particle and an *n*-hexadecane droplet have been measured by Snyder et al.[109] To simulate collision efficiencies in a vigorous reactor, extremely high scan rates were used, which inevitably complicates deconvolution of equilibrium force–distance relationships. A repulsive force was observed in water, which then yielded to particle engulfment at higher pressures/times. Addition of a cationic starch solution resulted in the removal of engulfment, presumably by electrosteric stabilization of the droplet.

11.12   SUMMARY AND FUTURE DIRECTIONS

From early studies of tip–sample interactions undertaken to enhance AFM imaging, the technique of AFM force measurement has expanded to allow the quantitative assessment of surface forces in an array of simple and complex experimental systems. DLVO theory has been tested in many of these and successfully used to describe the interactions between a large number of similar and dissimilar solid surfaces. In addition, the use of colloidal probes has allowed direct comparison between force measurements and other techniques, such as electrophoresis.

The zero distance problem imposes limitations on the study of compressible surfaces, but much qualitative and some quantitative information has been gained from studies of polymer, surfactant, and biomolecule adsorption. Perhaps the major instrumental challenge is to incorporate an independent measurement of tip–surface separation to allow further advances in the understanding of this important process.

Studies of the unusual forces between surfaces in the presence of nonadsorbing solutes are now underway. In addition, experiments involving the complex interactions between solid–liquid, liquid–liquid, and liquid–gas interfaces are yielding important information concerning the role of deformation in colloidal interactions.

In the future, the ability of the AFM to perform rapid kinetic surface force measurements in different conditions is likely to be the focus of much interest, while the possibility of surface force measurements between biological surfaces, including those of individual living cells, is an exciting prospect.

REFERENCES

1. G. Binnig and H. Rohrer, *Helv. Phys. Acta,* 726 (1982).

2. G. Binnig, C.F. Quate, and C. Gerber, *Phys. Rev. Lett.* **56**(9), 930.

3. P.F. Luckham and B.A.D. Costello, *Adv. Colloid Interface Sci.* **44**, 183 (1993).

4. P.K. Hansma, J.P. Cleveland, M. Radmacher, D.A. Walters, P.E. Hilner. M. Bezanilla, M. Fritz, D. Vie, and H.G. Hansma, *Appl. Phys. Lett.* **64**(13), 1738 (1994).

5. T.J. Senden and W.A. Ducker, *Langmuir* **8**(2), 733 (1992).

6. Y.I. Rabinovich and R.H. Yoon, *Colloids Surfaces A: Physicochem. Eng. Aspects* **93**, 263 (1994).

7. J.E. Sader, I. Larson, P. Mulvaney, and L.R. White, *Rev. Sci Instrum.* **66**(7), 3789 (1995).

8. J.P. Cleveland, S. Manne, D. Bocek, and P.K. Hansma, *Rev. Sci. Instrum.* **64**(2), 403 (1993).

9. B.V. Derjaguin, B.V., *Kolloid Zeitschrift* **69**, 155 (1934).

10. J.L. Hutter and J. Bechhoefer, *Rev. Sci. Instru.* **64**(7), 1868 (1993).

11. C.J. Drummond and T.J. Senden, *Colloids Surfaces A; Physicochem. Eng. Aspects* **87**(3), 217 (1994).

12. T. Arai and M. Fujihira, *J. Electroanal. Chem.* **374**(1–2), 269 (1994).

13. W.A. Ducker, T.J. Senden, and R.M. Pashley, *Nature* **353**, 239 (1991).

14. S.B. Johnson, C.J. Drummond, P.J. Scales, and S. Nishimura, *Langmuir* **11**(7), 2367 (1995).

15. S.B. Johnson, C.J. Drummond, P.J. Scales, and S. Nishimura, *Colloids Surfaces A: Physicochem. Eng. Aspects* **103**, 195 (1995).

16. P.G. Hartley, I. Larson, and P.J. Scales, *Langmuir* **13**(8), 2207 (1997).

17. W.A. Ducker, Z.G. Xu, and J.N. Israelachvili, *Langmuir* **10**(9), 3279 (1994).

18. A.J. Milling, P. Mulvaney, and I. Larson, *J. Colloid Interface Sci.* **180**, 460 (1996).

19. W.A. Ducker and E.J. Wanless, *Langmuir* **12**, 5915 (1996).

20. A.L. Weisenhorn, P.K. Hansma, T.R. Albrecht, and C.F. Quate, *Appl. Phys. Lett.* **54**(26), 2651 (1989).

21. U. Hartmann, *Phys. Rev. B* **43**(3), 2404 (1991).

22. A.L. Weisenhorn, P. Maivald, H.-J. Butt, and P.K. Hansma, *Phys. Rev. B* **45**(19), 11226 (1992).

23. R.L. Alley, K. Komvopoulos, and R.T. Howe, *J. Appl. Phys.* **76**(10 Part 1), 5731 (1994).

24. C.A. Johnson and A.M. Lenhoff, *J. Colloid Interface Sci.* **179**(2), 587 (1996).

25. C.D. Frisbie, L.F. Rozsnyai, A. Noy, M.S. Wrighton, and C.M. Lieber, *Science* **265**, 2071 (1994).

26. D.B. Hough and L.R. White, *Adv. Colloid Interface Sci.* **14**, 3 (1980).

27. T.J. Senden and C.J. Drummond, *Colloids Surfaces A: Physicochem. Eng. Aspects* **94**(1), 29 (1995).

28. C. Girard, D.V. Labeke, and J.M. Vigoureux, *Phys. Rev. B* **40**(18), 12133 (1989).

29. N. Garcia and V.T. Binh, *Phys. Rev. B* **46**(12), 7946 (1992).

30. C. Argento and R. French, *J. Appl. Phys.* **80**(11), 6081 (1996).

31. W.A. Ducker, R.F. Cook, and D.R. Clarke, *J. Appl. Phys.* **67**(91), 4045 (1990).

32. N.A. Burnham and R.J. Colton, *J. Vac. Sci. Tech. A* **7**, 2906 (1989).

33. N.A. Burnham, D.D. Dominguez, R.L. Mowery, and R.J. Colton, *Phys. Rev. Lett.* **64**, 1931 (1990).

34. N.A. Burnham, R.J. Colton, and H.M. Pollock, *Nanotechnology* **4**, 64 (1993).

35. G.S. Blackman, C.M. Mate, and M.R. Philpott, *Phys. Rev. Lett.* **65**(18), 2270 (1990).

36. B. Gauthier-Manuel, *Europhys. Lett.* **17**(3), 195 (1992).

37. W.A. Ducker and D.R. Clarke, *Colloids Surfaces A: Physicochem. Eng. Aspects* **93**, 275 (1994).

38. J.L. Hutter and J. Bechhoefer, *J. Vacuum Sci. Tech. B* **12**(3), 2251 (1994).

39. A. Meurk, P. Luckham, and L. Bergstrom, *Langmuir* **13**(14), 3896 (1997).

40. B. Gady, D. Schleef, R. Reifenberger, D. Rimai, and L.P. Demejo, *Phys. Rev. B-Condensed Matter* **53**(12), 8065 (1996).

41. C.J. Drummond, G. Georgaklis, and D.Y.C. Chan, *Langmuir* **1**(11), 2617 (1996).

42. S. Biggs and P. Mulvaney, *J. Chem. Phys.* **100**(11), 8501 (1994).

43. A.C. Hillier, S. Kim, and A.J. Bard, *J. Phys. Chem.* **100**(48), 18808 (1996).

44. L. Meagher, *J. Colloid Interface Sci.* **152**(1), 293 (1992).

45. I. Larson, C.J. Drummond, D.Y.C. Chan, and F. Grieser, *J. Am. Chem. Soc.* **115**(25), 11885 (1993).

46. D.T. Atkins and R.M. Pashley, *Langmuir* **9**(8), 2232 (1993).

47. G. Toikka, R.A. Hayes, and J. Ralston, *Langmuir* **12**(16), 3783 (1996).

48. G. Toikka, R.A. Hayes, and J. Ralston, *J. Colloid Interface Sci.* **180**, 329 (1996).

49. J.N. Israelachvili and G.E. Adams, *J. Chem. Soc. Faraday Trans.* **74**(1), 975 (1978).

50. E.J.W. Verwey and J.T.G. Overbeek, *Theory of the Stability of Lyophobic Colloids*, Elsevier, Amsterdam, 1948.

51. B.V. Derjaguin and L. Landau, *Acta Physiciochim. URSS* **14**, 633 (1941).

52. D. McCormack, S.L. Carnie, and D.Y.C. Chan, *J. Colloid Interface Sci.* **169**, 177 (1995).

53. L. Ip, D.Y.C. Chan, and S. Venters, *AFM Analysis Version 2.0*, Mathematics Department, University of Melbourne, 1995. Contact E-mail: D.Chan@ms.unimelb.edu.au.

54. X.Y. Lin, F. Creuzet, and H. Arribart, *J. Phys. Chem.* **97**(28), 7272 (1993).

55. H.-J. Butt, *Biophys. J.* **60**, 1438 (1991).

56. H.-J. Butt, *Biophys. J.* **63**, 578 (1992).

57. W.A. Ducker, T.J. Senden, and R.M. Pashley, *Langmuir* **8**, 1831 (1992).

58. M. Prica, S. Biggs, F. Grieser, and T.W. Healy, *Colloids Surfaces A: Physicochem. Eng. Aspects* **119**(2–3), 205 (1996).

59. P. Kekicheff, S. Marcelja, T.J. Senden, and V.E. Shubin, *J. Chem. Phys.* **99**(8), 6098 (1993).

60. Y.Q. Li, N.J. Tao, J. Pan, A.A. Garcia, and S.M. Lindsay, *Langmuir* **9**(3), 637 (1993).

61. D.Y.C. Chan, R.M. Pashley, and L.R. White, *J. Colloid Interface Sci.* **77**(1), 283 (1980).

62. R. Hogg, T.W. Healy, and D.W. Fuerstenau, *Trans. Faraday. Soc.* **62**, 1638 (1965).

63. P.G. Hartley, *Colloids Surfaces A: Physicochem. Eng. Aspects* **77**, 191 (1993).

64. I. Larson, C.J. Drummond, D.Y.C. Chan, and F. Grieser, *J. Phys. Chem.* **99**(7), 2114 (1995).

65. I. Larson, C.J. Drummond, D.Y.C. Chan, and F. Grieser, *Langmuir* **13**(7), 2109 (1997).

66. L. Meagher and R.M. Pashley, *Langmuir* **11**(10), 4019 (1995).

67. L. Meagher and R.M. Pashley, *J. Colloid Interface Sci.* **185**, 291 (1997).

68. R.G. Horn and J.N. Israelachvili, *J. Chem. Phys.* **75**, 1400 (1981).

69. S.J. O'Shea, M.E. Welland, and T. Rayment, *Appl. Phys. Lett.* **60**(19), 2356 (1992).

70. R.M. Pashley, *Adv. Colloid Interface Sci.* **16**, 57 (1982).

71. J.P. Cleveland, T.E. Schaffer, and P.K. Hansma, *Phys. Rev. B-Condensed Matter* **52**(12), R8692 (1995).

72. H.K. Christenson, M.E. Schrader and G. Loeb, eds., in *Modern Approaches to Wettability: Theory and Applications*, Plenum, New York, 1992.

73. R.M. Pashley, P.M. McGuiggan, B.W. Ninham, and D. Fennell Evans, *Science* **229**, 1088 (1985).

74. P. Kekicheff and O. Spalla, *Phys. Rev. Lett.* **75**(9), 1851 (1995).

75. M.W. Rutland and T.J. Senden, *Langmuir* **9**(2), 412 (1993).

76. R.H. Yoon, D.H. Flinn, and Y.I. Rabinovich, *J. Colloid Interface Sci.* **185**(2), 363 (1997).

77. Y.I. Rabinovich and R.H. Yoon, *Langmuir* **10**(6), 1903 (1994).

78. J.L. Parker, P.M. Claesson, and P. Attard, *J. Phys. Chem.* **98**(34), 8468 (1994).

79. L. Meagher and V.S.J. Craig, *Langmuir* **10**(8), 2736 (1994).

80. S.J. Miklavic, D.Y.C. Chan, L.R. White, and T.W. Healy, *J. Phys. Chem.* **98**, 9022 (1994).

81. S. Nishimura, P.J. Scales, S.R. Biggs, and T.W. Healy, *Colloids Surfaces A: Physicochem. Eng. Aspects* **103**(3), 289 (1995).

82. S. Biggs, P. Mulvaney, C.F. Zukoski, and F. Grieser, *J. Am. Chem. Soc.* **116**(20), 9150 (1994).

83. I. Larson, D.Y.C. Chan, C.J. Drummond, and F. Grieser, *Langmuir* **13**(9), 2429 (1997).

84. A.S. Lea, J.D. Andrade, and V. Hlady, *Colloids Surfaces A: Physicochem. Eng. Aspects* **93**, 349 (1994).

85. G.J.C. Braithwaite and P.F. Luckham, *J. Chem. Soc.-Faraday Trans.* **93**(7), 1409 (1997).

86. S. Biggs and T.W. Healy, *J. Chem. Soc.-Faraday Trans.* **90**(22), 3415 (1994).

87. S. Biggs, *Langmuir* **11**, 156 (1995).

88. G.J.C. Braithwaite, A. Howe, and P.F. Luckham, *Langmuir* **12**(17), 4224 (1996).

89. P.G. Hartley, I. Larson, and P.J. Scales, *Langmuir* **13**(8), 2207 (1998).

90. T. Ishino, H. Hieda, K. Tanaka, and N. Gemma, *Jpn. J. Appl. Phys. Part 1-Regular Papers Short Notes Rev. Papers* **33**(8), 4718 (1994).

91. S. Manne, J.P. Cleveland, H.E. Gaub, G.D. Stucky, and P.K. Hansma, *Langmuir* **10**(12), 4409 (1994).

92. L.M. Grant and W.A. Ducker, *J. Phys. Chem.* **101**, 5337 (1997).

93. B.G. Sharma, S. Basu, and M.M. Sharma, *Langmuir* **12**(26), 6506 (1996).

94. G.J. Fleer, M.A. Cohen Stuart, J.M.H.M. Scheutjens, T. Cosgrove, and B. Vincent, *Polymers at Interfaces*, Vol. 1, Chapman and Hall, London, 1993.

95. A. Milling and S. Biggs, *J. Colloid Interface Sci.* **170**(2), 604 (1995).

96. A.J. Milling, *J. Phys. Chem.* **100**(21), 8986 (1996).

97. P.F. Luckham and P.G. Hartley, *Adv. Colloid Interface Sci.* **49**, 341 (1994).

98. M. Ludwig, V.T. Moy, M. Rief, E.L. Florin, and H.E. Gaub, *Microscopy Microanaly. Microstructures* **5**(4–6), 321 (1994).

99. V.T. Moy, E.L. Florin, and H.E. Gaub, *Colloids Surfaces A: Physicochem. Eng. Aspects* **93**, 343 (1994).

100. E.L. Florin, V.T. Moy, and H.E. Gaub, *Science* **264**(5157), 415 (1994).

101. A. Chilkoti, T. Boland, B.D. Ratner, and P.S. Stayton, *Biophys. J.* **69**(5), 2125 (1995).

102. J.K. Stuart and V. Hlady, *Langmuir* **11**, 1368 (1995).

103. U. Dammer, M. Hegner, D. Anselmetti, P. Wagner, M. Dreier, W. Huber, and H.J. Guntherodt, *Biophys. J.* **70**(5), 2437 (1996).

104. U. Dammer, O. Popescu, P. Wagner, D. Anselmetti, H.-J. Guntherodt, and G.N. Misevic, *Science* **267**, 1173 (1995).

105. M.L. Fielden, R.A. Hayes, and J. Ralston, *Langmuir* **12**, 3721 (1996).

106. H.-J. Butt, *J. Colloid Interface Sci.* **166**, 109 (1994).

107. P. Mulvaney, J.M. Perera, S. Biggs, F. Grieser, and G.W. Stevens, *J. Colloid Interface Sci.* **183**(2), 614 (1996).

108. P.G. Hartley, F. Grieser, P.A. Mulvaney, and G. Stevens, in preparation.

109. B.A. Snyder, D.E. Aston, and J.C. Berg, *Langmuir* **13**(3), 590 (1997).

12 Surface Forces Apparatus: Studies of Polymers, Polyelectrolytes, and Polyelectrolyte–Surfactant Mixtures at Interfaces

PER M. CLAESSON

Laboratory for Chemical Surface Science, Department of Chemistry, Physical Chemistry, Royal Institute of Technology, SE-100 44 Stockholm, Sweden, and Institute for Surface Chemistry, Box 5607, SE-114 86 Stockholm, Sweden

12.1 INTRODUCTION

Direct measurement of surface forces is nowadays a rather common way to investigate interactions between solid surfaces,[1,2] between particles and surfaces,[3,4] in single-foam lamellae,[5] and in liquid crystalline phases.[6] For this purpose many different types of techniques may be used, all with their own advantages and disadvantages.[7] Nevertheless, it is clear that most data concerned with interactions between solid surfaces coated with polymers have been obtained with the interferometric surface force apparatus developed by Israelachvili.[1] The first studies of this type dealt with adsorption of homopolymers and their influence on surface interactions as reported by Israelachvili and co-workers[8] and by Klein.[9] Later, many reports have shown how the solvent quality, the segment–surface affinity, and the surface coverage affect the interfacial properties of homopolymers[10–17] and block copolymers.[18–22] The first investigations of the surface forces induced by adsorbed polyelectrolytes were reported by Klein and Luckham,[23,24] which soon were followed by others.[25–28] It is clear that the direct measurement of the forces induced by polymers and polyelectrolytes together with the parallel development of theoretical models for the structure of adsorbed layers and the interaction between polymer coated

Colloid–Polymer Interactions: From Fundamentals to Practice, Edited by Raymond S. Farinato and Paul L. Dubin
ISBN 0-471-24316-7   © 1999 John Wiley & Sons, Inc.

surfaces[29–37] have been of utmost importance and that the results obtained have a direct relevance to colloidal stability and particle deposition/removal.

In this chapter we first describe the basic principle of the interferometric surface force apparatus (SFA) and point out some advantages and disadvantages of the SFA compared to noninterferometric surface force techniques. More detailed descriptions of SFA can be found in the original literature.[1,38] In the next section we give examples of the information that can be obtained when studying aqueous polymer systems with SFA. First, aqueous nonionic ethyleneoxide-containing polymers at interfaces are considered, and the effects of changing the solvency (by changing the temperature), the molecular structure, and the polymer–surface affinity are discussed. The relevance of the findings for enhanced steric stabilization (steric stabilization above the cloud point temperature of the polymer), and for the protein rejecting properties of the surface layer are pointed out. Some recent data obtained for cellulose surfaces, with implications for papermaking, are also mentioned before we focus the attention on the forces acting between surfaces coated with cationic polyelectrolytes. Here the results obtained depend on polyelectrolyte concentration and linear charge density, ionic strength of the solution, and in some cases on the order of adding the polyelectrolyte and the inorganic salt. This has bearings on flocculation of particles with polyelectrolytes as in, for example, wastewater treatment and papermaking. Finally, very recent data for surface interactions in polyelectrolyte–surfactant mixtures are discussed. The results obtained here are not yet fully understood, but due to the technological importance of aqueous polyelectrolyte–surfactant mixtures it seems appropriate to include them here to provide an idea of the properties of polyelectrolyte–surfactant complexes at solid–liquid interfaces. We have chosen to take the examples from results obtained in our own laboratory rather than reviewing the many very rewarding surface force studies of polymers and polyelectrolytes at interfaces that has been carried out around the world. Hence, this chapter is not a review of the field, but it should rather be regarded as an introduction to what can be learned by studying polymers and polymer–surfactant mixtures employing the interferometric surface force technique. For a review of the data obtained for polymer systems up to 1990 the reader is referred to the report by Luckham.[39]

12.2   SURFACE FORCE APPARATUS

There are many different designs of the interferometric surface force apparatus.[1,12,38,40] The principle behind the different versions of the instrument is similar; see Figure 12.1. The preferred substrate in the SFA is muscovite mica, an aluminosilicate mineral that can easily be cleaved into large molecularly smooth sheets. However, other substrates have also been used.[41–43]

The main advantage of the interferometric surface force apparatus over noninterferometric techniques is that the use of optical interferometry allows the determination of absolute surface separations whereas the other techniques only measure distances relative to a "hard wall." The possibility to measure absolute

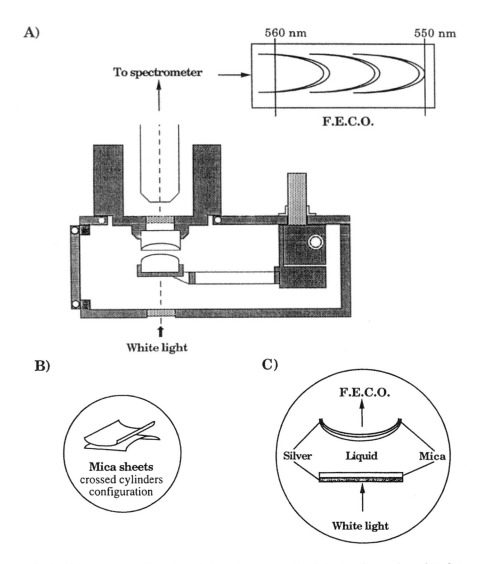

FIGURE 12.1 Schematic illustration of the main components of the interferometric surface force apparatus is provided in part (*a*). The stainless-steel measuring chamber contains the two interacting surfaces. In a typical experiment one mica surface is glued to a silica disk that is attached to a piezoelectric crystal (topmost part). The other surface, also glued to a silica disk, is mounted on a double cantilever force measuring spring. The surfaces are oriented in a crossed-cylinder configuration [see part (*b*)]. White light enters through the window in the bottom of the chamber. It is multiply reflected between the silver layers and a standing-wave pattern, fringes of equal chromatic order (FECO), is generated [see part (*c*)]. The standing waves exit through the top window and the wavelength and fringe shape are analyzed in a spectrometer.

distances with SFA means that adsorbed layer thicknesses and orientation of asymmetric molecules (e.g., proteins) on surfaces can be determined. The optical interference technique also allows studies of a range of phenomena such as surface deformation, local radius of curvature of the surfaces, phase separation, and measurements of refractive index. For these reasons one may state that this technique provides the most detailed information of all available surface force techniques. The main drawbacks with SFA are that it is a slow technique and that relatively few types of substrate surfaces fulfill the requirements of being sheetlike, transparent, and molecularly smooth. For this reason much effort is going into developing methods to chemically modify mica surfaces. Techniques that have been used for this purpose include Langmuir–Blodgett deposition,[44,45] plasma activation[46] sometimes followed by silanation,[47] plasma polymerization,[48] vacuum deposition of metals[49] or oxides,[50] and spin-coating.[51]

12.2.1 Determination of Surface Separation

The substrate surfaces are silvered on their backside and glued onto cylindrical silica disks. These surfaces are mounted in a crossed cylinder configuration inside the SFA. Collimated white light is directed perpendicularly to the surfaces. It passes through the lower surface and becomes multiply reflected between the silver layers. Due to constructive interference for certain wavelengths, multiple-beam interference fringes [or fringes of equal chromatic order (FECO)] are transmitted through the upper surface. The FECO wavelengths depend on the thickness (T) of the mica surfaces and their separation (D). The surface separation is determined by comparing the FECO wavelengths when the surfaces are in contact and apart. For a symmetric three-layer system the separation can be calculated with an accuracy of 0.1 to 0.2 nm from[52]

$$\tan\left(\frac{2\pi\mu D}{\lambda_n^D}\right) = \frac{2\bar{\mu}\sin(\phi\pi)}{(1+\bar{\mu}^2)\cos(\phi\pi) \pm (\bar{\mu}^2 - 1)} \tag{12.1}$$

where μ = refractive index of the medium
$\bar{\mu} = \mu_m/\mu$
μ_m = refractive index of mica at the wavelength λ_n^D
n = fringe order that equals the number of antinodes in the standing wave
+ and − = refers to n being odd and even, respectively
λ_n^D = wavelength of fringe n when the surfaces are separated a distance D

$$\phi = \frac{1 - (\lambda_n^0/\lambda_n^D)}{1 - (\lambda_n^0/\lambda_{n-1}^0)} \tag{12.2}$$

where λ_n^0 = wavelength of fringe n when the surfaces are in contact
λ_{n-1}^0 = wavelength of fringe $n - 1$ when the surfaces are in contact

The wavelengths of the standing waves that correspond to zero surface separation are determined with the surfaces in contact in either dry air or a dilute aqueous salt solution. Under these conditions a strong van der Waals force pulls the surfaces firmly together, and the glue holding the surfaces to the silica disks flattens locally.

Odd and even fringes depend in a different way on the refractive index of the medium between the surfaces. Hence, by measuring the wavelength of fringe n, $n -$ 1, and $n - 2$ with the surfaces in contact and the wavelength of fringe n and $n - 1$ with the surfaces apart, it is possible to determine both the surface separation and the refractive index. Knowing this it is possible to calculate the adsorbed amount on each surface (Γ) by using the relation[53]

$$\Gamma = \frac{D(n - n_b)}{2dn/dc} \tag{12.3}$$

where n = refractive index of the adsorbed layer
  n_b = refractive index of the bulk solution
  dn/dc = refractive index increment

12.2.2 Determination of Force

The distance between the two interacting surfaces is normally changed stepwise by either a synchronous motor or, more accurately, by a stepwise change in the voltage applied to a piezoelectric crystal. The movement of the piezoelectric crystal is calibrated at large surface separations using the FECO. During this calibration procedure, it is assumed that the expansion of the piezoelectric crystal (ΔD) is linear with the applied voltage (ΔV);

$$\Delta D = c\,\Delta V \tag{12.4}$$

where c is the proportionality constant. An alternative way to change the surface separation, which so far has been used only in a few cases, is to utilize a magnetic device.[54]

The force (F_c) between the two surfaces in crossed-cylinder configuration is measured by expanding or contracting the piezoelectric crystal by a known amount (ΔD) and then measuring interferometrically the actual distance that the two surfaces have moved relative to one another (ΔD_0). Any difference in the two distances when multiplied by the spring constant (k) gives, using Hooke's law, the difference between the forces at the initial and the final separations (ΔF_c):

$$\Delta F_c = k(\Delta D - \Delta D_0) = k(c\,\Delta V - \Delta D_0) \tag{12.5}$$

Spring deflections of the order of 1 nm can be determined in this way. Hence, using a spring with a spring constant of about 100 N/m and surfaces with a radius of about 2 cm gives an accuracy in force of 10^{-7} N, corresponding to a force normalized by radius of about 0.005 mN/m.

The mechanical spring system is unstable in the regions of the force–distance curve where the gradient of the force (dF_c/dD) exceeds the spring constant. In these regions the force cannot be measured in the same manner as discussed above. Instead springs of different stiffnesses may be used, and for each spring the position of the instability, "the jump distance," is determined. In this way it is possible to construct a graph of the force gradient as a function of surface separation. The drawback with this method is that it is very time consuming (only one point on the graph is obtained for each spring stiffness), and we have found that for slowly varying forces it is difficult to determine the instability point to better than 0.5 to 1 nm.

12.2.3   Determination of Radius

With the interferometric surface force apparatus it is very easy to determine the local radius, R, of the interacting surfaces from the shape of the interference fringes. Basically, one measures the width of the fringe $(2x)$ at a distance ΔD from the tip of the fringe having the surfaces close to contact (see Fig. 12.1).

$$R = \frac{(x/M)^2}{2\Delta D} \qquad (12.6)$$

where M is the magnification of the optical system.

The data obtained with the SFA is presented as force (F) normalized by radius (R) as a function of separation. This quantity is related to the free energy of interaction per unit area between flat surfaces (G_f)[55]:

$$\frac{F_c}{2\pi R} = \frac{F_{sf}}{2\pi R} = \frac{F_s}{\pi R} = G_f \qquad (12.7)$$

where subscript c stands for crossed-cylinder geometry, sf for a sphere against a flat surface, and s for two spherical surfaces. This relation, known as the Derjaguin approximation,[55] is valid provided the range of the interaction is much smaller than the radius of the surfaces. This is the case in the SFA where the typical radius of the surfaces employed is 2 cm. Another requirement is that the surfaces should not deform due to the action of the surface forces. This condition is not always fulfilled. In the interferometric SFA, surface deformations affect the value of F/R, whereas the distance between the surfaces remains correct. This is because it is the glue, which is located outside the optical cavity, that is compressed rather than the mica sheets. In contrast, when noninterferometric methods are used, such as the atomic force microscope (AFM) colloidal probe[3] or the measurement and analysis of surface interface forces (MASIF),[56] both the F/R value and the determined surface separation are affected by any surface deformation.

12.3 RESULTS AND DISCUSSIONS

12.3.1 Nonionic Polymers

The properties of nonionic polymers in solution and at interfaces are of immense practical importance. One powerful way to investigate the properties of polymers in bulk and at interfaces theoretically is by means of lattice mean-field theories,[31] which are based on the fundamental Flory–Huggins treatment of polymer solutions. An alternative is the scaling theory introduced by de Gennes.[57] Important properties of the polymer that influence the behavior of polymer systems are molecular weight, blockiness (homopolymer, random copolymers, or block copolymer), chain branching, and stiffness. The effect of the solvent is commonly discussed in terms of the segment–segment interaction across the solvent, expressed in terms of the χ parameter. We note that in an athermal solvent where the segment–segment, segment–solvent, and solvent–solvent interactions are identical, $\chi = 0$ by definition, and excluded volume effects result in a strong swelling of the polymer chain. When the segments attract each other in the solvent, the swelling of the polymer coil is decreased and $\chi > 0$. In a theta solvent, where $\chi = 0.5$, the net segment–segment attraction just balances the excluded volume effect and the polymer chains behave ideally. A further increase in the χ parameter results in poor solvency conditions and phase separations occur readily. The effective surface–segment interaction, expressed in terms of the χ_s parameter describes the energy change when a surface–solvent contact is replaced by a surface–segment contact. A positive χ_s value means that the segment–surface affinity is larger than the surface–solvent affinity. A polymer loses conformational entropy close to a surface, and therefore no adsorption occurs until the χ_s value has become sufficiently positive to compensate the entropy loss.

In the next section we focus on some surface force experiments that have been conducted to study the effects of adsorbed nonionic polymers on interactions between surfaces across aqueous solutions. First, the forces acting between an ethyleneoxide-based surfactant will be discussed to highlight the temperature dependence of the interactions between ethyleneoxide and water. Next we move on to ethyleneoxide-based polymers to illustrate the effect of the segment–solvent and effective segment-surface interaction on the adsorbed layer structure and the forces acting between polymer-coated surfaces.

12.3.2 Pentaoxyethylene Dodecyl Ether

The forces acting between hydrophobized mica surfaces coated with the surfactant pentaoxyethylene dodecyl ether $[C_{12}H_{25}(OCH_2CH_2)_5OH, C_{12}E_5]$ have been investigated with the interferometric surface force apparatus. To ensure a maximum packing of the surfactant at the surface a surfactant concentration of 6×10^{-5} M, that is, just above the critical micelle concentration (cmc), was used. The χ_s parameter describing the interaction between the alkyl chain and the hydrophobic surface is much larger than the χ_s parameter for the ethyleneoxide groups. Hence, the orientation is such that the alkyl chains are directed toward the surface with the ethyleneoxide groups being exposed to water. The force–distance profiles obtained at various temperatures are shown in Figure 12.2. The zero separation in this figure is defined as the position of the surfaces obtained under a high load ($F/R \approx 100$ mN/m). This hard-wall separation increases with temperature from 1.4 ± 0.4 nm at 20 °C to 2.6 ± 0.4 nm away from the hydrophobic surface at 37 °C.[58] Hence, the $C_{12}E_5$ layer thickness increases with temperature, which shows that the adsorbed amount increases as a consequence of the worsening of the solvency (increasing χ parameter) for the ethyleneoxide chain. The molecular mechanism behind this change in solvency with temperature will be discussed below.

The forces measured at 15 °C are purely repulsive and increase monotonically with decreasing separation. Purely repulsive forces are observed also at 20 °C, but now a local minimum in the force curve is present 3 nm from the hard wall. At higher temperatures a small part of the force curve is attractive. The magnitude of the attraction increases and the position of the force minimum moves closer to the hard wall at higher temperatures. This is also a consequence of the reduced solvency of the ethyleneoxide chain at higher temperatures.

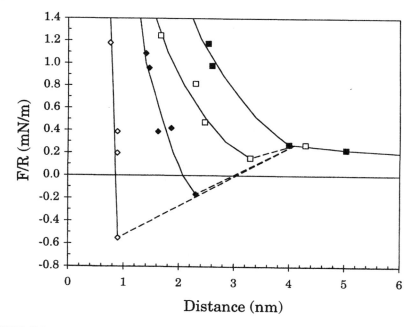

FIGURE 12.2 Force normalized by radius as a function of surface separation between hydrophobized mica surfaces across a 6×10^{-5} M $C_{12}E_5$ solution. The zero separation is defined as the position of the surfaces under a high force (≈ 100 mN/m). The forces were measured at 15 °C (■), 20 °C (□), 30 °C (♦), and 37 °C (◇). The lines are guides for the eye.

The fact that the χ parameter for the ethyleneoxide unit increases with temperature does affect, as seen above, the adsorption and interaction. It also affects the phase diagram for poly(ethyleneoxide)–water systems that display a closed miscibility gap[59] and the phase diagram for alkyl ethoxylate surfactants that phase separate on heating.[60] The cloud point of the $C_{12}E_5$–water system is about 27 °C,[61] and already below this temperature the size of the $C_{12}E_5$–micelles increase with increasing temperature.[62] The temperature dependence of the interactions between ethylene oxide chains in aqueous solutions has also great practical importance. For instance, oil-in-water emulsions stabilized by ethyleneoxide-based surfactants often change to water-in-oil emulsions at elevated temperatures.[63] This is a consequence of a decrease in head-group repulsion and preferred head-group area at the oil-water interface at higher temperatures.[64] Surfaces coated with poly(ethyleneoxide) (PEO) are used in other applications to reduce protein adsorption. It has been found that the protein rejecting ability of PEO coatings are most efficient when the molecular weight is large and the graft density is not too small.[65,66] It is also most efficient at low temperatures[67] where the water content of the layer is high and the chains extend relatively far out into solution. This means that the van der Waals attraction between the surface coating and the protein will be low and that when the protein approaches the underlying surface it experiences a strong steric repulsion due to the compression of the PEO chains.[65] At higher temperatures the PEO layer contracts and water is expelled. As a consequence the van der Waals attraction between the protein and PEO layer increases, and the steric repulsion due to compression of the PEO becomes less long ranged.[68] Based on similar considerations, it has been argued that for optimal protein rejection the density of the PEO chains at the surface should not be too low or too high.[66,69]

One may ask what is the origin of the temperature dependence of the solvency for ethyleneoxide-based surfactants and polymers. This has been discussed frequently in the literature,[70–75] and basically three types of mechanisms have been suggested.

Kjellander[70] argued that the hydration of the PEO chain is similar to that around hydrophobic solutes with water molecules forming a clathrate-like structure around it. However, a difference is that the ether oxygen in the ethyleneoxide group may participate in hydrogen bonds with water molecules. The model predicts that the association of two PEO chains in water results in release of water molecules and a reduction in number of hydrogen bonds, which results in an increase in enthalpy (ΔH, counteracts association) and entropy (ΔS, favors association). Both terms are large at room temperature, and they nearly cancel each other in the expression for the free energy, which also contains a small entropy term due to the ideal entropy of mixing (ΔS_i, counteracts association).

$$\Delta G = \Delta H - T \, \Delta S - T \, \Delta S_i$$

(12.8)

At low temperatures the free energy of association is positive and dominated by the positive enthalpy term. As the temperature is increased, the $T \, \Delta S$ term becomes increasingly important, and at a given temperature the free energy of association becomes negative and phase separation occurs. Observe that in this model the phase separation of PEO systems is a consequence of the hydration as such and not a consequence of the fact that the hydration of a single PEO chain is diminished close to the phase separation point. Instead, this model assumes that the hydration of the PEO chain is decreased only at very high temperatures, and when this occurs, both the ΔH and the $T\,\Delta S$ terms in the free energy expression are reduced, allowing the ideal entropy of mixing to become dominant, thus explaining the increased solubility found at very high temperatures.

The second type of models also focus on the hydrogen bonding between water and the ether oxygen in the PEO.[71,73-75] Common to these models is that they assume that the hydration of the PEO chain decreases with increasing temperature and that this reduces the solvent quality, thus causing the phase separation at elevated temperatures. Goldstein[71] proposes that the water molecules next to the PEO chain could exist in two states, nonbonded and bonded. At low temperatures most water molecules next to the PEO chain form hydrogen bonds to it and a hydration repulsion dominates. At higher temperatures the nonbonded state predominates, and the PEO chains effectively attract each other, explaining the phase separation. Similar arguments are used by Matsuyama and Tanaka,[73] but they consider that the PEO and water form polydisperse complexes, the hydration of which decreases with increasing temperature. Along a similar line, Bekiranov et al.[75] use an extended Flory–Huggins description with the assumption that a water molecule can form only one hydrogen bond to PEO segments, and that the hydration decreases with temperature and pressure. They show that this model predicts that the single-chain PEO can exist in two conformations, one strongly swelled conformation that is favored by low temperature and pressure, and a more collapsed state favored by a high temperature and pressure. De Gennes[74] approaches the problem in a different way by assuming that two PEO segments repel each other, whereas in a cluster the segments attract each other due to attractive higher-order virial coefficients. This model predicts that under certain conditions two different swelled states of the polymer solution coexist, and under other conditions a swelled state is in equilibrium with a collapsed state. Note the difference between these latter models and Kjellander's model. Kjellander[70] proposes that the phase separation is a consequence of the hydration, whereas in the other models the phase separation is due to a reduction of the hydration of the PEO chain at elevated temperatures.

A different approach was taken by Karlström,[72] who focused on the conformation of the ethyleneoxide chain. He proposed that the segments in the chain may exist in two main types of conformations. The conformation with the lowest energy has the oxygen atoms in gauche conformation around the C–C bond and a trans conformation is preferred around the C–O bond. This conformation gives rise to a large dipole moment and thus a strong interaction with water. At higher temperatures other less polar conformations with a larger χ parameter become, for entropic reasons, more common. Hence, the PEO chain becomes more hydrophobic at elevated temperatures. This two-state model of the PEO chain has been utilized by Linse and Björling in lattice mean-field calculations[76,77] that predict the structure of adsorbed and grafted

PEO chains and the interaction forces between surfaces coated with PEO. A good agreement between measured and calculated force curves has been reported.[21]

The advantage with Karlström's model is that it also can rationalize the temperature dependence of the solubility of ethyleneoxide-based polymers in nonaqueous solutions. It also has some support from nuclear magnetic resonance (NMR) spectroscopy.[78,79] The advantage of the other models is that they can be applied to aqueous solutions of other polymers and surfactants that phase separate on heating.

12.3.3 Block Copolymers

Diblock copolymers with a hydrophobic part and a hydrophilic part behave similarly to a surfactant in the sense that they associate in aqueous solutions and in that it is the hydrophobic interaction between the block copolymers and the hydrophobic surface that is the driving force for adsorption. A typical case is block copolymers of butyleneoxide (B) and ethyleneoxide (E). When comparing diblock and triblock copolymers with the same overall chemical composition, it has been found both experimentally and theoretically that the diblock copolymer adsorbs considerably more than the triblock.[21] This is because the diblock copolymers, for geometrical reasons, are able to pack more efficiently at the surface than any of the triblocks. Diblock copolymers are often very efficient as steric stabilizers since the anchor block (the adsorbing block) adsorbs strongly to the surface whereas the buoy block (the nonadsorbing block), which experience good solvency conditions, extends far out into the solution. It has been suggested that the most efficient steric stabilization is obtained when the anchor block constitutes about 10 to 20% of the diblock copolymer.[31]

The surface force apparatus was used for studying the interactions between uncharged hydrophobized mica surfaces across an aqueous 50 ppm B_8E_{41} solution. The results are displayed in Figure 12.3. The measured forces are purely repulsive and the same on approach and separation. This shows that the layers are in (restricted) equilibrium, that is, the adsorbed polymer chains adopt their equilibrium conformation at each surface separation. The force is of steric origin and essentially of the same type as for the $C_{12}E_5$ surfactant discussed above. The main difference is that the longer chain of the block copolymer results in a larger loss in conformational entropy when the surfaces are brought together, which reduces the relative importance of hydration effects. However, the same type of temperature dependence is expected for B_8E_{41} and $C_{12}E_5$, that is, an increased adsorbed amount and a more compact layer structure at elevated temperatures. The temperature dependence of the forces in the case of B_8E_{41} has not been studied, but the expected changes with temperature have been observed for PEO chains electrostatically adsorbed to mica surfaces via a cationic anchor group.[80]

The data shown in Figure 12.3 are obtained at two surface-to-volume-ratios. The interaction forces obtained across a small droplet (volume 0.05 to 0.1 mL) are considerably less long-ranged than those obtained with the whole measuring chamber (volume $\approx$ 350 mL) filled with the polymer solution. We have shown, using lattice

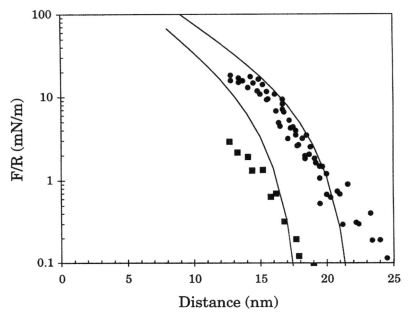

FIGURE 12.3 Force normalized by radius as a function of surface separation between hydrophobized mica surfaces across a 50-ppm B_8E_{41} solution. The forces were determined with the whole measuring chamber filled with the solution ($V \approx 350$ mL) (●), and with only a small droplet between the surfaces ($V \leq 0.1$ mL) (■). The solid lines are theoretically calculated force curves.

mean-field calculations,[21] that the reason for this is that the polymer sample used was slightly polydisperse, $M_w/M_n = 1.1$, according to the manufacturer. It is the components with high molecular weight and high B/E ratio that preferentially adsorb to the surface. Even when these components have a very low bulk concentration, they will dominate at the surface.[21] In the drop there is not enough of these components to cover the whole surface, which results in a shorter range force compared to the case when the whole measuring chamber is filled with the polymer solution. A consequence of this is that the difference between the polymer composition in the bulk and in the adsorbed layer is smaller in the droplet experiment than in the full box experiment.

In the original study[21] we compared the measured interaction with force curves calculated using a lattice mean-field theory, and found a good qualitative agreement. Here (Fig. 12.3), we instead compare the measured interaction with predictions based on the scaling approach for polymer brushes.[32] According to this theory, the pressure $P(D)$ between two flat polymer-coated surfaces is given by:

$$P(D) = \frac{kT}{s^3}\left[\left(\frac{D^*}{D}\right)^{9/4} - \left(\frac{D}{D^*}\right)^{3/4}\right]$$

(12.9)

where Eq. (12.9) is valid provided the separation, D, is less than D^* (where D^* is twice the length of the polymer tail), and s is the linear distance between the anchored chains on the surface. For the interactions between two crossed cylinders this relation is, according to the Derjaguin approximation, modified to:

$$\frac{F(D)}{R} = -2\pi \int_{D^*}^{D} P(D') \, dD' \qquad (12.10)$$

The parameters needed in order to calculate the force are the length of the extended polymer chain and the separation between the polymer chains on the surface. The latter parameter can be calculated from the adsorbed amount (obtained from ellipsometry measurements that gave $s = 1.89$ nm), whereas the length of the polymer chains enters as a fitting parameter. The values of the fitting parameters used here were 9 nm for the droplet experiment and 11 nm for the full box experiment. The radius of gyration for the ethyleneoxide chain in bulk solution is about 1.8 nm, that is, much less than the extended length of the polymer chain. The reason is the high adsorption density that forces the chains to stretch out from the surface, and the brush-model, assumed by Eq. (12.9), is thus expected to be applicable. We see that the forces calculated using scaling theory are in qualitative agreement with the measured interaction. The main difference is that the gradient of the force at large separations is less than predicted by theory, which can be rationalized by the polydispersity of the polymer which results in chains with different lengths. The deviation at small separations is most likely due to the fact that at high compressive forces some of the adsorbed polymers are being desorbed from the surface.

12.3.4   Ethyl(hydroxyethyl)cellulose

Ethyl(hydroxyethyl)cellulose (EHEC) is a nonionic polymer that has found use in a range of applications as rheology modifiers, steric stabilizers, and as inert coatings in medical technology; EHEC has a cellulose backbone that has been modified by grafted ethyl groups and short ethyleneoxide chains. This prevents an effective crystalline packing, and as a consequence the polymer becomes more soluble in water despite the fact that the grafted groups are rather hydrophobic. Aqueous EHEC solutions phase separate on heating due to the temperature dependence of the interaction between ethyleneoxide units and water.[81] The cloud point temperature depends on the degree of modification.

It has been shown that EHEC adsorbs strongly to hydrophobic surfaces and that only a small fraction of the EHEC molecules desorb when the polymer-coated surface is immersed in pure water.[82] The layer appears to be "irreversibly" adsorbed. The very slow desorption rate of most polymers is a consequence of their high-affinity adsorption isotherm. This means that when such a layer is immersed in solution, the free polymer concentration outside the surface is very low. Thus, the polymer concentration gradient away from surface, which is the driving force for transport to bulk solution, is extremely small even when the bulk polymer concentration is essentially zero. It has been shown by Cohen Stuart and Fleer[83] that this leads to extremely slow desorption, and for practical purposes the layer can be regarded as irreversibly adsorbed even though the adsorption is not a truly irreversible process.

In one set of experiments an EHEC fraction with mean molecular weight of 2.5×10^5 g/mol having a cloud point temperature of 39 °C was allowed to adsorb onto hydrophobic surfaces at 25 °C. The forces acting between the hydrophobic EHEC-coated surfaces were then determined across a polymer-free aqueous solution at three different temperatures; see Figures 12.4 and 12.5. At 25 °C, the repulsive steric force extends out to about 80 nm, and it becomes progressively stronger as the surface separation is decreased. The forces measured on approach and separation are the same, showing that the system is in restricted equilibrium. The layer contracts significantly when the temperature is increased to 44 °C. At this temperature a steric force is observed below about 25 nm. On further compression the force increases strongly. The compaction of the preadsorbed layer with increasing temperature is a consequence of the decreasing solvency of the EHEC molecule. This is well understood from a theoretical point of view, and a similar reduction in hydrodynamic thickness of preadsorbed polymer layers with decreasing solvency has been documented.[31]

One noticeable feature is that the forces measured at 44 °C are purely repulsive, although this is 5 °C above the cloud point. For a homopolymer above the cloud point,

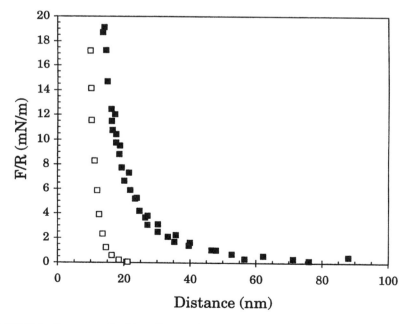

FIGURE 12.4 Force normalized by radius as a function of surface separation between hydrophobized mica surfaces precoated with an adsorbed layer of EHEC. The forces were measured at 25 °C (■) and 44 °C (□).

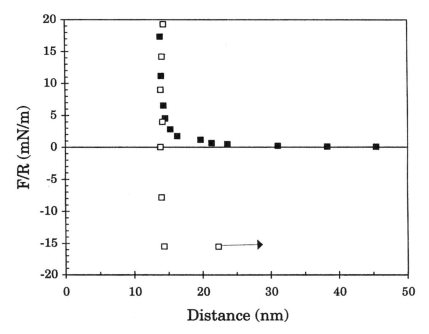

FIGURE 12.5   Force normalized by radius as a function of surface separation between hydrophobized mica surfaces precoated with an adsorbed layer of EHEC. The temperature was 55 °C. Filled and unfilled symbols represent forces measured on approach and separation, respectively.

there is an effective attraction between the polymer segments, which results in an attraction in the outermost part of the force curve. This has also been confirmed experimentally.[9] A plausible explanation to why this is not observed for EHEC is that the segments along the chain are not identical, but they have different substitutions of short ethyleneoxide chains and ethyl groups. The different segments therefore have a different interaction with water. The most hydrophobic segments of the polymer will be accumulated close to the hydrophobic surface and within the adsorbed layer, whereas the more hydrophilic segments will be concentrated in the outer part of the layer. This effect has been demonstrated theoretically for random (statistical) copolymers.[31] When the two EHEC-coated surfaces are brought together, it is the outermost segments that first interact with each other. If these segments experience good solvency conditions ($\chi < 0.5$) at 44 °C, they will repel each other, explaining the absence of any attraction in the measured force curve.

When the solution temperature is increased further to 55 °C, a strong attraction is observed when the adsorbed layers are close together; see Figure 12.5. Hence, at this high temperature even the outermost segments experience poor solvency conditions.

The experiment described above illustrates how a preadsorbed layer is affected when the solvency of the polymer is changed by changing the temperature. It is interesting to compare this with what happens in the corresponding experiment when

EHEC is present in solution during the temperature variation. This has been studied using a different EHEC fraction (M_w 475,000 g/mol, cloud point 35 °C).[84] Some results, obtained at 20 and 41 °C, are shown in Figure 12.6. At the lower temperature the range of the steric force is about 140 nm. The repulsion increases comparatively slowly with decreasing separation. At the higher temperature the range of the force is slightly larger, and, more significantly, it increases more rapidly with decreasing separation. Hence, for a given separation the repulsive force increases with increasing temperature. This is opposite to the situation with preadsorbed layers because with EHEC present in solution the adsorbed amount of the polymer increases when the temperature is raised.[81] This is a consequence of the reduced solvency. We note that in the EHEC system we used, it typically took a few hours before adsorption equilibrium had been established after changing the temperature.

The variation in the range of the steric force with temperature is illustrated in Figure 12.7 both for the case of preadsorbed layers and for the case with EHEC in solution. Note that when EHEC is present in solution the range of the steric force is rather independent of temperature. Similarly, it has been predicted[31] that the hydrodynamic thickness of a polymer layer should be insensitive to the solvency when the adsorbed amount is allowed to vary. This comparison should, however, not be taken too far since the hydrodynamic thickness is more sensitive to the tails of the polymer than the steric force.

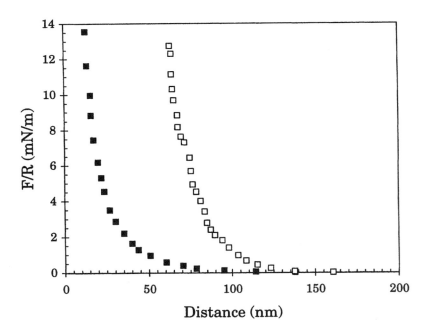

FIGURE 12.6 Force normalized by radius as a function of surface separation between hydrophobized mica surfaces across a 0.25% EHEC solution at 20 °C (■) and 41 °C (□).

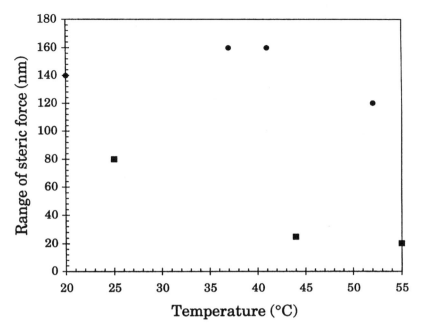

FIGURE 12.7   Measurable range of the steric force between EHEC-coated hydrophobic surfaces as a function of temperature. Data is shown for the situation with preadsorbed EHEC layers (■), and for the situation with EHEC present in solution (●).

Let us also consider how the surface segment affinity (the χ_s parameter) affects the adsorbed layers and the interaction forces. This can be illustrated by comparing the results discussed above for EHEC on uncharged hydrophobized mica surfaces (high segment–surface affinity) with those obtained for EHEC on negatively charged mica surfaces (low segment–surface affinity). Before carrying out this comparison it may be valuable to recapitulate some facts about the mica surface. Mica is a layered aluminosilicate mineral (Figure 12.8). The negative surface charge is due to isomorphous substitution where some silicon atoms have been replaced by aluminum. In the crystal these charges are compensated by potassium ions located between the aluminosilicate layers. When the surface is immersed in aqueous solutions the potassium ions located on the basal plane $(2.1 \times 10^{18} \text{ m}^{-2})$ are completely dissolved. Due to adsorption of protons and other cations present in solution, the charge of the mica surface in aqueous electrolyte solutions is, however, much less than expected from the crystal structure.

The forces acting between mica surfaces across aqueous EHEC solutions are displayed in Figure 12.9. At 20 °C the long-range force is dominated by an electrostatic repulsion originating from the mica surface charge. A weak steric force is observed only at distances below 5 nm. By increasing the compressive force, the polymer is easily squeezed out from between the surfaces, which is a convincing evidence that the polymer–surface affinity is very low indeed. This contrasts sharply to the

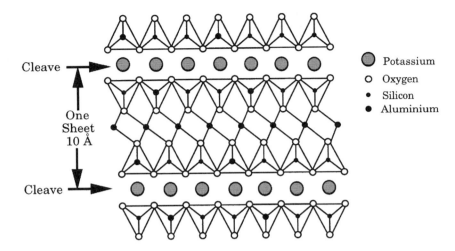

FIGURE 12.8   Schematic illustration of the mica lattice. The mica crystal is easily cleaved between the aluminosilicate layers. After cleavage the potassium ions located on the surface are exchanged for other cations present in solution. The potassium ions in the crystal compensate the negative charge; thus about 25% of the silicon atoms in the crystal are replaced by aluminum atoms.

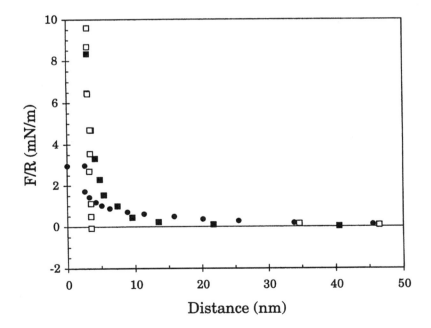

FIGURE 12.9   Force normalized by radius as a function of surface separation between hydrophilic negatively charged mica surfaces across a 0.1% EHEC solution at 20 °C (●) and 45 °C on approach (■) and separation (□).

long-range nature of the steric force between hydrophobic surfaces across EHEC solutions (compare Figures 12.6 and 12.9). When the temperature is raised to 40 to 45 °C, the force curve changes significantly due to an enhanced EHEC adsorption. A steep steric force is now encountered at distances below about 10 nm, and it is no longer possible to desorb the polymers by applying a strong compressive force. Clearly, the decreased solvency of the polymer results in an increased adsorption and steric repulsion when the segment–surface affinity is low. Consistent with the findings above, it has been shown by ellipsometric measurements that the adsorbed amount of EHEC is much larger on hydrophobized silica compared to on bare silica.[81] It was also found that on both surfaces the adsorbed amount increased with increasing temperature, and more dramatically so on the hydrophobized silica.

12.3.5 Cellulose

Cellulose, the most abundant polysaccharide on earth, is not soluble in water because the extended conformation of the polysaccharide is ideal for formation of intrachain and interchain hydrogen bonds that promote crystalline structures. It is of great interest to investigate the forces acting between cellulose surfaces and between

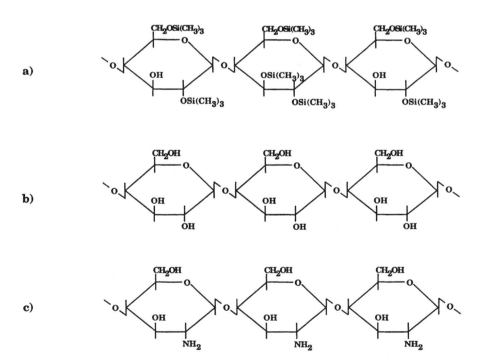

FIGURE 12.10 Structure of (*a*) trimethylsilyl cellulose. Ideally each segment should have 3 trimethylsilyl substituents, but in most cases the degree of substitution is less, (*b*) cellulose, and (*c*) chitosan.

cellulose and other materials since it has direct technological relevance for papermaking and the deposition and removal of particles from textiles. Complications in the technical processes include the fact that the cellulose fibers do not consist of pure cellulose, but also contain hemicellulose (papermaking), and that the cellulose fibers might be oxidized. In order to investigate the interactions between model cellulose surfaces we have applied a Langmuir–Blodgett method[85] to prepare well-defined and smooth surfaces.[86] In the studies presented below 10 layers of trimethylsilyl cellulose (Fig 12.10) were deposited on hydrophobized mica. The cellulose was regenerated in humid HCl atmosphere after deposition. It was found that the thickness of each cellulose layer in air was 0.4 nm, increasing to 0.65 nm when the humidity was increased to close to 100%. Crystalline cellulose takes up a very limited amount of water,[87] and hence the rather large uptake of water seen here indicates that our sample was mainly composed of amorphous cellulose. The swelling in liquid water is even larger, and it is seen that under a high compression $F/R \approx 10$ mN/m; each layer is about 0.8 nm thick. The interaction between cellulose surfaces across water, at pH 5.5 to 6.0, is shown in Figure 12.11. No long-range double-layer force is observed, which is as expected for unoxidized cellulose considering that the pK_a of glucose is around 12.4.[88] The range of the repulsive force is about 30 nm. It has been

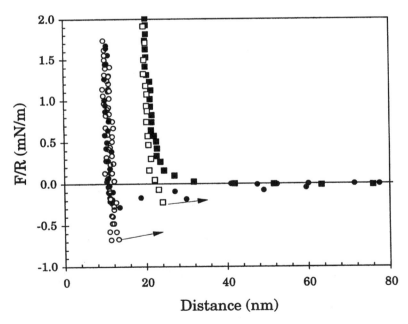

FIGURE 12.11   Force normalized by radius as a function of surface separation between two cellulose surfaces prepared by a Langmuir–Blodgett deposition method. The forces were measured across a 0.1 mM KBr solution. Forces were measured on approach (■) and separation (□). The forces measured between one cellulose surface and one chitosan-coated surface on approach (●) and separation (○) are also shown. The arrows indicate outward jumps.

identified as a steric force arising from compression of some dangling tails extending away from the surface.[86] A weak attraction is observed on separation. The origin of the attraction is most likely a van der Waals force between the hydrated cellulose layers. The Hamaker constant for the hydrated cellulose layer was determined from Lifshitz theory using the measured refractive index and an estimated value of the dielectric constant. It was found to be 0.7 to 0.9×10^{-20} (J) in water and 8.4×10^{-20} (J) in dry air.[86]

In the papermaking process, polyelectrolytes are often used as retention agents and wet and dry strength additives. Hence, it is relevant to learn more about interactions between cellulose and polyelectrolytes and how the polyelectrolytes affect the interaction between cellulose and other materials. Two surface force studies aiming in this direction have been reported. In one set of experiments the interactions between one cellulose surface and one mica surface coated with chitosan (Fig. 12.10), a weak cationic polyelectrolyte with $pK_a \approx 6.5$, were studied at pH 5.5. The coating of the mica surface with chitosan changed the surface potential from about -100 to $+30$ mV. It was found that at distances below 40 nm an attraction was present between the chitosan-coated mica surface and the cellulose surface (Fig. 12.11). The data were

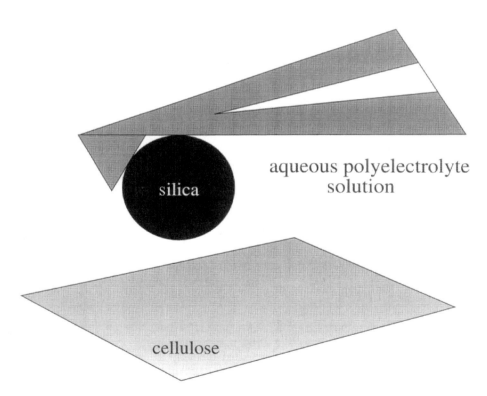

FIGURE 12.12   Experimental geometry used when studying interactions between silica and cellulose using the AFM colloidal probe technique.

interpreted in terms of a bridging mechanism where extended cellulose tails formed bridges to the chitosan-coated surface.[86] The normalized adhesion force was, however, relatively weak, below 1 mN/m.

In another set of experiments the colloidal probe technique (Fig. 12.12) was used for studying interactions between a flat uncharged cellulose surface and a negatively charged colloidal silica particle.[89] The long-range interaction was dominated by a repulsive double-layer force (Fig. 12.13) originating from the confinement of the counterions to the negatively charged silica probe. Addition of the cationic polyelectrolyte poly(2-acryloxyethyltrimethylammonium chloride) (PCMA) (the structure is provided in Fig. 12.15) to a concentration of 20 ppm decreased the repulsive double-layer force, thus lowering the barrier for deposition of silica onto the cellulose surface. Surprisingly, the presence of the polyelectrolyte did not result in any large increase in the adhesion force between the silica particle and the cellulose surface. Hence we conclude that for this system the main effect of the polyelectrolyte is to lower the energy barrier preventing the particle deposition, whereas the polyelectrolyte does not significantly affect the energy barrier for removing the particle. The data obtained so far is too limited to allow any conclusion about how general these features are, and one may suspect that for different particles and polyelectrolytes other effects (e.g., changes in adhesion forces) may be of importance.

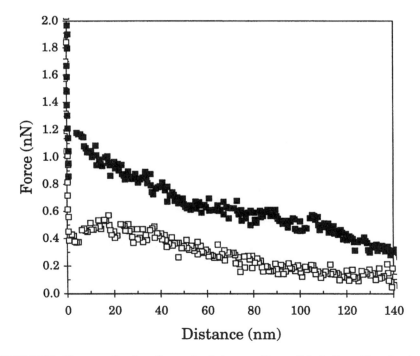

FIGURE 12.13 Force as a function of separation between a silica particle (radius ≈ 13 μm) and a flat cellulose surface across 0.1 mM KBr before (■) and after (□) adding the cationic polyelectrolyte PCMA to a concentration of 20 ppm. The measurements were done on approach.

In particular, for the case of negatively charged cellulose fibers, one may expect that addition of cationic polyelectrolytes will increase the adhesion between negatively charged particles and the fiber. This will undoubtedly be an important field for future research.

12.3.6 Polyelectrolytes

The interactions between polyelectrolytes and surfaces are obviously strongly affected by electrostatic forces between the polyelectrolyte and the surface and between the polyelectrolyte segments. Hence, all parameters that affect the electrostatic interactions (e.g., polyelectrolyte charge density, surface charge density, and ionic strength) are of prime importance for the adsorbed layer structure and the interactions between polyelectrolyte-coated surfaces.[90–93] Adsorption occurs readily when the polyelectrolyte and the surface are oppositely charged, and one important parameter is the ratio of the distance between charges on the polyelectrolyte and on the surface (Fig. 12.14). Charge neutralization can be achieved with a very thin polyelectrolyte layer when this ratio is close to unity. On the other hand, when the distance between charges on the polyelectrolyte is significantly larger than that between the charged surface sites, one may expect that the uncharged stretches of the polyelectrolyte form loops on the surface in order to allow a high adsorbed amount. When highly charged polyelectrolytes adsorb to an oppositely charged surface of low charge density, a local charge reversal may occur. The repulsion between the excess charges of the polyelectrolyte will in this case favor an extended polymer conformation, and one can visualize that negatively charged patches on one particle may interact with positively charged patches on another particle, which is the mechanism behind the so-called patch-wise flocculation. Extended polyelectrolyte chains also results from association between a polyelectrolyte and a colloidal particle (or a surfactant micelle), which is small compared to the polyelectrolyte. The problem of polyelectrolyte adsorption on curved interfaces has been studied in detail by von Goeler and Muthukumar.[94] The structure of adsorbed polyelectrolyte layers under various conditions has been investigated theoretically using mean-field theories,[31,90–93] scaling arguments,[36,37] and simulation methods.[33]

Adsorption of polyelectrolytes on surfaces with the same sign of charge may also occur provided the unfavorable electrostatic interaction is sufficiently reduced and a favorable nonelectrostatic interaction between the surface and the polymer exists.[31] This situation arises most easily when the polyelectrolyte has a low charge density and under high ionic strength conditions. One example is the adsorption of hemicellulose on mica.[95] Adsorption of polyelectrolytes on similarly charged surfaces is also facilitated by the presence of multivalent counterions, and the surface forces in such a system have been investigated.[96]

Below we discuss some surface force data obtained for cationic polyelectrolytes having different charge densities, ranging from one charge per segment to one charge per hundred segments, on highly negatively charged surfaces, that is, for the situations

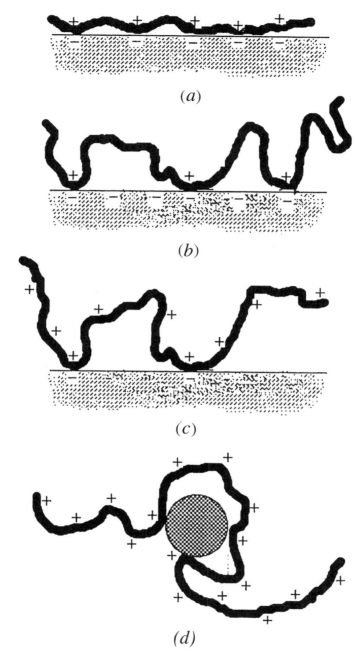

FIGURE 12.14   Illustration of how one may expect that the distance between charges on the surface and along an oppositely charged polyelectrolyte may influence the structure of adsorbed layers. In (a) the two distances are similar; (b) illustrates the case of a low charge density polyelectrolyte on a high charge density surface; (c) a high charge density polyelectrolyte on a low charge density surface; and (d) the situation with a polyelectrolyte and a small colloidal particle.

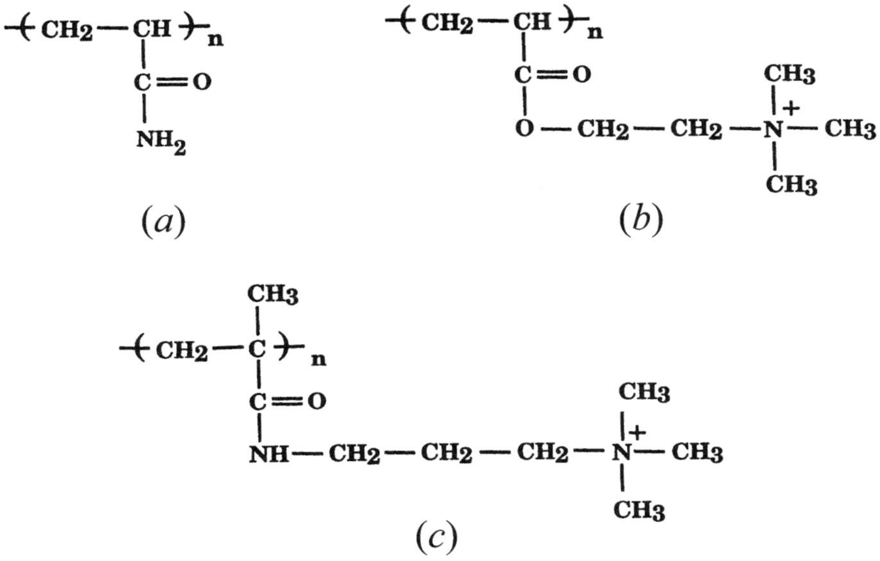

$$-(CH_2-CH)_n$$
$$|$$
$$C=O$$
$$|$$
$$NH_2$$

(a)

$$-(CH_2-CH)_n$$
$$|$$
$$C=O$$
$$|$$
$$O-CH_2-CH_2-N^+-CH_3$$

with CH_3 groups on the N^+

(b)

$$-(CH_2-\overset{CH_3}{\underset{|}{C}})_n$$
$$|$$
$$C=O$$
$$|$$
$$NH-CH_2-CH_2-CH_2-N^+-CH_3$$

with CH_3 groups on the N^+

(c)

FIGURE 12.15 Chemical structures of the segments in the cationic polyacrylamides studied. (a) AM, acrylamide; (b) CMA, 2-acryloxyethyltrimethylammonium chloride; (c) MAPTAC, 3-methacrylamido-propyltrimethylammonium chloride.

illustrated by part (a) and (b) of Figure 12.14. The structure of the cationic monomers (MAPTAC and CMA) and the uncharged monomer (AM) are shown in Figure 12.15.

12.3.7 Effect of Polyelectrolyte Adsorption Density

The forces acting between two mica surfaces across a 0.1 mM KBr solution containing different concentrations of the polyelectrolyte PCMA, which has one positive charge per segment and a mean molecular weight of about 1.5×10^6 g/mol, are illustrated in Figures 12.16 and 12.17.[97] The data shown in Figure 12.16 are obtained at PCMA concentrations at or below 20 ppm. The long-range repulsion, which has a decay length consistent with a double-layer force, decreases with polyelectrolyte concentration. Hence, we identify this repulsion as originating from overlap of the diffuse ionic clouds associated with the net (negatively) charged surface partially coated with polyelectrolytes. With increasing polyelectrolyte concentration the adsorbed amount increases and the net charge and thus the electrostatic double-layer force decreases. At smaller separations an attractive force component, which increases in range and magnitude with increasing polyelectrolyte concentration is observed. Under the action of this attractive force the surfaces reach a separation of about 1 nm. Hence, the adsorbed layer is extremely thin considering the high molecular weight of this polymer. The reasons for this is the high polymer–surface affinity and the similarity of the distance between charges on the polymer and in the surface lattice.

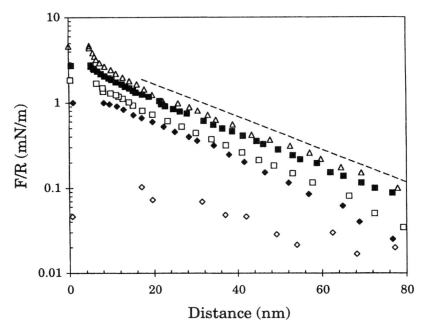

FIGURE 12.16 Force normalized by radius as a function of surface separation between mica surfaces across a 0.1 mM KBr solution containing the cationic polyelectrolyte PCMA at concentrations of 0 ppm (△), 1 ppm (■), 5 ppm (□), 10 ppm (♦), and 20 ppm (◇). The dashed line illustrates the slope of a double layer in 0.1 mM 1:1 electrolyte.

The very thin polyelectrolyte layer observed in low ionic strength solutions is consistent with theoretical predictions.[90]

When the polyelectrolyte concentration is increased further, the long-range repulsive force increases somewhat (Fig. 12.17). This indicates that the charges of the adsorbed polyelectrolytes slightly overcompensate the charge of the substrate surface, and that the polymer-coated surfaces now have a net positive charge. It is interesting to note that under these conditions the decay length of the force is larger than the decay length of a double-layer force at the given electrolyte concentration. Hence, the observed force has also a steric origin, and it seems plausible that it is due to a few polyelectrolyte tails extending far away from the surface. We may term this force electrosteric since it has contributions from both electrostatic and steric interactions.

The attractive part of the force curve between surfaces coated with a similar polyelectrolyte, MAPTAC, has been investigated in more detail and compared with Monte Carlo simulations (Fig. 12.18).[98] It is found that the measured attraction is about 1 order of magnitude larger than the van der Waals force between mica surfaces. From the Monte Carlo simulations it could be concluded that the attractive force originates from a bridging mechanism where different parts of the polyelectrolytes are attracted to different surfaces.[98] Due to the long-range nature of electrostatic forces, such a bridging force may arise even though the polyelectrolytes are not directly

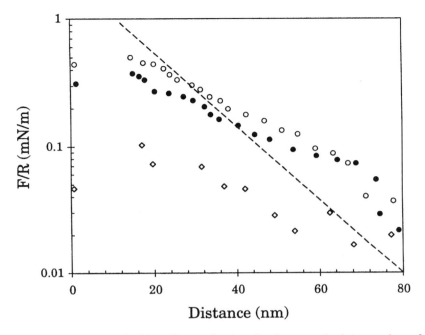

FIGURE 12.17 Force normalized by radius as a function of surface separation between mica surfaces across a 0.1 mM KBr solution containing the cationic polyelectrolyte PCMA at concentrations of 20 ppm (◇), 50 ppm (●), and 100 ppm (○). The dashed line illustrates the slope of a double layer in 0.1 mM 1:1 electrolyte.

bound to both surfaces. This is an important difference compared to bridging forces induced by uncharged polymers that interact with surfaces via short-range forces.

12.3.8 Effect of Ionic Strength

An increase in ionic strength of the solution results in a screening of the repulsion between the charges on the polyelectrolyte chain. This means that the electrostatic persistence length decreases and the polyelectrolyte conformation in solution becomes less extended and more like that of an uncharged polymer, which, for instance, affects the electrophoretic mobility of the polyelectrolyte.[99] A further consequence is that the attraction between the polyelectrolyte and the surface decreases, which results in dramatic changes in the structure of the adsorbed layer and the interaction between polyelectrolyte-coated surfaces. This was studied by allowing PCMA to adsorb onto mica surfaces from 20 ppm solutions of different ionic strengths, and then measuring the resulting surface forces. The results obtained are illustrated in Figure 12.19. When the adsorption occurs from a 1 mM KBr solution, a strong double-layer force dominates the long-range interaction. This is in sharp contrast to the very weak double-layer force observed in 0.1 mM KBr. The thickness

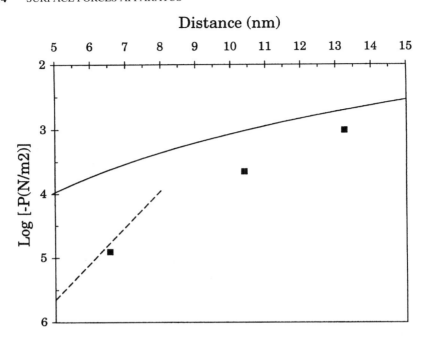

FIGURE 12.18 Attractive pressure between flat surfaces as a function of surface separation. The points are recalculated from the slope of the attractive force between mica surfaces neutralized by adsorption of the cationic polyelectrolyte MAPTAC. The solid line corresponds to the nonretarded van der Waals force between bare mica surfaces. The dashed line corresponds to the bridging attraction obtained from Monte Carlo simulations for a system mimicking MAPTAC-coated mica surfaces. For details see Dahlgren et al.[98]

of the adsorbed layer under a maximum compression is larger in 1 mM salt solution than in 0.1 mM, demonstrating the somewhat more extended conformation of the layer when the electrostatic segment–surface attraction is screened by addition of a small amount of electrolyte. The bridging attraction observed in 0.1 mM KBr has also decreased significantly. At even higher ionic strengths the range and magnitude of the repulsive force increases further. Clearly, the coiled structure of the polyelectrolyte in high ionic strength solutions is only perturbed upon adsorption. The distance dependence of the force is no longer consistent with that of a double-layer force. From this we concluded that the force is of steric origin due to tails and loops extending far out into the solution. However, the electrostatic interactions remain important since it drives the adsorption to the surface and strongly affect the conformation of the adsorbed polyelectrolyte. It should also be noted that when the ionic strength is 10 mM or higher the repulsive force measured on the first approach is significantly larger than on subsequent approaches.[100] Hence, the act of compressing the adsorbed layers results in an apparent "irreversible" compaction of the adsorbed layer.

In high ionic strength solutions it has been found that the forces measured depend on the history of the system.[100] The very long-range forces shown in Figure 12.19 are

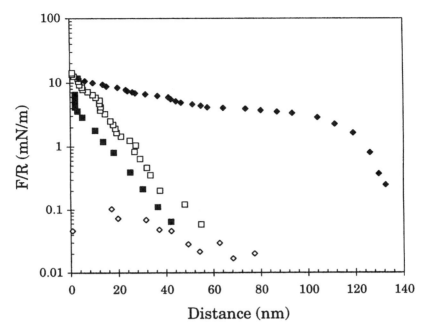

FIGURE 12.19 Force normalized by radius as a function of surface separation between mica surfaces across a 20-ppm PCMA solution also containing 0.1 mM KBr ($\diamond$), 1 mM KBr ($\blacksquare$), 10 mM KBr ($\square$), and 100 mM KBr ($\blacklozenge$). The adsorption was carried out from the respective electrolyte solution.

obtained when the polyelectrolyte is adsorbed from a solution with the indicated ionic strength. On the other hand, forces of much smaller range are observed when the polyelectrolyte first is adsorbed from a low ionic strength solution, which results in formation of a very thin adsorbed layer, and the ionic strength is increased afterwards. This indicates that the equilibrium structure of the adsorbed layer is reached extremely slowly (a week's equilibration time is not enough) due to the fact that the polyelectrolyte is bound to the surface by many segments and the adsorption energy per segment is large. The lateral interactions between adsorbed polyelectrolytes also contribute to the slow reequilibration. Another way of saying that the structure of the adsorbed layer changes very slowly when the ionic strength is changed is to say that the rate of spreading/despreading of each polyelectrolyte molecule in the layer is very slow. For this reason it is difficult to compare the forces measured at high ionic strength solutions with theoretical predictions based on equilibrium models.[100,101] The slow rearrangement of adsorbed polyelectrolytes on surfaces have implications for various applications. It has for instance been seen that when particles are flocculated with polyelectrolytes the floc properties depend on how the polyelectrolyte is dosed.[102]

12.3.9 Effect of Polyelectrolyte Charge Density

The concentration of polyelectrolytes needed to neutralize the mica surface charge can be determined by measuring the forces as a function of polyelectrolyte concentration (see, e.g., Figs. 12.16 and 12.17). At a given polyelectrolyte concentration no, or a very weak, repulsive double-layer force is observed (at 20 ppm in the case of PCMA, Fig. 12.16). This concentration we denote the charge neutralization concentration (cnc), which depends on the surface, the polyelectrolyte, and the ionic strength. The long-range forces acting between polyelectrolyte-coated mica surfaces at the cnc are shown for various polyelectrolytes in Figure 12.20. All data were obtained in low ionic strength solutions (≤ 0.1 mM). The results obtained for the 100% charged polyelectrolytes have been discussed above. We just note that the force curve determined for a polyelectrolyte with mean molecular weight 10^5 g/mol is nearly identical to the one measured for a higher molecular weight (1.5×10^6 g/mol). This is

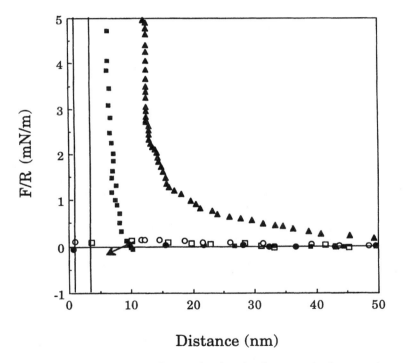

Distance (nm)

FIGURE 12.20 Force normalized by radius as a function of surface separation between mica surfaces across a 0.1 mM KBr solution also containing polyelectrolytes. The data shown are for two polyelectrolytes where each segment contains one charge. (●) are data for a MAPTAC solution, and (○) are for a PCMA solution; (□) represents data for a polyelectrolyte, AM-MAPTAC-30, having 30% of the segments charged, (■) represents data for a polyelectrolyte, AM-CMA-10, having 10% of the segments charged, and (▲) represents data for a polyelectrolyte, AM-MAPTAC-1, having 1% of the segments charged. In no case was any electrostatic double-layer force observed, indicating that the polyelectrolyte together with small ions within the layer neutralize the mica surface charge. Only force curves measured on approach are shown.

a consequence of the very thin adsorbed layer and in agreement with theoretical predictions.[101]

The forces between mica surfaces coated with a polyelectrolyte with 30% of the segments charged are in several respects similar to the ones observed for the 100% charged case. No long-range repulsion is observed whereas a bridging attraction dominates at distances below 10 to 15 nm. The position of the attractive minimum is located at a separation of about 3 to 4 nm, that is, about 2 nm further out than for the 100% charged case. No further compression of the adsorbed layer occurs under a high load. The magnitude of the pull-off force (normalized by surface radius) needed to separate the surfaces is between 2 and 5 mN/m, which is at least a factor of 30 lower than for the 100% charged case.

A further reduction in polyelectrolyte charge density to 10% of the segments results in a further increase in adsorbed layer thickness and decrease in the magnitude of the normalized pull-off force. A bridging attraction is still present in the distance range 10 to 20 nm. At very low polyelectrolyte charge densities (1% of the segments being charged), no bridging attraction is observed. Instead, a long-range steric force is the dominating feature. Some of the important features of the force curves shown in Figure 19.20 are summarized in Table 12.1. Clearly at the charge neutralization concentration in low ionic strength solutions, the layer thickness, layer compressibility, and range of the steric/bridging force decreases whereas the magnitude of the normalized pull-off force increases with polyelectrolyte charge density. These trends are also predicted from Monte Carlo simulations.[98]

An attempt has been made to correlate surface force data with the flocculation properties of latex particles.[103] When carrying out such a comparison, one has to take into account the difference in surface charge density of mica, 2.1×10^{18} charges m^{-2} (the surface used in the surface force measurements) and the latex particles, 0.7 to 0.5 $\times 10^{18}$ charges m^{-2}. With this in mind the following correlations can be found. When the distance between charges on the surface is similar to the distance between charges along the polyelectrolyte, thin adsorbed layers are obtained in low ionic strength solutions. This gives rise to strong attractive forces and flocs that are open and rigid.

TABLE 12.1 Effect of Polyelectrolyte Charge Density on Interaction Between Mica Surfaces at Charge Neutralization Concentration[a]

Polyelectrolyte Charge Density (%)	Adhesion Force (F/R) (mN/m)	Distance at Force Minimum (nm)	Distance at Maximum Compression (nm)	Range of Steric/Bridging Force (nm)
100	150–250	1–1.5	1–1.5	10–15
30	2–5	3–4	3–4	10–15
10	0.5–1	10–15	6–7	20
1	0	—	12–20	60–90

[a]Data was obtained in dilute electrolyte solutions ($C \leq 0.1$ mM). The adhesion force is the depth of the attractive force minimum.

These flocs are relatively easily broken by a high shear, but they also readily reflocculate when the shearing is stopped.[103] The use of less strongly charged polyelectrolytes results in thicker adsorbed layers and weaker attractive forces. This gives denser flocs where the particles, due to the thick adsorbed layer, have some freedom to move relative to each other. For these reasons the flocs are less easily broken down by shear. However, if the flocs are broken down they reflocculate less efficiently.

12.3.10 Aqueous Polyelectrolyte–Surfactant Mixtures

Both polyelectrolytes and surfactants are present in various technical applications for rheology control, solubilization of hydrophobic drugs or perfume, wetting, and particle deposition/removal.[104] For this reason polyelectrolytes mixed with oppositely charged surfactants have been extensively studied during the last decades. It is mainly the bulk association that has been the focus of the research and much has been learned about the initial association, the phase separation occurring at slightly higher surfactant concentrations, and the redispersion of the complexes that sometimes occur when the surfactant is present in large excess. There are several excellent reviews on the subject.[104–107] Here we will only consider aqueous solutions containing mixtures of a negatively charged surfactant and a highly charged cationic polyelectrolyte without any strongly hydrophobic groups. The important interactions to consider are electrostatic and hydrophobic. An electrostatic repulsion is present between the polyelectrolyte segments and between the surfactant head groups, whereas the surfactant head groups are attracted to the charged segments of the polyelectrolyte. The hydrophobic interaction acts mainly between the surfactant tails. Not surprisingly it has been shown by numerous studies that polyelectrolytes and oppositely charged surfactants associate strongly at surfactant concentrations well below, typically 1 to 3 orders of magnitude below, the critical micellar concentration and that the association step is highly cooperative.[105] Further, it has been shown that the changes in free energy, enthalpy, and entropy occurring during association between polyelectrolytes and oppositely charged surfactants are similar to the changes observed during micelle formation in bulk solution.[107–110] One reason why the polyelectrolyte–surfactant association occurs at such a low surfactant concentration is the high electrostatic potential just outside the polyelectrolyte chain that makes the local surfactant concentration much higher than in bulk solution.[110] It is fruitful to view the association process as a polyelectrolyte-induced micelle formation, where the polyelectrolyte is very efficient in screening the repulsion between the surfactant head groups.[111–113] The cooperativity in the association process is due to the hydrophobic interaction between the surfactant tails. From the many studies carried out at low surfactant concentrations, below the cmc of the surfactant, one may draw the following conclusions. The critical association concentration (cac) between polyelectrolytes and oppositely charged surfactants decreases with the surfactant chain length[114] and the polyelectrolyte charge density,[115] whereas the cac increases with the ionic strength of the solution.[116,117] These observations can be rationalized in terms of electrostatic and hydrophobic interactions.[111-113] It has also been observed that the cac is insensitive to the molecular weight of the polyelectrolyte (above a critical value),[107] and it decreases with polyelectrolyte chain flexibility[111] and hydrophobicity.[107] The binding isotherm for surfactants to polyelectrolytes can often be fitted by theoretical models taking into account nearest neighbor interactions.[107,118,119] However, a drawback is that the models contain several adjustable parameters.

Interactions between polyelectrolytes and surfactants well above the cmc have been much less studied. However, Dubin et al. have addressed this point in a series of studies.[106,120,121] One main finding was that only micelles with a charge density above a critical value interacted with the polyelectrolytes, demonstrating the importance of electrostatic forces.

Here we will discuss mixtures containing 20 ppm of the cationic polyelectrolyte PCMA and various amounts of the anionic surfactant sodium dodecyl sulfate (SDS). The turbidity difference between the polyelectrolyte–surfactant mixture and the corresponding polyelectrolyte-free surfactant solution (which always has a very low turbidity) is shown in Figure 12.21 as a function of total (i.e., bound and free) SDS concentration. We have chosen to plot the surfactant concentration in units of cmc in order to emphasis that the association between the surfactant and the polyelectrolyte occurs at concentrations much below the cmc for SDS (8.3×10^{-3} M). The measurements were carried out 10 and 30 min after mixing the solutions. The same results were obtained at both these times. Longer equilibration times (up to 24 h) changes the absolute turbidity values somewhat, but the results remain similar to the data shown in Figure 12.21.

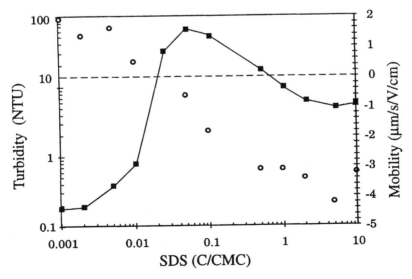

FIGURE 12.21 Turbidity (■) of solutions containing 20 ppm PCMA, 0.1 mM KBr, and various concentrations of SDS, expressed relative to the cmc. The electrophoretic mobility (○) of the aggregates are also shown as measured with a ZetaSizer 4. The mobility values obtained for solutions having a low turbidity are less reliable than those obtained at higher turbidities due to a lower particle counting rate.

There is a small but significant turbidity difference at an SDS concentration of 0.005 cmc, showing that some association has taken place. An increase in total SDS concentration to above 0.01 cmc results in a sharp increase in turbidity. The turbidity maximum is reached at an SDS concentration close to 0.05 cmc. At higher SDS concentrations the turbidity difference decreases again. Measurements of the electrophoretic mobility of the aggregates formed show that they are positively charged at low surfactant concentrations, uncharged close to the turbidity maximum, and negatively charged at higher SDS concentrations (Fig. 12.21). In the initial state of binding the electrophoretic mobility is hardly affected by the surfactant. This can be rationalized by considering the ion condensation next to highly charged polyelectrolytes,[122] and thus the initial binding of the surfactant can be viewed as an ion-exchange process, as also pointed out by Wei and Hudson.[107] At slightly higher surfactant concentrations the association of the polyelectrolyte and the surfactant results in a charge neutralization, and when this occurs the polyelectrolyte chains with the associated micelles form large aggregates. Up to this point the association is driven by both hydrophobic and electrostatic forces. Once the aggregates are uncharged, further incorporation of SDS occurs despite the unfavorable increase in electrostatic free energy of the aggregates. Hence, the further uptake of SDS is not favorable from an electrostatic point of view if one considers the aggregate as a whole. Instead, we propose that it is due to the same factors that drives the association between uncharged polymers and ionic surfactants where the polymer chain makes the surfactant self-association more favorable than in bulk by reducing the repulsion between the surfactant head groups and/or reducing the interfacial tension between the micelle core and the aqueous phase. In the present case it may also be possible that a local favorable electrostatic interaction between the surfactants and the polyelectrolyte chain contributes to the recharging of the aggregate as a whole. The decrease in turbidity at high surfactant concentrations (above 0.05 cmc) indicates that the recharging results in a decrease in size of, but not in a complete disintegration of, the polyelectrolyte–surfactant aggregates.

The association between polyelectrolytes and surfactants in bulk solution is rather extensively studied.[104–107] However, the association of polyelectrolytes and surfactants at a solid surface is a more novel research topic with only a few surface force studies being reported so far.[123–127] Most of these have been carried out with preadsorbed polyelectrolyte layers where the effects on the adsorbed layer and the surface forces of adding surfactants to the polyelectrolyte-free solution have been investigated. The technically more important situation where the polyelectrolytes and surfactants first are allowed to associate in bulk solution and then the aggregates are allowed to adsorb onto solid surfaces have been studied to a very limited degree.[128] Below we will consider some recent data concerned with the forces acting between negatively charged mica surfaces across solutions containing various PCMA/SDS mixtures. Now we have to bear in mind that the polyelectrolytes experience an electrostatic attraction to the surface whereas the surfactants are electrostatically repelled.

Even at very low surfactant concentrations the forces measured across the polyelectrolyte–surfactant mixture are rather different to the ones measured across the surfactant-free polyelectrolyte solution, and, unlike in the surfactant-free case, they show a clear dependence on the equilibration time (Fig. 12.22). In the SDS concentration range 0.001 cmc to 0.01 cmc a strong steric force dominates the long-range interaction immediately after introducing the polyelectrolyte–surfactant mixture into the measuring chamber. Once the surfaces are 10 nm apart, a bridging attraction pulls them together to a separation of about 1 nm, similar to the contact position found in the surfactant-free case. A typical result, obtained at 0.01 cmc SDS is displayed in Figure 12.22. It was also noted that, just as for the surfactant-free case at high ionic strength, the repulsion experienced on the first approach was more long-ranged than on subsequent approaches. Another finding was that the range of the repulsive force decreased with the equilibration time, and after more than 12 h the forces observed were only moderately stronger than in the surfactant-free case (Fig. 12.22). We interpret these observations such that even at these low SDS concentrations the surfactants associate with the polyelectrolyte and induce a more compact conformation in bulk solution, which prevents the formation of a very flat adsorbed layer. This interpretation is supported by light-scattering studies for similar systems, that show a decrease in hydrodynamic radius for the polyelectrolyte chain as a result of the initial binding of the surfactant.[129] The relation between bulk association, as studied with dynamic light scattering, and structures of adsorbed layers

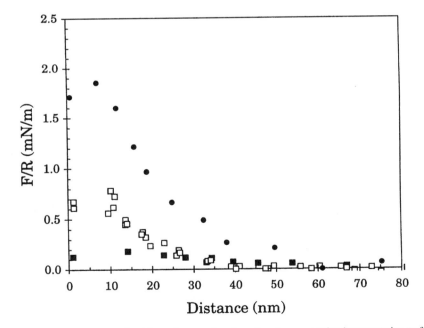

FIGURE 12.22 Force normalized by radius as a function of surface separation between mica surfaces immersed in a solution containing 20 ppm PCMA, 0.1 mM KBr and no SDS (■), 0.083 mM (0.01 cmc) SDS after a few hours adsorption (●), and after more than 12 hours adsorption (□). For the latter case the data points are from three different force measurements.

as studied with the surface force technique is discussed in some detail in a recent publication using a 30% charged polyelectrolyte and an anionic surfactant.[128]

When the layers are compressed, or when they are left undisturbed on the surface for a long time, their structures change and the polyelectrolyte tails become less extended normal to the surface. It seems likely that the change in layer structure is accompanied by an expulsion of SDS molecules, but this should at this stage be regarded as a hypothesis. We also note that the data obtained immediately after introducing the polyelectrolyte-surfactant mixture showed a rather poor reproducibility, whereas the data obtained after prolonged adsorption were reproducible (Fig. 12.22).

In the reminder of this section we do not consider kinetic effects but only discuss data obtained after more than 12 h of equilibration. When the surfactant concentration is above 0.02 cmc, the aggregates formed in solution are nearly uncharged. The data obtained are qualitatively reproducible, whereas we see slight differences in range and magnitude of the forces measured in different experiments.[130] Hence, below we only discuss the qualitative features of the force curves. The forces measured at an SDS concentration of 0.043 cmc is shown in Figure 12.23. Here a repulsive force is encountered once the surfaces are less than about 350 nm apart, demonstrating that the adsorbed layer on each surface is about 170 to 180 nm thick. In this situation it is

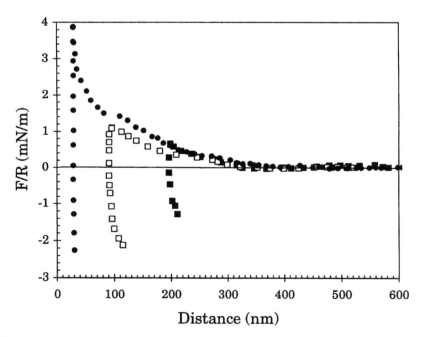

FIGURE 12.23   Force normalized by radius as a function of surface separation between mica surfaces immersed in a solution containing 20 ppm PCMA, 0.1 mM KBr, and 0.36 mM (0.043 cmc) SDS. The different symbols represent force data where the surfaces have been brought to different distances before being separated. The adsorbed layers rearranged very slowly when the surface separation was changed, preventing measurements of (quasi)equilibrium forces.

extremely difficult to measure equilibrium forces. The reason is that the adsorbed layer approaches its equilibrium structure very slowly after the surface separation has been changed. We note, however, that once the layers are brought into contact an attractive force develops between them as noticed upon separation. We propose that this attraction is due to formation of interlayer polyelectrolyte–micelle crosslinks. In Figure 12.23 three consecutive force curves are shown. First, the surfaces were brought to a separation of 200 nm and separated again. The next time they were brought to 90 nm before separation and in the final measurement to 25 nm. The attraction measured on separation increases somewhat with decreasing distance, which can be interpreted as a slight increase in the number of interlayer crosslinks.

Similar, but less long-range, forces are observed at an SDS concentration of 0.022 cmc and up to an SDS concentration of about 0.1 cmc. Hence, in this range of SDS concentrations it seems that polyelectrolyte–SDS aggregates adsorb to the surface, and that the thickness of the adsorbed layer increases when the net charge of the aggregates decreases. It is interesting that also aggregates with a net negative charge adsorb firmly to the negatively charged surface. This implies that during the adsorption process some SDS is leaving the aggregates, allowing the formation of strong surface–polyelectrolyte segment contacts.

At even higher SDS concentrations the character of the measured forces changes again. The measured interaction when the total SDS concentration equals the cmc is illustrated in Figure 12.24. At this high SDS concentration (quasi)equilibrium forces

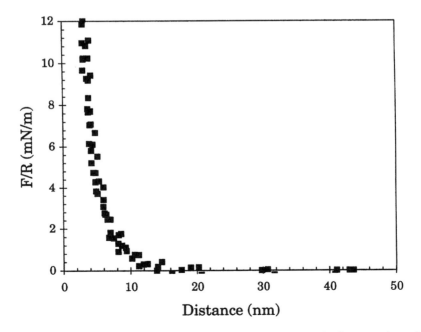

FIGURE 12.24 Force normalized by radius as a function of surface separation between mica surfaces immersed in a solution containing 20 ppm PCMA, 0.1 mM KBr, and 8.3 mM (1 cmc) SDS.

are again measured, and the force curve is purely repulsive. The most plausible interpretation of this finding is that a few single polyelectrolyte chains with associated SDS micelles adsorb to the negatively charged surface. Again, to facilitate the adsorption, some SDS molecules have to be released from the single polyelectrolyte chain–SDS complex. The part of the polyelectrolyte chain that is not in direct contact with the surface is associated with SDS and extends out into the solution giving rise to an electrosteric repulsion. We note that the high turbidity of the solution at these high SDS concentrations indicates that large aggregates (with a net negative charge) are present in the bulk solution. However, the short range of the measured forces shows that the large aggregates do not adsorb to the negatively charged surface.

One way of summarizing the results is to plot the thickness of the undisturbed layer, estimated as half the range of the steric repulsion or bridging attraction, as a function of SDS concentration; see Figure 12.25. At low surfactant concentrations, region 1, the polyelectrolyte adsorbs to the surface, and the surfactant mainly affects the range of the interaction through its effect on the polyelectrolyte conformation. Under a high compression the adsorbed layers become very flat, and an attraction due to polyelectrolytes bridging between the surfaces is important. In the middle region, region 2, aggregates adsorb to the surface. In this region the thickness of the layer increases with decreasing net aggregate charge in bulk solution, and attractive forces due to formation of interlayer polyelectrolyte–micelle crosslinks are of importance.

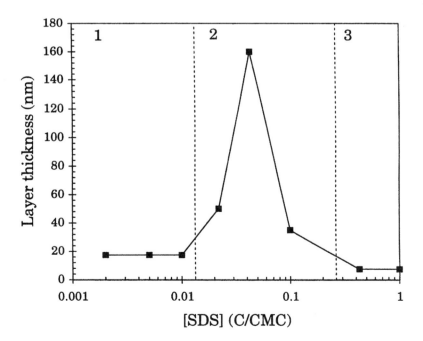

FIGURE 12.25 Undisturbed layer thickness defined as half the range of the steric or bridging force measured in solutions containing 20 ppm PCMA, 0.1 mM KBr, and various amounts of SDS.

At high surfactant concentrations, region 3, it seems that a few individual polyelectrolyte chains loaded with surfactants adsorb to the surface. The layers are relatively thin, and the interaction is characterized by an electrosteric repulsion. No or very weak attractive forces are observed in this case, suggesting that each polyelectrolyte chain is saturated with surfactants.

12.4 SUMMARY

Experimental and theoretical considerations of interacting surfaces and particles have during the last 30 years considerably advanced our understanding of colloidal systems and interactions. The (Derjaguin, Landau, Verwey, and Overbeek) DLVO theory, which takes into account additive contributions of van der Waals and double-layer forces, has been found to be a good starting point when discussing forces in polymer-free systems. However, even for such systems it does not give the complete picture, but short-range hydration or steric/protrusion forces are often of importance.[131] Further, more advanced theories that take into account ion–ion correlation effects have demonstrated that it is not strictly correct to treat van der Waals and double-layer forces as additive.[132] The importance of specific ion adsorption effects has also been recognized.[133]

Lattice mean-field and scaling theories together with Monte Carlo and molecular dynamics simulations have significantly improved our understanding of surface interactions in the presence of homopolymers, block copolymers, and polyelectrolytes. Experiments are often in qualitative agreement with theoretical predictions but quantitative comparisons are few. One inherent problem in such comparisons is that the exact values of the interaction parameters in bulk and at the surface are unknown. Another problem is that it is very difficult to obtain polymers with a sufficiently low polydispersity. This is important since there is a strong preferential adsorption of higher molecular weight species and of polymers that for other reasons have a higher affinity for the surface than the average polymer. To advance this field further and allow quantitative comparisons between experiments and theories, there is a need for systematic studies employing extremely well-defined polymers and polyelectrolytes.

Very few studies of surface interactions induced by polyelectrolyte–surfactant mixtures have so far been carried out. Here we have the situation that not even the main features are well established experimentally or theoretically. Perhaps the most rapid way to advance this area of research is to study both bulk and interfacial properties of the mixtures and elucidate the relation between bulk association and association at interfaces. In the next step, qualitative and quantitative comparisons with theoretical predictions of layer structures and surface interactions in such complex mixtures may be possible.

In this chapter we have not discussed dynamic interactions between surfaces coated with polymers, polyelectrolytes, or polymer–surfactant mixtures. Nevertheless this topic deserves extensive studies in the future. There is certainly a lack of knowledge about how systems that are trapped in a nonequilibrium situation change as they approach the equilibrium state. Also comparatively little is known about hydrodynamic and frictional forces between polymer coated surfaces, as well as the viscoelastic properties of polymer and polymer–surfactant coatings.

REFERENCES

1. J.N. Israelachvili and G.E. Adams, *J. Chem. Soc. Faraday Trans. 1* **74**, 975 (1978).

2. J.N. Israelachvili, *Intermolecular and Surface Forces*, Academic, London, 1991.

3. W.A. Ducker, T.J. Senden, and R.M. Pashley, *Langmuir* **8**, 1831 (1992).

4. D.L. Sober and J.Y. Walz, *Langmuir* **11**, 2352 (1995).

5. V. Bergeron and C.J. Radke, *Langmuir* **8**, 3020 (1992).

6. V.A. Parsegian, N. Fuller, and R.P. Rand, *Proc. Natl. Acad. Sci. USA* **76**, 2750 (1979).

7. P.M. Claesson, T. Ederth, V. Bergeron, and M.W. Rutland, *Adv. Colloid Interface Sci.* **67**, 119 (1996).

8. J.N. Israelachvili, R.K. Tandon, and L.R. White, *J. Colloid Interface Sci.* **78**, 430 (1980).

9. J. Klein, *Nature* **288**, 248 (1980).

10. J. Klein, *Adv. Colloid Interface Sci.* **16**, 101 (1982).

11. J. Klein and P.F. Luckham, *Nature* **300**, 429 (1982).

12. J. Klein, *J. Chem. Soc. Faraday Trans. 1* **79**, 99 (1983).

13. J. Klein and P.F. Luckham, *Nature* **308**, 836 (1984).

14. J.N. Israelachvili, M. Tirrell, J. Klein, and Y. Almog, *Macromolecules* **17**, 204 (1984).

15. J. Marra and M.L. Hair, *Macromolecules* **21**, 2349 (1988).

16. J. Marra and M.L. Hair, *Macromolecules* **21**, 2356 (1988).

17. K. Ingersent, J. Klein, and P. Pincus, *Macromolecules* **23**, 548 (1990).

18. G. Hadziioannou, S. Patel, S. Granick, and M. Tirrell, *J. Am. Chem. Soc.* **108**, 2869 (1986).

19. J. Marra and M.L. Hair, *Colloids Surf.* **34**, 215 (1988/89).

20. H.J. Taunton, C. Toprakcioglu, and J. Klein, *Macromolecules* **21**, 3333 (1988).

21. K. Schillén, P.M. Claesson, M. Malmsten, P. Linse, and C. Booth, *J. Phys. Chem.* **101**, 4238 (1997).

22. G.F. Belder, G. ten Brinke, and G. Hadziioannou, *Langmuir* **13**, 4102 (1997).

23. J. Klein and P.F. Luckham, *Colloids and Surf.* **10**, 65 (1984).

24. P.F. Luckham and J. Klein, *J. Chem. Soc. Faraday Trans. 1* **80**, 865 (1984).

25. T. Afshar-Rad, A.I. Bailey, P.F. Luckham, W. MacNaughtan, and D. Chapman, *Colloids Surf.* **25**, 263 (1987).

26. L.R. Dix, C. Toprakcioglu, and R.J. Davies, *Colloids Surf.* **31**, 147 (1988).

27. J. Marra and M.L. Hair, *J. Phys. Chem.* **92**, 6044 (1988).

28. P.M. Claesson and B. Ninham, *Langmuir* **8**, 1406 (1992).

29. J.M.H.M. Scheutjens and G.J. Fleer, *Adv. Colloid Interface Sci.* **16**, 361 (1982).

30. J.M.H.M. Scheutjens and G.J. Fleer, *Macromolecules* **18**, 1882 (1985).

31. G.J. Fleer, M.A. Cohen Stuart, J.M.H.M. Scheutjens, T. Cosgrove, and B. Vincent, *Polymers at Interfaces*, Chapman & Hall, London, 1993.

32. P.G. de Gennes, *Adv. Colloid Interface Sci.* **27**, 189 (1987).

33. T. Åkesson, C. Woodward, and B. Jönsson, *J. Chem. Phys.* **91**, 2461 (1989).

34. F. von Goeler and M. Muthukumar, *Macromolecules* **28**, 6608 (1995).

35. F. von Goeler and M. Muthukumar, *J. Chem. Phys.* **105**, 11335 (1996).

36. P. Pincus, *Macromolecules* **24**, 2912 (1991).

37. R.S. Ross and P. Pincus, *Macromolecules* **25**, 2177 (1992).

38. J.L. Parker, H.K. Christenson, and B.W. Ninham, *Rev. Sci. Instum.* **60**, 3135 (1989).

39. P.F. Luckham, *Adv. Colloid Interface Sci.* **34**, 191 (1991).

40. J.N. Israelachvili and P.M. McGuiggan, *J. Mater. Res.* **5**, 2223 (1990).

41. R.G. Horn, D.R. Clarke, and M.T. Clarkson, *J. Mater. Res.* **3**, 413 (1988).

42. R.G. Horn, D.T. Smith, and W. Haller, *Chem. Phys. Lett.* **162**, 404 (1989).

43. W.A. Ducker, *J. Am. Ceram. Soc.* **77**, 437 (1994).

44. J. Marra and J. Israelachvili, *Biochemistry* **24**, 4608 (1985).

45. P.M. Claesson, C.E. Blom, P.C. Herder, and B.W. Ninham, *J. Colloid Interface Sci.* **114**, 234 (1986).

46. P.M. Claesson, J.L. Parker, and J. Fröberg, *J. Disp. Sci. Techn.* **15**, 375 (1994).

47. J.L. Parker, D.L. Cho, and P.M. Claesson, *J. Phys. Chem.* **93**, 6121 (1989).

48. D. Cho, P.M. Claesson, C.G. Gölander, and K. Johansson, *J. Appl. Polymer Sci.* **41**, 1373 (1990).

49. J.L. Parker and H.K. Christenson, *J. Chem. Phys.* **88**, 8013 (1988).

50. G. Vigil, Z. Xu, S. Steinberg, and J. Israelachvili, *J. Colloid Interface Sci.* **165**, 367 (1994).

51. R. Neuman, J. Berg, and P.M. Claesson, *Nordic Pulp Paper Res. J.* **8**, 96 (1993).

52. J.N. Israelachvili, *J. Colloid Interface Sci.* **44**, 259 (1973).

53. J.A. de Feijter, J. Benjamins, and F.A. Veer, *Biopolymers* **17**, 1759 (1978).

54. A.M. Stewart and H.K. Christensson, *Meas. Sci. Tech.* **1**, 1301 (1990).

55. B. Derjaguin, *Kolloidnyi Zhurnal* **69**, 155 (1934).

56. J. Parker, *Prog. Surf. Sci.* **47**, 205 (1994).

57. P.-G. de Gennes, *Scaling Concepts in Polymer Physics*, Cornell University Press, Ithaca, NY, 1979.

58. P.M. Claesson, R. Kjellander, P. Stenius, and H.K. Christenson, *J. Chem. Soc. Faraday Trans. 1* **82**, 2735 (1986).

59. S. Saeki, N. Kuwahara, M. Nakata, and M. Kaneko, *Polymer* **17**, 685 (1976).

60. G.J.T. Tiddy, *Phys. Rep.* **57**, 3 (1980).

61. D.J. Mitchell, G.J.T. Tiddy, L. Waring, T. Bostock, and M.P. McDonald, *J. Chem. Soc. Faraday Trans. 1* **79**, 975 (1983).

62. P.-G. Nilsson, H. Wennerström, and B. Lindman, *J. Phys. Chem.* **87**, 1377 (1983).

63. B. Bergenståhl and P.M. Claesson, *Surface Forces in Emulsions* Marcel Dekker, New York, 1990.

64. A. Kabalnov and H. Wennerström, *Langmuir* **12**, 276 (1996).

65. S.I. Jeon, H.J. Lee, J.D. Andrade, and P.G. de Gennes, *J. Colloid Interface Sci.* **142**, 149 (1991).

66. S.I. Jeon and J.D. Andrade, *J. Colloid Interface Sci.* **142**, 159 (1991).

67. C.G. Gölander, J.N. Herron, K. Lim, P.M. Claesson, P. Stenius, and J.D. Andrade, in J.M. Harris, ed., *Poly(Ethylene Glycol) Chemistry: Biotechnical and Biomedical Applications*, Plenum, New York, 1992, Chap. 15.

68. K. Bergström, E. Österberg, K. Holmberg, A.S. Hoffman, T.P. Schuman, A. Kozlowski, and M.J. Harris, *J. Biomater. Sci. Polymer Edn.* **6**, 123 (1994).

69. K. Lim and J.N. Herron, in J.M. Harris, ed., *Poly(ethylene glycol) Chemistry. Biotechnical and Biomedical Applications*, Plenum, New York, 1992, pp. 29–56.

70. R. Kjellander, *J. Chem. Soc. Faraday Trans. 2* **78**, 2025 (1982).

71. R.E. Goldstein, *J. Chem. Phys.* **80**, 5340 (1984).

72. G. Karlström, *J. Phys. Chem.* **89**, 4962 (1985).

73. A. Matsuyama and F. Tanaka, *Phys. Rev. Lett.* **65**, 341 (1990).

74. P.-G. de Gennes, *C.R. Acad. Sci. Paris* **313**, 1117 (1991).

75. S. Bekiranov, R. Bruinsma, and P. Pincus, *Europhys. Lett.* **24**, 183 (1993).

76. P. Linse and B. Björling, *Macromolecules* **24**, 6700 (1991).

77. M. Björling, *Macromolecules* **25**, 3956 (1992).

78. M. Björling, G. Karlström, and P. Linse, *J. Phys. Chem.* **95**, 6706 (1991).

79. T. Ahlnäs, G. Karlström, and B. Lindman, *J. Phys. Chem.* **91**, 4030 (1987).

80. P.M. Claesson and C.G. Gölander, *J. Colloid Interface Sci.* **117**, 366 (1987).

81. M. Malmsten and B. Lindman, *Langmuir* **6**, 357 (1990).

82. M. Malmsten, P.M. Claesson, E. Pezron, and I. Pezron, *Langmuir* **6**, 1572 (1990).

83. M.A. Cohen Stuart and G.J. Fleer, *Ann. Rev. Mater. Sci.* **26**, 463 (1996).

84. M. Malmsten and P.M. Claesson, *Langmuir* **7**, 988 (1991).

85. M. Schaub, G. Wenz, G. Wegner, A. Stein, and D. Klemm, *Adv. Mater.* **5**, 919 (1993).

86. M. Holmberg, J. Berg, S. Stemme, L. Ödberg, J. Rasmusson, and P.M. Claesson, *J. Colloid Interface Sci.* **186**, 369 (1997).

87. C. Fringant, I. Tvaroska, K. Mazeau, M. Rinaudo, and J. Desbrieres, *Carbohydrate Res.* **278**, 27 (1995).

88. G.G. Fasman, *Handbook of Biochemistry and Molecular Biology. Physical and Chemical Data, Vol. 1*, CRC, Cleveland, 1976.

89. M. Holmberg, R. Wigren, R. Erlandsson, and P.M. Claesson, *Colloids Surfaces A* **129–130**, 175 (1997).

90. M.R. Böhmer, O.A. Evers, and J.M.H.M. Scheutjens, *Macromolecules* **23**, 2288 (1990).

91. V. Shubin and P. Linse, *J. Phys. Chem.* **99**, 1285 (1995).

92. P. Linse, *Macromolecules* **29**, 326 (1996).

93. V. Shubin and P. Linse, *Macromolecules* **30**, 5944 (1997).

94. F. von Goeler and M. Muthukumar, *J. Chem. Phys.* **100**, 7796 (1994).

95. P.M. Claesson, H.K. Christenson, J.M. Berg, and R.D. Neuman, *J. Colloid Interface Sci.* **172**, 415 (1995).

96. J.M. Berg, P.M. Claesson, and R.D. Neuman, *J. Colloid Interface Sci.* **161**, 182 (1993).

97. M.A.G. Dahlgren, P.M. Claesson, and R. Audebert, *J. Colloid Interface Sci.* **166**, 343 (1994).

98. M.A.G. Dahlgren, Å. Waltermo, E. Blomberg, P.M. Claesson, L. Sjöström, T. Åkesson, and B. Jönsson, *J. Phys. Chem.* **97**, 11769 (1993).

99. M. Muthukumar, *Electrophoresis* **17**, 1167 (1996).

100. M.A.G. Dahlgren, H.C.M. Hollenberg, and P.M. Claesson, *Langmuir* **11**, 4480 (1995).

101. M.A.G. Dahlgren and F.A.M. Leermakers, *Langmuir* **11**, 2996 (1995).

102. G. Khosa, M.Sc. Thesis, Imperial College of Science, Technology and Medicine, London, 1990.

103. P.M. Claesson, M.A.G. Dahlgren, and L. Eriksson, *Colloids Surf.* **93**, 293 (1994).

104. E.D. Goodard, in K.P. Ananthapadmanabhan and E.D. Goodard, eds. *Interactions of Surfactants with Polymers and Proteins*, CRC, Boca Raton, FL, 1993, pp. 395–414.

105. B. Lindman and K. Thalberg, in K.P. Ananthapadmanabhan and E.D. Goodard, eds. *Interactions of Surfactants with Polymers and Proteins*, CRC, Boca Raton, FL, 1993, pp. 203–276.

106. Y. Li and P.L. Dubin, *ACS Symposium Series*, **578**, 320 (1994).

107. Y.-C. Wei and S.M. Hudson, *J.M.S.-Rev. Macromol. Chem. Phys.* **C35**, 15 (1995).

108. A. Malovikova, K. Hayakawa, and J.C.T. Kwak, *J. Phys. Chem.* **88**, 1930 (1984).

109. J.P. Santerre, K. Hayakawa, and J.C.T. Kwak, *Colloids Surf.* **13**, 35 (1985).

110. J. Skerjanc, K. Kogei, and G. Vesnaver, *J. Phys. Chem.* **92**, 6382 (1988).

111. T. Wallin and P. Linse, *Langmuir* **12**, 305 (1996).

112. T. Wallin and P. Linse, *J. Phys. Chem.* **100**, 17873 (1996).

113. T. Wallin and P. Linse, *J. Phys. Chem.* **101**, 5506 (1997).

114. H. Okuzaki and Y. Osada, *Macromolecules* **27**, 502 (1994).

115. I. Satake, T. Takahashi, K. Hayakawa, T. Maeda, and M. Aoyagi, *Bull. Chem. Soc. Jpn.* **63**, 926 (1990).

116. K. Hayakawa and J.C.T. Kwak, *J. Phys. Chem.* **86**, 3866 (1982).

117. K. Hayakawa and J.C.T. Kwak, *J. Phys. Chem.* **87**, 506 (1983).

118. I. Satake and J.T. Yang, *Biopolymers* **15**, 2263 (1976).

119. Y.C. Wei and S.M. Hudson, *Macromolecules* **26**, 4151 (1993).

120. P.L. Dubin and R. Oteri, *J. Colloid Interface Sci.* **95**, 453 (1983).

121. D.R. Rigsbee and P.L. Dubin, *Langmuir* **12**, 1928 (1996).

122. G.S. Manning, *J. Phys. Chem.* **88**, 6654 (1984).

123. J.-F. Argillier, R. Ramachandran, W.C. Harris, and M. Tirell, *J. Colloid Interface Sci.* **146**, 242 (1991).

124. V. Shubin, P. Petrov, and B. Lindman, *Colloid Polym. Sci.* **272**, 1590 (1994).

125. P.M. Claesson, A. Dedinaite, E. Blomberg, and V.G. Sergeyev, *Ber. Bunsenges. Phys. Chem.* **100**, 1008 (1996).

126. P.M. Claesson, A. Dedinaite, M. Fielden, U.R.M. Kjellin, and R. Audebert, *Prog. Colloid Polymer Sci.* **106**, 24 (1997).

127. U.R.M. Kjellin, P.M. Claesson, and R. Audebert, *J. Colloid Interface Sci.* **190**, 476 (1997).

128. P.M. Claesson, M. Fielden, A. Dedinaite, W. Brown, and J. Fundin, *J. Phys. Chem. B* **102**, 1270 (1998).

129. J. Fundin, W. Brown, and M.S. Vethamuthu, *Macromolecules* **29**, 1195 (1996).

130. P.M. Claesson, A. Dedinaite, M. Fielden, M. Kjellin, and R. Audebert, *Prog. Colloid Polym. Sci.* **106**, 24 (1997).

131. J.N. Israelachvili and H. Wennerström, *J. Phys. Chem.* **96**, 520 (1992).

132. R. Kjellander and S. Marcelja, *J. Phys. Chem.* **90**, 1230 (1986).

133. B.W. Ninham and V. Yaminsky, *Langmuir* **13**, 2097 (1997).

13 Scanning Angle Reflectometry and Its Application to Polymer Adsorption and Coadsorption with Surfactants

ROBERT D. TILTON

Department of Chemical Engineering, Colloids, Polymers and Surfaces Program, Carnegie Mellon University, Pittsburgh, Pennsylvania 15213

13.1 INTRODUCTION

The surface excess concentration, perhaps the most rudimentary quantity pertaining to polymer adsorption, must be known in order to fully appreciate the microscopic behavior of adsorbed polymer layers and the role adsorption plays in the control of colloidal fluid properties.[1] For a basic introduction to the concept of surface excess concentration, see, for example, Adamson and Gast.[2] The hydrodynamic and electrokinetic properties of polymer-decorated interfaces are intimately linked to the surface excess concentration. Colloid stability in polymer solutions is similarly dependent on the surface excess concentration, particularly when steric stabilization, bridging flocculation, or depletion flocculation could each potentially compete to control stability. In matters of colloid stability, *adsorption kinetics* can be crucial. Whereas a rapidly evolved steric layer promotes stabilization, slow adsorption may lead to bridging flocculation. This is one of the main motivations for quantifying polymer adsorption kinetics and hence for applying optical reflectometry techniques to problems of polymers at interfaces. On solid surfaces, adsorbed macromolecular layers very often experience slow structural relaxations over hours or days, and their notoriously slow *desorption kinetics* lead to effectively irreversible adsorption. These adsorbed layers reside in nonequilibrium states over time scales relevant to the stability of colloidal fluids. Since the macroscopic consequences of polymer–colloid

Colloid–Polymer Interactions: From Fundamentals to Practice, Edited by Raymond S. Farinato and Paul L. Dubin

ISBN 0-471-24316-7   © 1999 John Wiley & Sons, Inc.

interactions, for example, colloid stability, may be governed by transient or kinetically trapped adsorbed layer states, rather than the thermodynamic equilibrium state, it is important to measure not only the extent of adsorption but also the *kinetics* and reversibility of adsorption.

Optical reflectometry, a noninvasive technique requiring no external probes, is well suited for measuring adsorption and desorption kinetics over time scales ranging from seconds to days. Reflectometry has been applied in different configurations to measure polymer, protein, and surfactant adsorption, as well as coadsorption, at both solid–liquid and liquid–vapor interfaces; see, for example, various studies available.[3–16] It has also been used to measure colloid deposition on solid surfaces.[17,18] A closely related but usually nonquantitative technique, Brewster angle microscopy, is increasingly popular for nonfluorescent imaging of monolayers at air–water interfaces.[19,20]

This chapter begins with a brief discussion of the reflection of polarized light from multilayer, striated interfaces, the basis of reflectometry measurements. The discussion follows well-known treatises on the topic, for example, Azzam and Bashara.[21] This presentation will motivate a discussion of several modes for data collection and analysis and also the subject of uncertainty in the data analysis. The chapter concludes with applications of reflectometry to the measurement of polymer adsorption and coadsorption with low-molecular-weight amphiphiles. These are intended to demonstrate some capabilities of the technique, the most important being in situ measurements of adsorption kinetics.

13.2   REFLECTION OF POLARIZED LIGHT FROM STRIATED ISOTROPIC INTERFACES

Reflectometry is based on the reflectivity of a striated interface illuminated by a polarized light source. The reflectivity depends on the thickness, d, and refractive index, n, of each individual interfacial layer, and also on the angle of incidence. The following discussion refers to the schematic interface diagram in Figure 13.1 which also provides basic definitions and vocabulary.

The basic principle of reflectometry is quite simple. Since adsorption changes the refractive index profile near a material interface, it will change the reflectivity of that interface. The change in reflectivity is most pronounced when parallel (p) polarized light is incident to the surface at the Brewster angle. So, by measuring changes in reflectivity at the Brewster angle, it is possible to extract sufficient information about the optical properties of the adsorbed layer to calculate a surface excess concentration.

The Brewster angle is that angle of incidence where the reflectivity of p-polarized light is zero for a sharp or Fresnel interface between two semi-infinite media. The Brewster angle,

$$\theta_B = \tan^{-1}(n_t/n_i)$$

$$(13.1)$$

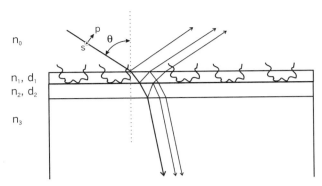

FIGURE 13.1 Schematic illustration of a striated interface comprising two intermediate layers 1 and 2 separating two semi-infinite media 0 and 3. Here layer 1 is the adsorbed polymer layer having an average refractive index n_1 and effective thickness d_1. Layer 2 might be an oxide layer on a silicon wafer, for example. The two semi-infinite media have refractive indices n_0 and n_3. The medium of refractive index n_0 contains the incident and reflected beams, while the medium of refractive index n_3 contains the transmitted beam. Interference of rays reflected at the various interfaces determines the reflectivity of the striated interface. The dotted line is normal to the interface. The laser beam is incident from the left side of the figure. The plane of incidence is defined as that plane containing the incident beam and the normal to the interface—that is, lying in this page. If the electric field vector of a linearly polarized laser beam is parallel to the plane of incidence, it is said to be parallel (p) polarized, as shown by the arrow. If it is perpendicular to the plane of incidence, it is perpendicular (s) polarized, shown by the circle. The angle of incidence, θ, is measured from the surface normal.

depends on the refractive indexes of the media containing the transmitted (n_t) and incident (n_i) light. In standard reflectometry methods, the reflectivity of p-polarized light, R_p, is measured as a function of the incident angle over a small range of angles, typically $0.5°$ to $4°$, centered on θ_B. The intensity reflectivity is defined as

$$R_p(\theta) \equiv \frac{I_p(\theta)}{I_{0p}(\theta)} \tag{13.2}$$

where $I_p(\theta)$ and $I_{0p}(\theta)$ are the reflected and incident intensities, respectively. Depending on how the experiment is performed, I_{0p} may or may not be a function of the incident angle, as will be discussed in Section 13.3.

Following Azzam and Bashara,[21] the reflectivity of a striated interface consisting of M layers of thickness d_i and refractive index n_i between two semi-infinite bulk media of refractive index n_0 and n_{M+1} can be calculated by the matrix method of Abelès. The electric field at any position z is represented vectorially as

$$\mathbf{E}(z) = \begin{bmatrix} E^+(z) \\ E^-(z) \end{bmatrix} \tag{13.3}$$

where the superscripts $+$ and $-$ indicate light propagating in the transmitted and reflected directions, respectively, and z is counted in the direction normal to the interface. In Figure 13.1, the $+$ or transmission direction is "downward and to the right" and the $-$ or reflection direction is "upward and to the right ." Thus, $E^+(z)$ is the electric field amplitude for light traveling in the direction of transmission, at a particular position z. Note that $\mathbf{E}(z)$ is not the electric field vector per se; it is merely a vector whose elements are electric field amplitudes.

The electric field amplitudes at any two different planes j and k are related by a 2×2 scattering matrix, $\mathbf{S}$, such that $\mathbf{E}(z_j) = \mathbf{S}\mathbf{E}(z_k)$. Thus, the electric field in the semi-infinite medium containing the incident laser beam (0) is related to the electric field in the other semi-infinite medium ($M + 1$) as

$$\begin{bmatrix} E^+(0) \\ E^-(0) \end{bmatrix} = \mathbf{S} \begin{bmatrix} E^+(M+1) \\ 0 \end{bmatrix} \tag{13.4}$$

The second element of the vector on the right-hand side is zero because there is no light traveling in the direction of reflection in medium $M + 1$; $\mathbf{S}$ includes the effects of refraction and reflection at the interfaces between all adjoining layers, described by "interface matrices" $\mathbf{I}$ as well as the effects of phase changes as the light passes through each layer. The latter are described by the "layer matrices" $\mathbf{L}$. In the following, it suffices to use z to refer to some particular medium, that is, either of the two semi-infinite media or any of the M layers. Each interface between layers z and $z + 1$ has an associated interface matrix $\mathbf{I}_{z(z+1)}$, and each layer z has an associated layer matrix $\mathbf{L}_z$. The following expressions are used to construct $\mathbf{S}$:

$$\mathbf{S} = \mathbf{I}_{01} \prod_{z=1}^{M} \mathbf{L}_z \mathbf{I}_{z(z+1)} \tag{13.5}$$

$$\mathbf{I}_{z(z+1)} = \begin{bmatrix} 1 & r_{z(z+1)} \\ r_{z(z+1)} & 1 \end{bmatrix} \frac{1}{t_{z(z+1)}} \tag{13.6}$$

$$\mathbf{L}_z = \begin{bmatrix} \exp(j\beta_z) & 0 \\ 0 & \exp(-j\beta_z) \end{bmatrix} \tag{13.7}$$

where $r_{z(z+1)}$ and $t_{z(z+1)}$ are the Fresnel reflection and transmission coefficients, respectively, for the interface between layers of refractive index n_z and n_{z+1}. For parallel polarized light, these are

$$r_{z(z+1)} = \frac{\tan(\theta_z - \theta_{z+1})}{\tan(\theta_z + \theta_{z+1})} \tag{13.8}$$

$$t_{z(z+1)} = \frac{2\sin\theta_{z+1}\cos\theta_z}{\sin(\theta_z + \theta_{z+1})\cos(\theta_z - \theta_{z+1})} \tag{13.9}$$

while for perpendicular polarized light, the reflection and transmission coefficients are

$$r_{z(z+1)} = \frac{-\sin(\theta_z - \theta_{z+1})}{\sin(\theta_z + \theta_{z+1})} \tag{13.10}$$

$$t_{z(z+1)} = \frac{2\sin\theta_{z+1}\cos\theta_z}{\sin(\theta_z + \theta_{z+1})} \tag{13.11}$$

The angles θ_z and θ_{z+1} are the angles of incidence in layers z and $z + 1$, respectively. These are dictated by Snell's law $n_z\sin\theta_z = n_{z+1}\sin\theta_{z+1}$, thereby accounting for each of the refractive indices in the striated interface. The phase thickness of layer z, β_z is defined as

$$\beta_z = 2\pi\left(\frac{d_z}{\lambda_0}\right)n_z\cos\theta_z \tag{13.12}$$

where λ_0 is the wavelength in vacuum. The amplitude reflection coefficient of the entire striated interface, r, is defined as the ratio of the reflected and incident electric field amplitudes in medium 0. This is readily shown to be equal to the ratio of two elements from the scattering matrix $S_{(\text{row/column})}$

$$r = \frac{E^-(0)}{E^+(0)} = \frac{S_{21}}{S_{11}} \tag{13.13}$$

Note that all of the transmission coefficients included in Eq. (13.6) cancel in Eq. (13.13) and hence do not have to be calculated. Intensity, not electric field amplitude, is the experimentally measurable quantity. Thus, the intensity reflectivity

$$R = \frac{I}{I_0} = r \cdot r^* \tag{13.14}$$

where the asterisk denotes the complex conjugate, is the quantity to be calculated from the raw intensity data. To calculate reflectivities for either s or p polarization, one

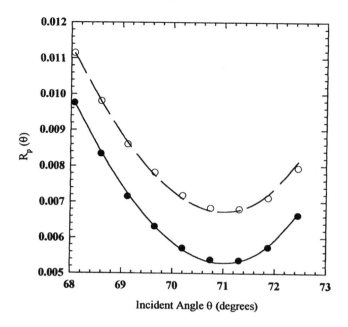

FIGURE 13.2   Reflectivity profiles for p-polarized light before (closed symbols) and after (open symbols) adsorption of cetyltrimethylammonium bromide surfactant on silica (oxide layer thermally grown on silicon wafer) from aqueous solution. Curves are regression results generated by the Abelès matrix method: d_{oxide} = 29.1 nm; adsorbed layer results: n_{av} = 1.380, d_{av} = 8.95 nm, Γ = 2.80 mg/m^2, corresponding to approximately 20 Å^2/molecule as expected for a defective (i.e., interpenetrating) bilayer of adsorbed surfactants. Data from Pagac.[15]

simply chooses the appropriate Fresnel reflection coefficient expression, either Eq. (13.8) or (13.10).

Examples of reflectivity profiles are shown in Figure 13.2. These correspond to a layer of silicon oxide formed on a silicon wafer and to a cetyltrimethylammonium bromide (CTAB) surfactant layer adsorbed on the same oxide layer.[17] Water is the incident medium in both cases. The refractive index of the oxide is 1.46, while the underlying silicon has a complex refractive index. At the HeNe laser wavelength 632.8 nm, n_{si} = 3.882 + 0.019i, where i is $\sqrt{-1}$. The refractive index of silicon is tabulated for different wavelengths by Palik.[22]

13.3   REFLECTOMETRY MEASUREMENTS

A simple instrument used in the author's laboratory to study adsorption to solid–liquid interfaces is illustrated schematically in Figure 13.3.[11] To study adsorption in situ, a slit-flow cell is constructed where the upper wall is formed by one face of a fused-silica prism while the lower surface is a silicon wafer. The advantages of a flow cell are (1) adsorption kinetics can be measured under well-defined hydrodynamic

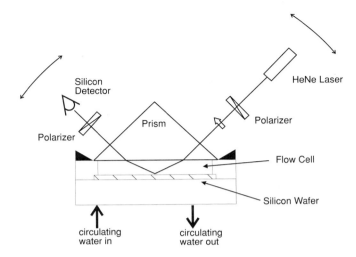

FIGURE 13.3 Schematic diagram of a scanning angle reflectometer for solid–liquid interfaces. The HeNe laser and detector are mounted on two separate optical rails that are rotated by equal but opposite angles via a gear mechanism to change the angle of incidence. The polarization vector is indicated by the arrow superimposed on the beam. The flow cell housing is thermostatted by circulating water. The direction of flow in the rectangular flow cell ($2.47 \times 1.27 \times 0.145$ cm) is perpendicular to the plane of incidence.

conditions and (2) the solution contents can be changed rapidly by pumping from one of any number of different feed reservoirs. The latter makes it possible to examine layer response to compositional perturbations. Constant temperature is maintained by circulating thermostatted water through the flow cell housing. The HeNe laser beam propagates through the prism, refracts according to Snell's law at the prism–solution interface and strikes the surface of the wafer at a prescribed angle close to θ_B.

Reflected light then propagates through the solution. into the prism and out to a silicon detector (laser power meter). It is important that the prism angles be cut so that the angle of incidence to the prism–solution interface is far removed from its own Brewster angle. Otherwise, adsorption to the prism would decrease the transmission of light from the prism into the solution.

The detector signal $I_p(\theta)$ is collected by a computer data acquisition system. The laser and the detector are mounted on two rails that may be rotated simultaneously by equal but opposite angles to vary the angle of incidence. The beam and the detector lie on concentric radii. Because a prism with flat surfaces is used rather than a hemispherical prism, a small amount of beam displacement occurs while sweeping the full range of incident angles. The beam displacement depends on the depth of the flow cell and the full range of angles sampled. Typically, in the author's laboratory, the beam displacement is less than one beam radius.

The incident intensity does not depend on angle in this experiment. To measure I_{0p}, the two rails are fully rotated to the horizontal position, allowing the beam to bypass the prism and strike the detector directly. The sample $R_p(\theta)$ profiles shown in Figure 13.2 were obtained by this procedure.

Since one scanning angle measurement provides one value for the surface excess concentration, this procedure is not well suited for kinetics. This is because mechanically changing the angle of incidence to measure the full reflectivity profile takes several minutes. This time resolution would be unacceptable for kinetic studies, and would be tiresome over a long adsorption experiment. To measure kinetics, one sets the angle of incidence to θ_B and continuously monitors the reflectivity at that single angle.[5] The analysis required to convert the reflectivity to surface excess concentration will be described later in Section 13.4.

For the instrument configuration shown in Figure 13.3, I_{0p} is independent of angle. This is not the case in the dynamic scanning angle reflectometry method. This was developed by Leermakers and Gast[8] to enable rapid, repetitive acquisition of full reflectivity profiles. Their instrument records the reflectivity profile by focusing the incident beam in the plane of incidence and spatially resolving the reflected intensities with a linear photodiode array. Each pixel on the array detects light reflected at a slightly different angle of incidence. Since the incident beam is focused, I_{0p} is a function of θ. The photodiode array samples the entire angle range in milliseconds, so the time resolution of this instrument is limited only by the array scan time.

The photodiode array reports reflected intensity as a function of pixel number. To convert this $I_p(\theta)$ to $R_p(\theta)$, first the incident intensity profile $I_{0p}(\theta)$ must be measured and stored on the computer. In the Leermakers and Gast[8] experiments, polymer adsorption was measured directly at a prism–solution interface (unlike the instrument in Fig. 13.3 where the incident beam propagates through the solution before striking the surface of interest), so $I_{0p}(\theta)$ could be measured precisely by totally internally reflecting the incident focused beam at the prism–air interface before filling the flow cell. The final step is calibration of the array to convert pixel number to angle. This is done by noting the pixel of minimum intensity that corresponds to the Brewster angle for a series of liquids of varying refractive index, applying Eq. (13.1) for θ_B.

This dynamic scanning angle reflectometry method was modified in the author's laboratory to study adsorption at the air–water interface.[10,14] See Figure 13.4. The modifications were motivated mainly by difficulties associated with the free air–water interface, particularly water-level control and water condensation on optics housed inside a humid enclosure. One problem was the difficulty of measuring $I_{0p}(\theta)$ without the benefit of a total internal reflection. Another more serious problem was the potential instability of the incident profile over the course of a long experiment.

To circumvent these difficulties the modified procedure calls for separately measuring the reflected intensities of both p- and s-polarized light instead of measuring only the parallel polarized reflectivity profile. From these one calculates the normalized parallel reflectivity (NPR).

$$\text{NPR}(\theta) \equiv \frac{I_p(\theta)}{I_p(\theta) + I_s(\theta)} \tag{13.15}$$

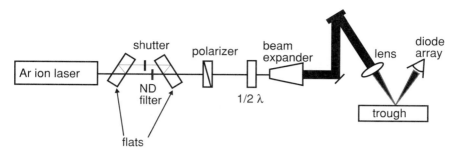

FIGURE 13.4 Schematic of a scanning angle reflectometer used for the air–water interface. Back reflections in the optical flats are exploited to separate an argon ion laser beam into a bright beam and a highly attenuated beam. The intensity of the bright beam may be adjusted by inserting a neutral density filter. A shutter slides between the two to select either the bright or attenuated beam for reflectivity measurements. The beam is initially p polarized. The polarization may be changed from p to s polarization by rotating the half-wave plate 45°. Because $R_s \gg R_p$ near the Brewster angle, the attenuated and bright beams are used when measuring I_s and I_p, respectively. The beam is expanded and focused by a cylindrical lens, such that the reflected beam is diverging when it strikes the photodiode array. Each pixel on the array detects light at a slightly different angle of incidence (centered approximately on the Brewster angle for the air–water interface).

where I_p and I_s are the reflected intensities of parallel and perpendicular polarized light, respectively. Normalized parallel reflectivity can be expressed in terms of the theoretically calculated reflectivities:

$$\text{NPR}(\theta) = \frac{R_p(\theta)}{R_p(\theta) + fR_s(\theta)} \tag{13.16}$$

where f is a measurable quantity that accounts for the relative intensities of incident p- and s-polarized light as well as polarization-dependent optical losses. As shown in Figure 13.4 the incident beam is either perfectly s-polarized or perfectly p-polarized to measure I_s or I_p. It is not elliptically polarized or even linearly polarized at an offset angle.

NPR does not depend on the magnitude of the incident intensity and is therefore completely insensitive to changes in the laser output or beam profile. Because of the way in which p- and s-polarized incident beams are produced (see Fig. 13.4) f would not change, even if the overall laser power or Gaussian beam profile were to change during an experiment. In fact, a filter can be partly inserted into the beam with no ill effect on the NPR measurement. This is a convenient, if obvious, consequence of the self-normalizing character of NPR measurements. The NPR profiles may be predicted from Eq. (13.16) using the Abelès matrix method described above simply by inserting the appropriate Fresnel coefficients for s- or p-polarized light. Sample experimental NPR profiles obtained for polystyrene–poly(ethyleneoxide) *block copolymers* (PS–PEO) adsorbed to (penetrating, actually) insoluble *lipid monolayers* at the air–water interface are shown in Figure 13.5.

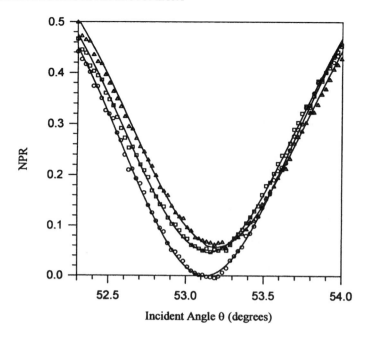

FIGURE 13.5 Normalized parallel reflectivity profiles for lipid monolayers penetrated by polystyrene-poly(ethyleneoxide) diblock copolymers at the air–water interface. Symbols are data; curves are regression results. The bottom profile is for the clean air–water interface; the middle profile is the unpenetrated DPPC monolayer prior to introducing polymer into the subphase; the top profile is after polymer penetrated the same monolayer from the subphase. The lipid monolayer is spread in the two-phase coexistence regime at 62 Å²/molecule. Γ_{PS-PEO} = 1.43 mg/m². After Charron and Tilton;[10] used by permission.

Dijt et al.[7,23] describe a related method employed to measure polymer adsorption to oxidized silicon wafers. Rather than focusing the beam and measuring a self-normalized angular reflectivity profile, they used an incident beam linearly polarized at some offset angle, and they split the reflected beam with a polarizing beam splitter, directing the s- and p-polarized components to two separate detectors. For analysis they normalized the data as

$$S = \frac{I_p}{I_s} \tag{13.17}$$

without varying the angle of incidence. To determine surface excess concentrations Γ, they calculated an approximate linear relationship between S and Γ. The proportionality constant depends on the oxide layer thickness (or whatever intermediate film there may be). Determination of that thickness requires a calibration against a reference technique (e.g., ellipsometry). Use of a scanning angle reflectometry method would eliminate the need for external calibration, since the scanning angle method can itself be used to measure the oxide layer thickness. Scanning angle reflectometry measurements of oxide layers on silicon wafers have been tested against ellipsometry measurements.[24] Agreement between the two is excellent.

Whether one uses the $R_p(\theta)$ profile or the NPR profile, one regresses the experimental reflectivity data to determine the average thickness and refractive index of the adsorbed layers. Typically a simple homogeneous film model is adopted for the nonlinear least-squares regression of the measured profiles. In the case of an intermediate film, such as the silica layer, one employs a two-film model, one for the oxide and one for the adsorbed layer. One could choose a more realistically detailed optical model, for example, a parabolic profile such as might be predicted by a mean-field polymer adsorption model, but there is usually no benefit to be gained by doing this. The simplest model, a homogeneous adsorbed film of uniform refractive index suffices, to determine the surface excess concentration. Unlike neutron reflectometry, optical reflectometry cannot distinguish between a simple homogeneous film and a more subtle layer profile, and it is probably pointless to attempt such a distinction. The surface excess concentration is the true objective of an optical reflectometry measurement. By choosing a single homogeneous film model, data regression will provide an optical average layer thickness (d_{av}) and refractive index (n_{av}) of the adsorbed layer. Since these parameters are model dependent, their physical significance is limited to a first-order approximation to the extension and density of the layer. Furthermore, the thickness and index obtained in this way are each subject to considerable experimental scatter. This occurrence is familiar in ellipsometry as well. Errors in index and thickness are mutually compensating, such that the quantity $(n_{av} - n_0)d_{av}$, the second refractive index n_0 referring to the bulk solution, is invariant.

Regarding the choice of optical model, one notes that the substrate material itself may be multilayered. This is the case for the oxidized silicon wafers often used for reflectometry. (Silicon offers the advantage of a large refractive index contrast to polymer solutions, dramatically improving resolution.) Molecules adsorb to an oxide layer that overlays the bulk silicon, rather than to the bare silicon. In fact, as will be discussed below, the sensitivity can be improved somewhat by thermally growing the oxide layer to larger thickness. In the presence of an intermediate layer (of known refractive index), its thickness prior to adsorption is measured both under water and in dry air. Equal film thicknesses must be obtained if the instrument is properly aligned; a good test before each experiment. After adsorption, an optical model of two homogeneous films is adopted in order to determine the unknown thickness and refractive index of the adsorbed layer.

The regression provides the average adsorbed layer optical properties. From these, one calculates the surface excess concentration Γ via:

$$\Gamma = \frac{(n_{av} - n_0)d_{av}}{dn/dc} \tag{13.18}$$

where dn/dc is the refractive index increment of the adsorbing solute. The surface excess concentration is invariant, that is, independent of the choice of optical model and insensitive to experimental scatter in the regressed thickness and refractive index. This is demonstrated in the Section 13.3.1.

Equation (13.18) derives from the mathematical form of the "effective" or optical average refractive index and thickness of an adsorbed layer. For any layer represented by an arbitrary refractive index profile $n(z)$, these quantities are given by[25,26]

$$n_{av} = \frac{\int\limits_0^\infty n(z)[n(z) - n_0]\, dz}{\int\limits_0^\infty [n(z) - n_0]\, dz} \tag{13.19}$$

$$d_{av} = \frac{\int\limits_0^\infty [n(z) - n_0]\, dz}{n_{av} - n_0} \tag{13.20}$$

The refractive index anywhere in the layer is related to the local concentration as

$$n(c) = n_{solvent} + c\,\frac{dn}{dc} \tag{13.21}$$

For most practical purposes, experimental bulk solutions are usually sufficiently dilute that n_0 is equal to $n_{solvent}$. The surface excess concentration of an adsorbing solute is by Gibbs's definition

$$\Gamma = \int\limits_0^\infty [c(z) - c_0]\, dz \tag{13.22}$$

where $c(z)$ and c_0 are, respectively, the solute concentrations at some distance z from the surface and the bulk concentration, that is, as z approaches infinity. Equation (13.18) is obtained by substituting Eqs. (13.19) to (13.21) into (13.22).

The refractive index increment, the dependence of refractive index on solution concentration, is an independently measured material property. Usually one uses a differential refractometer to measure the refractive indices of a series of concentrated solutions of the surface-active solute and assumes that the refractive index increment is constant, even at the very high local concentrations found in adsorbed layers. This, assumption has been tested and found acceptable for proteins.[26] For self-assembling solutes, a fortuitous consequence arises from the need to use concentrated solutions for differential refractometry measurements. These concentrations are usually well in excess of the critical micelle concentration. As a result, the refractive index increment is measured for surfactants assembled in micelles, that is, already present at local densities approximating those of adsorbed layers.

It is sometimes possible to use scanning angle reflectometry to measure the refractive index increment itself. This was done for the water-insoluble lipid dipalmitoylphosphatidylcholine (DPPC) by spreading it to a known surface excess concentration at the air–water interface. Instead of the usual reflectometry process of determining Γ from n_{av}, d_{av}, and the known dn/dc via Eq. (13.18) one now uses the same equation to calculate dn/dc from n_{av}, d_{av}, and the known Γ. The refractive index increment of DPPC is 0.119 ± 0.004 cm³/g, independent of the monolayer phase state.[10]

13.3.1 Insensitivity of the Surface Excess Concentration to the Choice of Optical Model

The adequacy of the single homogeneous film model to represent more complicated adsorbed layer profiles is summarized in Table 13.1. This table presents the single-film regression analyses of simulated reflectivity data for nonuniform layers. The layers were adsorbed on 20-nm-thick oxide layers atop silicon wafers, a typical oxide layer thickness for experiments conducted in the author's laboratory. The solution refractive index was 1.333 in all cases.

Several arbitrary refractive index profiles were considered, including a linear decay, a truncated linear decay, two Gaussian profiles with different standard deviations, and a linear "ramp" profile. To analyze these profiles, each one was

TABLE 13.1. Application of Single-Film Optical Model to Arbitrary Adsorbed Layer Refractive Index Profiles

	Profile	True Γ (mg/m²)	Single Film Γ (mg/m²)	Error (%)
Linear decay	$0 < z < 2$ nm: $n(z) = 1.500 - 0.084z$ $z \geq 2$ nm: $n = 1.333$	1.22	1.13	7
Truncated linear decay	$0 < z < 2$ nm: $n(z) = 1.500 - 0.042z$ $z \geq 2$ nm: $n = 1.333$	1.83	1.74	5
Gaussian	$n(z) = 1.333 + (1.500 - 1.333)e^{-z^2}$	1.10	1.00	9
Gaussian	$n(z) = 1.333 + (1.500 - 1.333)e^{-2z^2}$	0.81	0.73	10
Linear "ramp"	$0 < z < 2$ nm: $n(z) = 1.333 + 0.084z$ $z \geq 2$ nm: $n = 1.333$	1.23	1.11	10

discretized into 11 uniform slices of thickness δ to theoretically generate $Rp(\theta)$ at $0.2°$ increments over the angle range $\theta_B(Si/H_2O) \pm 2°$. Average thicknesses and refractive indexes obtained by regressing these profiles according to the single homogeneous film model produce the "single film" surface excess concentrations shown in the table. The true surface concentration corresponding to each discretized profile is generated by

$$\Gamma_{true} = \sum_i \frac{(n_i - n_0)\delta}{dn/dc} \tag{13.23}$$

taking dn/dc as 0.15 cm^3/g. The summation is over the 11 layers. As shown in Table 13.1 regression of the simulated data according to the single-film model is consistently within 10% of the correct surface concentration in all cases.

The profiles considered above were all assumed to be laterally homogeneous. If adsorption occurs in a laterally discontinuous or "patchwise" fashion, the effective layer refractive index n_{av} depends on the volume fraction of patches in the layer ϕ, on the refractive index of the patches n_f, and on the refractive index of the ambient solution separating the patches, n_0 according to[21]

$$\frac{n_{av}^2 - n_0^2}{n_{av}^2 + 2n_0^2} = \phi \frac{n_f^2 - n_0^2}{n_f^2 + 2n_0^2} \tag{13.24}$$

This typically will have little effect on the value of Γ calculated from reflectometry signals. One can simulate the effect of patchy adsorption by assigning a fixed number of patches per unit area N and allowing the surface excess concentration to increase by increasing the number of molecules per patch m. Let the cross-sectional area of each molecule in a patch be a (i.e., the number of adsorbed molecules per unit area in a patch is constant at $\Gamma_{patch} = 1/a$) and the thickness of the patch be d. During adsorption the patches simply grow in diameter at constant thickness and constant refractive index, n_f where

$$n_f = n_0 + \frac{dn/dc}{ad} \tag{13.25}$$

Since the cross-sectional area of each molecule in the patches is constant, the volume fraction of patches is independent of N, and grows simply as

$$\phi = Nma = \Gamma a \tag{13.26}$$

Table 13.2 compares the effective refractive index of a film developing by patchwise adsorption with the refractive index of a laterally homogeneous film grown at the same constant thickness d. For these calculations, $dn/dc = 0.18$ cm^3/g, $a = 1200$ Å^2, $d = 4$ nm, and molecular weight $= 14,000$ g/mol. These are reasonable values for

TABLE 13.2. Comparison of Patchy versus Homogeneous Adsorbed Layers

Γ (mg/m^2)	n_{av} (patchy)	n_{av} (homogeneous)	$R_p^{1/2}(\theta_B, \Gamma, d_{ox} = 20 \text{ nm})$ (patchy)	$R_p^{1/2}(\theta_B, \Gamma, d_{ox} = 20 \text{ nm})$ (homogeneous)
0	1.3330	1.3330	0.0497	0.0497
0.1	1.3374	1.3375	0.0501	0.0501
0.2	1.3419	1.3420	0.0505	0.0505
0.3	1.3464	1.3465	0.0509	0.0510
0.4	1.3508	1.3510	0.0514	0.0514
0.5	1.3553	1.3555	0.0518	0.0518
0.6	1.3598	1.3600	0.0522	0.0522
0.7	1.3643	1.3645	0.0526	0.0526
0.8	1.3687	1.3690	0.0529	0.0530
0.9	1.3732	1.3735	0.0533	0.0533
1.0	1.3777	1.3780	0.0537	0.0537

example, for proteins. For $0 < \Gamma < 1.0$ mg/m^2, the difference in layer refractive indices is in the third or fourth decimal place. Examination of the reflectivities calculated for these layers on top of a 20-nm-thick oxide layer, also presented in Table 13.2, demonstrates that patchiness has no significant effect on the determination of Γ.

13.3.2 Sensitivity

Increasing the oxide layer thickness from zero increases R_p and its sensitivity to adsorption. One manifestation of improved sensitivity is improved precision of regressed results. This effect can be simulated by theoretically generating noisy reflectivity profiles for a homogeneous adsorbed layer. For the current simulations, the refractive index was taken as 1.46 and the thickness as 1.417 nm, giving $\Gamma = 1.000$ mg/m^2 for $dn/dc = 0.180$ cm^3/g. To emulate experimental scatter the simulated reflectivity data contained Gaussian noise with a 2% standard deviation. For each oxide layer thickness, 30 simulated profiles were regressed according to the homogeneous film model and the results averaged. The results are shown in Table 13.3. There is no discernible trend in the magnitude of the regressed surface excess concentrations, all values being quite close to 1 mg/m^2, but the standard deviation is largest for the thinnest oxide layer. Thus precision is somewhat improved with thicker layers. Note that silicon exposed to air displays an approximately 2-nm-thick native oxide layer. For practical purposes, this sets a lower limit on the oxide layer thickness. Note, however, that the reflectivity function is periodic, so the improved precision afforded by thicker oxide layers will not increase monotonically without limit.

In a similar manner one can examine how the precision of the measurement depends on the range of angles used for a scanning angle measurement, by simulating reflectivity profiles with Gaussian noise with a 2% standard deviation for $\Gamma = 1.000$ mg/m^2. Reflectivities were generated every 0.2° for total angular ranges of 2.2°, 4.2°,

TABLE 13.3. Sensitivity Dependence on Oxide Layer Thickness

Oxide Layer Thickness (nm)	Regressed Refractive Index[a]	Regressed Thickness (nm)	Regressed Γ (mg/m^2)[b]
2	1.421 ± 0.065	4.2 ± 4.0	0.94 ± 0.13
3	1.440 ± 0.067	3.9 ± 4.2	1.00 ± 0.14
5	1.422 ± 0.069	6.8 ± 7.6	0.97 ± 0.10
10	1.442 ± 0.062	3.9 ± 5.1	0.99 ± 0.099
15	1.429 ± 0.060	7.1 ± 9.8	0.98 ± 0.077
20	1.446 ± 0.052	3.8 ± 7.0	0.97 ± 0.086
25	1.437 ± 0.054	7.0 ± 12	0.96 ± 0.067
30	1.442 ± 0.050	5.3 ± 12	0.95 ± 0.077
35	1.442 ± 0.051	11.4 ± 25	1.02 ± 0.088

[a]Error limits are $\pm$ one standard deviation.
[b]Γ should be 1.00 mg/m^2.

and 8.2°, each centered on θ_B. The oxide layer was 20 nm thick, and 30 independent simulated data sets were analyzed for each case. Surface excess concentrations obtained by regressing the simulated data were 0.99 ± 0.079, 0.97 ± 0.086, and 1.0 ± 0.098 mg/m^2, in order of increasing angular range. Thus, precision is somewhat improved by limiting measurements to a smaller range of angles around the Brewster angle.

These simulations pertain to oxide layers, but the trends are the same for other types of intermediate layers. While it is most convenient to measure adsorption on the oxide layer resident on silicon wafers, reflectometry is not limited to these surfaces. For example, polymer films may be spin-cast on the wafers to serve as adsorption substrates. The optical effect of the thin polymer film is equivalent to that of the oxide layer (accounting, of course, for the polymer refractive index) and, just as in the case of the oxide layer, it is necessary that the film thickness be measured prior to adsorption. Reflectometry has been used in this way to measure *protein adsorption* to spin-cast polystyrene and polydimethylsiloxane films.[9] Silica surfaces may also be readily modified by silane chemistry to produce a variety of surface functionalities for adsorption studies.[15] Reflectometry has also been applied to the fused silica–water interface[5] and to the interface between high index glasses and organic solutions.[8] In the last two applications there are no intermediate layers; only the single adsorbed layer separates the two semi-infinite media.

13.4 ADSORPTION KINETICS

13.4.1 Methodology and Data Interpretation

Adsorption kinetics are measured by setting the angle of incidence to θ_B and measuring the reflected intensity $I_p(\theta_B, t)$ during the adsorption process. Provided the oxide layer and surface excess concentration are within certain limits that must be calculated, the instantaneous surface excess concentration may be calculated to very good approximation from the proportionality

$$R_p(\theta_B, \Gamma)^{1/2} - R_p(\theta_B, 0)^{1/2} \propto \Gamma \qquad (13.27)$$

The proportionality constant may be obtained by performing a full scanning angle measurement to determine Γ at the adsorption plateau [and inserting it into Eq. (13.27)], or it may be calculated theoretically, using the Abelès matrix method to generate reflectivities for incrementally increasing values of the surface excess concentration. The surface excess concentration can either be increased by increasing refractive index at constant thickness or vice versa according to Eq. (13.18). This procedure requires a scanning angle measurement of the oxide layer thickness prior to adsorption. Examining Figure 13.6, one can see that this square root scaling is correct to within a few percent up to quite reasonable surface concentrations. In principle it might appear to be quite a simple matter to calculate the exact nonlinear relationship

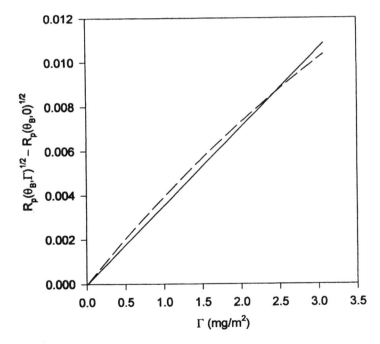

FIGURE 13.6   Square root of R_p at the Brewster angle is approximately proportional to the surface excess concentration. The solid line was generated by increasing Γ at constant refractive index (increasing thickness), while the dashed curve was generated by increasing at constant thickness (increasing refractive index). To generate the proportionality constant used for kinetic measurements, one fits both curves simultaneously by linear regression. The resulting error in doing so clearly depends on the range of surface excess concentrations encountered. This plot was generated for a 20-nm oxide layer on silicon, for an aqueous solution and HeNe laser light having a 632.8-nm wavelength.

between $R_p(\theta_B)$ and Γ in order to convert instantaneous reflectivities to instantaneous surface excess concentrations, but the benefit in improved accuracy will be of little consequence over the range of surface excess concentrations encountered in most polymer, protein, or surfactant adsorption experiments. Furthermore, discrepancies between the R_p behavior for layers grown at constant refractive index versus layers grown at constant thickness are also on the order of a few percent for reasonable oxide layer thicknesses and surface excess concentrations, making this additional effort unwarranted and unjustified.

Very often, the initial adsorption rate is of most interest in distinguishing between a transport limited or kinetic limited adsorption mechanism. Because of the curvature of the $R_p(\theta_B, \Gamma)^{1/2} - R_p(\theta_B, 0)^{1/2}$ versus Γ relationship, the largest error that would arise in the initial adsorption rate would occur when the proportionality constant is determined from the final scanning angle measurement. In other words, it is least satisfactory to determine the surface excess concentration in the limit of zero coverage by using a proportionality constant that was estimated at maximum coverage. This effect is shown in Figure 13.7, where the error in initial adsorption rate is plotted versus the plateau surface excess concentration used to calculate the proportionality

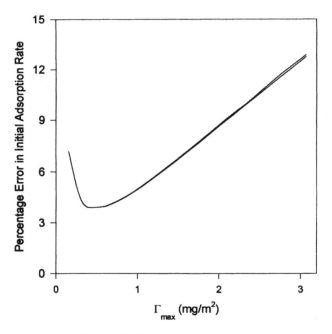

FIGURE 13.7 If instead of theoretically calculating the proportionality constant between $R_p^{1/2}(\theta_B, 0) - R_p^{1/2}(\theta_B, \Gamma)$ and Γ, one generates it experimentally via a scanning angle measurement of the final surface excess concentration at the end of an experiment, the initial adsorption rate will be overestimated. The magnitude of this error depends on how large the final surface excess concentration is, but it does not depend significantly on the oxide layer thickness. Two curves, generated for 2- and 20-nm-thick oxide layers, are virtually superimposed in this plot.

constant. For values of Γ shown by experience to be reasonable for adsorbing surfactants, proteins, and many polymers, calculating the proportionality constant from the final scanning angle measurement causes a 4 to 12% overestimate of the initial adsorption rate. The oxide layer thickness has no noticeable effect on this error. Thus, when the goal is to measure initial adsorption rates, it is preferable to generate the proportionality constant theoretically via the Abelès matrix method in the limit of low surface excess concentration.

13.4.2   Protein Adsorption Kinetics at the Solid–Liquid Interface

The initial steady-state rate of transport-limited adsorption from solution in fully developed laminar slit flow is described by the well-known Lévêque equation.[27]

$$\left.\frac{d\Gamma}{dt}\right|_0 = 0.538 \left(\frac{\gamma}{x}\right)^{1/3} D^{2/3} c_0 \tag{13.28}$$

where x is the distance from the flow cell inlet to the point of observation, γ is the wall shear rate, and D and c_0 are, respectively, the diffusion coefficient and bulk concentration of the surface-active species. The accuracy of reflectometry measurements, particularly at low surface excess concentrations, may be tested by

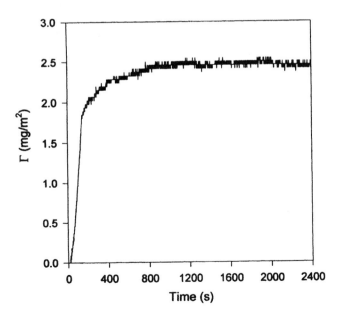

FIGURE 13.8   Adsorption kinetics for lysozyme adsorbing to silica from 0.005 M ionic strength aqueous solution.

checking for consistency with the Lévêque equation. Of course, not all adsorption experiments will be transport limited, and some experiments may not attain a steady-state adsorption rate before surface coverage effects start to dominate the kinetics. Experimental agreement with the Lévêque equation therefore should not always be expected, but quantitative confirmation of it in some cases lends credence to other adsorption kinetic measurements conducted with the same apparatus.

Kinetics for the adsorption of the protein lysozyme on thermally grown oxide layers on silicon wafers at 25°C are shown by the data in Figure 13.8 which represent part of a combined total internal reflection fluorescence and reflectometry study of crowding-induced protein layer reconfigurations.[13] The $\Gamma(t)$ values were obtained from the time-dependent reflectivity at the Brewster angle, $R_p(\theta_B, t)$, using the proportionality between Γ and $R_p(\theta_B, \Gamma)^{1/2} - R_p(\theta_B, 0)^{1/2}$. Initial steady-state adsorption rates for several experiments conducted at constant 10-ppm bulk lysozyme concentration are plotted in Figure 13.9 as a function of $\gamma^{1/3}$, as suggested by the Lévêque equation. The diffusion coefficient obtained from the slope of this plot is 1.2×10^{-6} cm²/s, in quantitative agreement with the literature value for lysozyme.[28] Adsorption is transport limited until $\Gamma \approx 1.8$ mg/m². The independent fluorescence data showed that the transition to surface-limited kinetics at that point correlates with a spontaneous lysozyme reorientation on the surface.

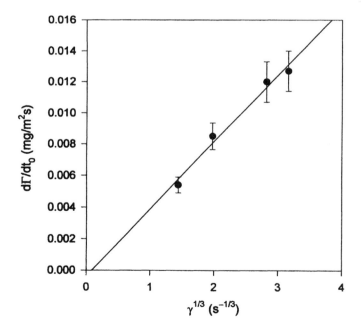

FIGURE 13.9 Initial steady-state adsorption rate scales with the one-third power of the wall shear rate for lysozyme adsorption from solutions in laminar slit flow, in quantitative agreement with the Lévêque equation. From Robeson and Tilton;[13] used by permission.

13.5 MULTICOMPONENT ADSORPTION

Many industrial and biomedical applications of macromolecular adsorption involve multicomponent mixtures containing both macromolecules and low-molecular-weight surfactants. There are many opportunities for interspecies interactions in these systems and, consequently, for unique adsorption behavior. The simplest multicomponent adsorption effects would be expected for noninteracting solutes that merely compete for limited interfacial area, that is, simple competitive adsorption. The situation is more complex and interesting when different solution components interact strongly in solution. In those cases, polymer–surfactant binding can be a critical factor in coadsorption mechanisms. It is well established that ionic surfactants bind with high affinity to oppositely charged polyelectrolytes or proteins in solution, and certain classes of surfactants will bind readily to nonionic polymers, for example, anionic surfactants binding to polyethyleneoxide (reviewed in Goddard and Ananthapadmanabhan[29]). In the presence of polymer–surfactant binding, coadsorption may be rather complex. Polymer–surfactant complexes may be more surface active than either component alone (synergy) or surfactant binding might solubilize a polymer and decrease its adsorption. It is also possible that binding may even lead to adsorption of one component (either the polymer or surfactant) that is by itself not attracted to the surface in question. In such a case, one species "delivers" the other to the surface in the form of a complex. This behavior depends both on the polymer–surfactant interaction and the affinity of the species for the surface.

While quantitative studies of polymer–surfactant coadsorption are in their infancy, kinetic limitations (i.e., nonequilibrium effects) are now recognizable as potentially dominant factors for determining the composition of mixed layers.[11,15]

13.5.1 Methodology and Analysis for Coadsorption Measurements

Reflectometry senses the average optical properties of adsorbed layers. As such it does not offer any chemical specificity. (An exception would be if a multiple wavelength spectroscopic reflectometry approach were taken for mixtures containing one chromophoric species.) If an adsorbed layer contains surface excesses of more than one species, reflectometry cannot discern the relative amounts of each adsorbed species; at best it can determine the total surface excess concentration of all species. Nevertheless, if these limitations are borne in mind, reflectometry can be quite useful for studying coadsorption.

How does the composition of the layer influence the reflectivity? Regardless of the layer composition, the effective or optical average refractive index and thickness of an adsorbed layer with arbitrary refractive index profile $n(z)$ remain as defined in Eqs. (13.19) and (13.20).

The surface excess concentration of any particular species i in a multicomponent adsorbed layer is

$$\Gamma_i = \int_0^\infty [c_i(z) - c_{i0}] \, dz \qquad (13.29)$$

where $c_i(z)$ and c_{i0} are now the concentrations of the particular species i at a distance z from the surface and in the bulk, respectively. Assuming that multicomponent interactions in the layer do not alter the refractive index increment of any component, then[11]

$$n(z) = n_{\text{solvent}} + \sum_i c_i(z) \frac{dn}{dc_i} \qquad (13.30)$$

where again $n_{\text{solvent}} \approx n_0$. It is readily shown from these equations that the reflectometric properties of a multicomponent adsorbed layer are described by

$$(n_{\text{av}} - n_0) \, d_{\text{av}} = \sum_i \Gamma_i \frac{dn}{dc_i} \qquad (13.31)$$

Thus, if all surface-active components have equal refractive index increments on a mass basis (e.g., if dn/dc were expressed in centimeters cubed per gram), reflectometry reports the total surface excess mass concentration of all species.

Might the refractive index increment depend on the total composition of the adsorbed layer? A solute's refractive index increment depends on the molar refractivities of its constituent chemical bonds and on its specific volume. If polymer–surfactant binding were to significantly alter specific volume, it could have a measurable effect on refractive index increment. The author's group used differential refractometry to investigate the effect of polymer–surfactant binding in solutions of polyethyleneoxide and micellar sodium dodecyl sulfate (SDS). These solutions display extensive binding and might therefore be expected to show composition-dependent refractive index increments. The refractive index increments of SDS and polyethyleneoxide at $\lambda = 632.8$ nm are 0.12 and 0.13 cm^3/g, respectively, in either single-component solutions or in mixtures. (SDS was above its critical micelle concentration in all cases.) So at least in one system where binding is known to occur, the assumption of constant, composition-independent refractive index increments would appear to be acceptable.

The following sections describe two applications of reflectometry to study multicomponent adsorbed layers. The first case, penetration of insoluble monolayers by bulk-soluble homopolymers and block copolymers, in some sense represents an ideal case where not only the total adsorbed amount but also the specific composition is directly obtained. In the second, coadsorption of bulk-soluble polyelectrolytes and surfactants, is an example of how reflectometry can directly provide the total adsorbed amount but requires additional work to infer the composition.

13.5.2   Penetration of Insoluble Monolayers at the Air–Water Interface

Lipids spontaneously self-assemble into oriented monolayers at the air–water interface. These monolayers often display well-defined two-dimensional phase transitions. The monolayer phase behavior, expressed via its surface pressure versus surface concentration isotherm, controls the Gibbs elasticity of these monolayers. These monolayers may be penetrated by soluble macromolecules. The resulting penetrated monolayers often display much different thermodynamic behavior, an effect of considerable importance for proper function of pulmonary surfactant in the lungs, for example.

Charron and Tilton[10,14] used scanning angle reflectometry to quantitatively examine the relationship between water-soluble polymer adsorption and the two-dimensional phase behavior of monolayers. In the context of this chapter, this problem represents a simpler application of Eq. (13.31). If both of the surface-active species in a two-component system are bulk soluble, one cannot determine the relative amounts of the two components in the adsorbed layer in a single experiment. If, however, one of the surface-active components is insoluble and introduced to the interface by spreading a known amount of material, the problem is alleviated. Such is the case for monolayer penetration, where Γ_{lipid} is known a priori.

The NPR profiles presented in Figure 13.5 represent part of a study of DPPC monolayer penetration by polystyrene-b-poly(ethyleneoxide) block copolymers (PS–PEO, $M_w = 379,000$, 10 wt % PS). The first step in these experiments was to measure the NPR profile for the unpenetrated monolayer. Regressing this monolayer data, one determines the quantity $d_{\text{av}}(n_{\text{av}} - n_0)$, from which the lipid's refractive index increment may be calculated via the single-component Eq. (13.18). This is because Γ_{DPPC} is known. The DPPC refractive index increment measured in this manner is extremely reproducible from one experiment to the next and is independent of Γ_{DPPC}.

Normalized parallel reflectivity profiles subsequently measured after penetration of the same monolayer are analyzed to determine the effective thickness and refractive index of the two-component layer. The polymer surface excess concentration then is obtained directly from $d_{\text{av}}(n_{\text{av}} - n_0)$, Γ_{DPPC}, and dn/dc_{DPPC} via the multicomponent Eq. (13.31). The polymer refractive index increment is known independently. By measuring the polymer surface excess concentration as a function of the area per lipid in the monolayer, plotted in Figure 13.10, it is evident that the polymer adsorbs by penetrating between DPPC molecules in the monolayer, as opposed to adsorbing "beneath" the monolayer via attractive interactions with the head groups. This mode of adsorption is responsible for the decrease in adsorption to monolayers in the liquid condensed regime. Most notable is that liquid-expanded monolayers and even monolayers in the phase coexistence regime do not diminish the extent of polymer adsorption at all relative to the bare air–water interface, in spite of their appreciable lateral densities. This notable tolerance for penetration is the result of an excluded area sink effect. This effect arises from the ability of polymeric penetrants to nucleate and drive the monolayer phase transition from the liquid-expanded to the liquid-condensed state.[14] Since the lipid-excluded area decreases approximately

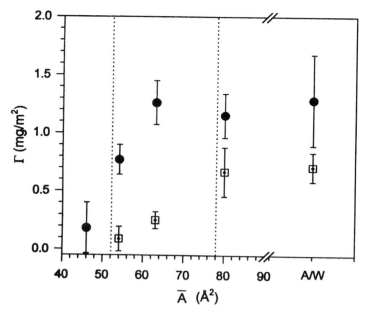

FIGURE 13.10 DPPC monolayer penetration by water-soluble polymers; Γ is the polymer surface excess concentration determined from the NPR profile after monolayer penetration; $\overline{A}$ is the average area per DPPC molecule spread at the air–water interface. A/W indicates the clean air–water interface, i.e., in the absence of DPPC. Filled circles represent PS–PEO diblock copolymers and open squares represent PEO homopolymer of comparable molecular weight. Dashed lines represent the monolayer phase boundaries between liquid expanded (at higher $\overline{A}$), liquid expanded–liquid condensed coexistence in the middle, and liquid condensed (at lower $\overline{A}$). After Charron and Tilton;[14] used by permission.

twofold as it experiences this phase transition, it has the effect of liberating available surface area for polymer segments to adsorb. Thus, monolayers that can experience this phase transition (i.e., monolayers below their critical temperature) do not resist penetration until they are in the liquid-condensed state.

PEO homopolymers of comparable molecular weight still benefit from the excluded area sink effect but are less effective monolayer penetrators than PS–PEO. The difference in their ability to penetrate denser layers is due to the ability of the diblock to assume an extended conformation at the interface and adjust its conformation in response to the lipid density.[14]

13.5.3 Coadsorption from Mixed Solutions

The author's group[11,15,30] studied coadsorption of the cationic polyelectrolyte, poly-L-lysine hydrobromide (PLL) and the cationic surfactant, cetyltrimethyl-ammonium bromide (CTAB) from aqueous solutions to negatively charged silica surfaces (thermally grown on silicon wafers). Both CTAB and PLL are water soluble.

Whereas the specific composition of the mixed layers was directly obtainable in the case of monolayer penetration described above, when both species are bulk soluble only the total surface excess concentration is directly obtainable. The mixed layer composition cannot be proven but, as will be shown, can be inferred from a series of closely related experiments.

Although both CTAB and PLL are bulk soluble and surface active on silica, they are mutually repellent and will not associate in solution. Regarding the interpretation of reflectometry data, note that the refractive index increments of CTAB and PLL are 0.15 and 0.16 cm^3/g at $\lambda = 632.8$ nm, respectively. Both are independent of solution ionic strength. Based on the near equality of refractive index increments, it is safe to say that reflectometry detects the total surface excess mass concentration for these coadsorbed layers.

The results of single-component PLL adsorption experiments will be presented before discussing coadsorption. Coadsorption will also be contrasted to sequential adsorption experiments in order to highlight the importance of kinetic traps as the dominant feature of the coadsorption mechanism.

Polylysine Adsorption without CTAB   Polyelectrolytes adsorb electrostatically to oppositely charged surfaces. Figure 13.11 shows the kinetics of PLL (degree of polymerization 860) adsorption to negatively charged silica from a 200-ppm aqueous solution in the absence of added electrolyte. The wall shear rate is 1.5 s^{-1}. The adsorption reaches its plateau in approximately 30 s. The *apparent* diffusion coefficient obtained by applying the Lévêque equation to the initial adsorption data would be 1.4×10^{-8} cm^2/s. This is one order of magnitude smaller than the 1.5×10^{-7} cm^2/s literature value[31] for PLL of this size. PLL adsorption was evidently not transport limited. In pure water, final surface excess concentrations of PLL on silica are 0.20 to 0.25 mg/m^2. This is slightly lower than the 0.25 to 0.5 mg/m^2 adsorption limit observed in 10 mM KBr. The influence of ionic strength is expected due to screening of polyelectrolyte conformation and solvation effects.

In either KBr or KBr-free solutions polylysine adsorption is entirely irreversible against rinsing for at least 24 h. This observation is particularly important for interpreting the coadsorption experiments described below.

Polylysine Coadsorption with CTAB   Here one can take advantage of the similarity of refractive index increments to interpret reflectometry results in terms of the total surface excess mass concentration during coadsorption. Only by performing additional reflectometry experiments, especially desorption measurements, may one *infer* the relative amounts of surfactant and polyelectryolyte in a mixed layer. To prove any conclusion about relative amounts in mixed adsorbed layers, one would need to apply other independent techniques, perhaps vibrational or fluorescence spectroscopies if appropriate, or solution depletion measurements on finely dispersed materials similar to those being examined by reflectometry. The following discussion is intended mainly to illustrate the applicability of reflectometry to mixtures. In particular, the advantages of being able to conduct kinetic measurements and to rapidly change solution composition will be highlighted.

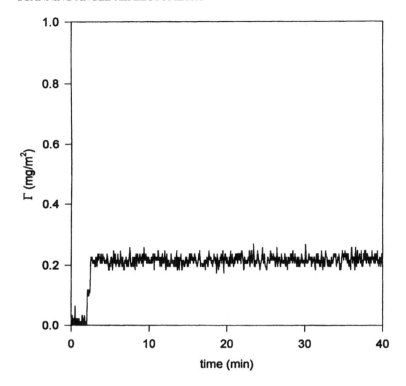

FIGURE 13.11. Polylysine adsorption kinetics on silica from 200 ppm aqueous solution with no additional electrolyte at 25°C.

In spite of their mutual repulsion, CTAB and PLL do coadsorb to form mixed layers on negatively charged silica surfaces. Coadsorption in the presence of 10 mM KBr is contrasted with CTAB adsorption from a single-component CTAB solution in Figure 13.12. In both cases, the CTAB concentration is 50 times the critical micelle concentration (0.20 mM in 10 mM KBr), and PLL concentration was 200 ppm. The total surface excess concentration attained by coadsorption is significantly larger than the amount of CTAB adsorbed from a single-component CTAB solution. This certainly implies that both CTAB and PLL are present in the adsorbed layer. The strongest evidence that the layer formed by the mixed solutions contains both CTAB and PLL is obtained from desorption experiments.

Desorption was induced by replacing the flowing solution of PLL and/or CTAB by flowing 10 mM KBr solution (containing neither CTAB nor PLL) with the results shown in Figure 13.13. Whereas all the adsorbed CTAB desorbs from the single-component CTAB layer, a significant fraction of the mixed layer remains irreversibly adsorbed. Most interesting is that the amount remaining adsorbed, 0.55 mg/m², is close to the amount of PLL that would adsorb irreversibly from a single-component PLL solution in 10 mM KBr. If one assumes that the irreversibly

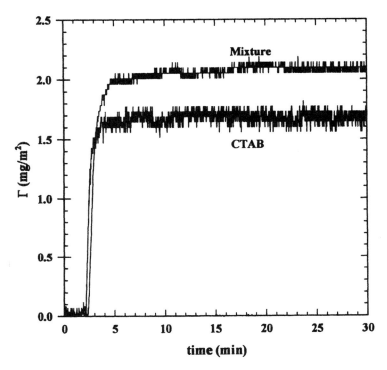

FIGURE 13.12. Comparison of adsorption kinetics on silica for single-component CTAB solutions and CTAB mixed with 200-ppm polylysine in the presence of 10 mM KBr. The CTAB concentration in both cases is 10 mM, i.e., 50 × cmc. Γ is the total surface excess concentration.

adsorbed material represents the entire amount of PLL adsorbed and subtracts this amount from the coadsorption plateau (Γ_{total} = 2.1 mg/m^2), one would conclude that the mixed layer contained 1.55 mg/m^2 CTAB. This is the same amount of CTAB that adsorbed from the single-component CTAB solution.

Similar observations have been made for mixtures containing CTAB at a concentration of 1 × cmc in 10 mM KBr. Mixed solutions produce mixed adsorbed layers containing approximately the same amounts of PLL and CTAB that would adsorb from either single-component solution, that is, adsorption was approximately additive in the presence of 10 mM KBr. One consequence of having mixed solutions that is not apparent in Figure 13.12 for highly concentrated solutions is that mixtures produce slower adsorption kinetics compared to single-component CTAB solutions. Although one might expect increased total rates of adsorption from the mixture, due to the higher total activity of all surface-active species, coadsorption kinetics are considerably slower. This is more easily seen at lower concentrations, as shown in Figure 13.14, where the concentrations of CTAB, PLL, and KBr are 1 × cmc, 200 ppm, and 10 mM, respectively. Coadsorption is clearly slower at early times. This early coadsorption deficit is made up eventually, and coadsorption ultimately

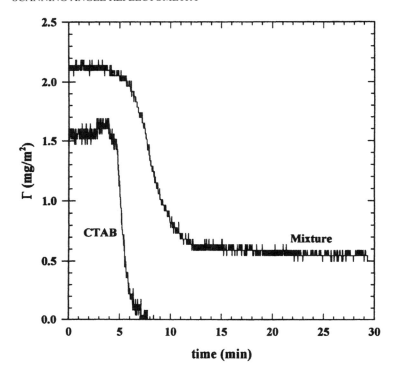

FIGURE 13.13. Same experiments shown in Figure 13.12 concluded by rinsing in 10 mM KBr solutions containing neither polylysine nor CTAB. CTAB desorbs rapidly and completely, while part of the coadsorbed layer remains bound (for over a day, not all data shown).

surpasses the single-component CTAB adsorption to display the same enhancement of total adsorption seen at $50 \times$ cmc.

In order to understand this effect, one should consider the relative adsorption rates of CTAB and PLL. Recall that PLL adsorption was not transport limited. It is especially important to note that (single-component) CTAB adsorption was not transport limited either. Apparent CTAB diffusion coefficients calculated from initial adsorption rates[11] were one to two orders of magnitude smaller than the literature values, for either monomeric or micellar CTAB. For example, at the cmc in 10 mM KBr, the apparent diffusion coefficient was 2.5×10^{-7} cm^2/s, which is far lower than the monomeric diffusion coefficient,[32] 5×10^{-6} cm^2/s, or even the micellar diffusion coefficient[33] 8×10^{-7} cm^2/s. Similar disagreements between apparent diffusion coefficients and literature values were found above the cmc as well, even if the decrease in monomeric surfactant ion concentration above the cmc[32,34–36] were taken into account.

In fact, the absolute PLL adsorption rate at 200 ppm was faster than the CTAB adsorption rate at the concentrations examined. Thus when PLL/CTAB mixtures contact a fresh silica surface, PLL dominates the surface at early times. Although the

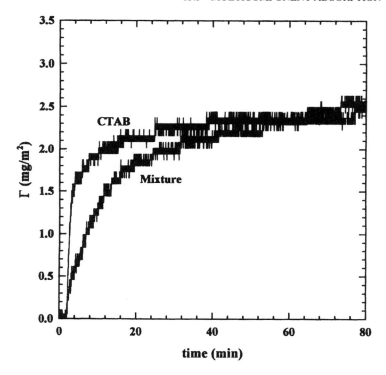

FIGURE 13.14 At CTAB concentrations of 0.2 mM (1 × cmc) in 10 mM KBr, it is evident that coadsorption kinetics are hindered relative to single-component CTAB adsorption on silica. The total surface excess concentration for the mixture does eventually surpass that attained for CTAB by itself.

amount of PLL adsorbed is fairly small, it is sufficient to reverse the net surface charge from negative to positive. (This was confirmed by electrophoresis measurements on colloidal quartz particles before and after adsorbing PLL.) The vast majority of CTAB that adsorbs does so against a net electrostatic repulsion from the surface.

This by itself does not satisfactorily explain why coadsorption kinetics lag behind single-component CTAB adsorption kinetics over such a large range of surface excess concentrations. It has been well established that CTAB adsorption reverses the silica surface charge at extremely low surface excess concentrations,[37] so most of the adsorbed CTAB layer is formed by molecules adsorbing against an electrostatic repulsion from the net positively charged interface, regardless of whether or not PLL is coadsorbed. The CTAB adsorption beyond the point of surface charge reversal is then driven by cooperative hydrophobic interactions among the surfactant tails.

Why then are the coadsorption kinetics hindered? Excluded area effects, enhanced by lateral electrostatic repulsions between adsorbing CTAB molecules and adsorbed PLL chains are obvious contributing factors. We have already noted, though, that in the presence of 10 mM KBr this inhibition is not sufficiently large to have a significant effect on the final amount of CTAB that adsorbs, and both CTAB and PLL adsorb from mixtures in amounts that are quite similar to those attained in experiments with their respective single-component solutions.

If one conducts coadsorption experiments in the absence of any added screening electrolyte, the results are quite different, and the total surface excess concentration attained by coadsorption is far less (by as much as 0.7 mg/m^2) than the amount of CTAB that adsorbs from single-component solutions. Desorption experiments similar to those described above indicate that the amount of polyelectrolyte that adsorbs is small but unaffected by the presence of the surfactant; on the other hand, the rapid polyelectrolyte adsorption dramatically decreases the amount of adsorbed surfactant. This is due to the stronger lateral electrostatic repulsions within the layer. These magnify the excluded area surrounding adsorbed PLL chains in the absence of KBr.

A more subtle effect could very well be that adsorbed PLL chains inhibit the rate of slow structural relaxations in the coadsorbed layer. Even in the absence of PLL, adsorbed CTAB layers undergo slow rearrangements as the surface excess concentration evolves toward its plateau.[15] These structural relaxations are apparently rate determining. It would not be surprising if tightly bound PLL chains were to slow the dynamics of these rate-determining rearrangements.

The irreversibility of polylysine adsorption (kinetic trapping) plays a critical role in coadsorption. This is best illustrated by the following presentation of sequential adsorption experiments for CTAB and PLL.

Sequential Adsorption   When one compares the coadsorption experiments described above to *sequential adsorption* experiments, it becomes apparent that kinetic limitations related to the irreversibility of PLL adsorption dominate the coadsorption process. Of course, if equilibrium were achieved in any of the coadsorption experiments, the total adsorbed amounts measured for any particular solution composition by definition should not depend on the path taken to that solution composition. The following paragraphs describe how preadsorption of either component prevents the subsequent adsorption of the other component introduced later.

In contrast to the mixture coadsorption experiments, preadsorption of PLL from a 200 ppm single component 10 mM KBr solution, followed by rinsing the adsorbed layer in water, and finally by introduction of CTAB, leads to no further adsorption. The total adsorbed amount remains identical to that attained by PLL even if the CTAB concentration is increased up to $5 \times$ cmc. Even though CTAB at this bulk concentration would adsorb from single-component solution to approximately 1.5 mg/m^2, the surface excess concentration remains stable at approximately 0.45 mg/m^2.[11] The preadsorbed polyelectrolyte does not desorb, and it prevents all subsequent surfactant adsorption.

Other sequential adsorption experiments have been conducted with no added electrolyte.[15] If one preadsorbs CTAB until a steady surface excess concentration is attained and then adds PLL to produce a two-component mixed solution, the adsorbed amount remains constant at the level produced by the CTAB preadsorption. Recall that coadsorption without any CTAB preadsorption would have produced a lower total surface excess concentration in the absence of KBr. Polylysine is unable to penetrate the preadsorbed CTAB layer or to displace any CTAB.

In addition to these experiments, one can take advantage of the ease of adjusting solution composition and arbitrarily switch between mixtures and single-component solutions to demonstrate more complex pathway-dependent surface excess concentrations. The final surface excess concentration is exquisitely path dependent. The conclusion from all of this is that kinetic traps dominate CTAB/PLL coadsorption.

The sequential adsorption experiments are most easily understood in terms of the charge reversal caused by either PLL or CTAB adsorption. CTAB can adsorb beyond the point of charge reversal only because of cooperative hydrophobic interactions between the tails. Surfactants adsorbed via electrostatic attraction provide a foothold for this hydrophobic adsorption. If PLL is irreversibly preadsorbed (reversing the net surface charge in the process), there can be no electrostatic driving force for CTAB to adsorb, and since PLL offers no hydrophobic tails, neither can there be any hydrophobic driving force. In a similar manner, if CTAB is preadsorbed (reversing the net surface charge), there is no electrostatic driving force for PLL to adsorb.

The observation that preadsorbed CTAB could not be displaced by PLL (from the subsequently introduced PLL + CTAB mixture), even though CTAB adsorption is indeed reversible, indicates that CTAB has the higher adsorption affinity of the two. Preadsorbed PLL exerts its effect in sequential adsorption experiments by virtue of its irreversibility, that is, via a kinetic trap.

The only reason that CTAB and PLL can coadsorb from mixtures (with no preadsorption step) is that there are enough charged sites on silica for both species to adsorb to some extent in the earliest stages. The future fate of the coadsorbed layer will depend exquisitely on the relative amounts of PLL and CTAB that share the surface at the earliest moments of adsorption. This is why the final adsorbed amount is so sensitive to changes in the path taken to achieve a given bulk composition.

13.6 SUMMARY

Optical reflectometry provides the capability for noninvasive measurement of adsorption and desorption kinetics. The property that can be reported with most confidence is the surface excess concentration. Although the analysis of reflectometry data provides an optical average thickness and an optical average refractive index for the adsorbed layer, these are best considered as nothing more than intermediate quantities obtained in the process of calculating the surface excess concentration. The thickness and refractive index are model dependent and prone to considerable experimental scatter. To within a few percent, the surface excess concentration is model independent and thus carries far greater significance.

Reflectometry as described here is not a spectroscopic technique in that it offers no molecular specificity. However, if one plans a series of related coadsorption and single-component adsorption experiments, this technique can be quite useful for measuring adsorption from multicomponent solutions. Multicomponent adsorbed layers are also amenable to study when one component forms an insoluble monolayer. Data interpretation is more straightforward in these cases.

Reflectometry is noninvasive and has excellent time resolution, making it particularly useful for kinetic measurements. The observations on the coadsorption mechanism presented here highlight the advantages of kinetic reflectometry experiments, and the significance of the ability to measure adsorbed layer responses to sudden changes in bulk compositions.

REFERENCES

1. D.H. Napper, *Polymeric Stabilization of Colloidal Dispersions*, Academic, San Diego, 1983.

2. A.W., Adamson and A.P. Gast, *Physical Chemistry of Surfaces*, 6th ed., Wiley, New York, 1997, pp. 71–77.

3. S. Welin, H. Elwing, H. Arwin, and I. Lundström, *Analytica Chim. Acta* **163**, 263 (1984).

4. H. Arwin and I. Lundström, *Anal. Biochem.* **145**, 106 (1985).

5. P. Schaaf, P. Déjardin, and A. Schmitt, *Langmuir* **3**, 1131 (1987).

6. P. Schaaf and P. Déjardin, *Colloids Surf.* **31**, 89 (1988).

7. J.C. Dijt, M.A. Cohen Stuart, J.E. Hofman, and G.J. Fleer, *Colloids Surf.* **51**, 141 (1990).

8. F.A.M. Leermakers and A.P. Gast, *Macromolecules* **24**, 718 (1991).

9. J.L. Robeson and R.D. Tilton, *Biophys. J.* **68**, 2145 (1995).

10. J.R. Charron and R.D. Tilton, *J. Phys. Chem.* **100**, 3179 (1996).

11. E.M. Furst, E.S. Pagac, and R.D. Tilton, *Ind. Eng. Chem Res.* **35**, 1566 (1996).

12. L. Heinrich, E.K. Mann, J.C. Voegel, G.J.M. Koper, and P. Schaaf, *Langmuir* **12**, 4857 (1996).

13. J.L. Robeson and R.D. Tilton, *Langmuir* **12**, 6104 (1996).

14. J.R. Charron and R.D. Tilton, *Langmuir* **13**(21), 55524 (1997).

15. E.S. Pagac, Ph.D. Dissertation, Carnegie Mellon University, Pittsburgh, 1997.

16. E.S. Pagac, D.C. Prieve, Y. Solomentsev, and R.D. Tilton, *Langmuir* **13**, 2993 (1997).

17. G.J.M. Koper and P. Schaaf, *Europhys. Lett.* **22**, 543 (1993).

18. E.K. Mann, E.A. van der Zeeuw, G.J.M. Koper, P. Schaaf, and D. Bedaux, *J. Phys. Chem.* **99**, 790 (1995).

19. S. Hénon and J. Meunier, *Rev. Sci. Instrum.* **62**, 936 (1991).

20. D. Hönig and D. Möbius, *J. Phys. Chem.* **95**, 4590 (1991).

21. R.M.A. and N.M. Bashara, *Ellipsometry and Polarized Light*, North-Holland, Amsterdam, 1987.

22. E.D. Palik, ed., *Handbook of Optical Constants of Solids*, Academic, London, 1985.

23. J.C. Dijt, M.A. Cohen Stuart, and G.J. Fleer, *Adv. Colloid Interface Sci.* **50**, 79 (1994).

24. N.S.B. Prasad, M.S. Thesis, Carnegie Mellon University, Pittsburgh, 1994.

25. F.L. McCrackin and J.P. Colson, in E. Passaglia, R.R. Stromberg, and J Kruger, eds., *Ellipsometry in the Measurement of Surfaces and Thin Films*, Natl. Bureau Standards Misc. Publ. **256**, 1964, p. 61.

26. J.A. de Feijter, J. Benjamins, and F.A. Veer, *Biopolymers* **17**, 1759 (1978).

27. B.K. Lok, Y.-L. Cheng, and C.R. Robertson, *J. Colloid Interface Sci.* **91**, 104 (1983).

28. T.E. Creighton, *Proteins: Structures and Molecular Properties*, 2nd ed., W.H. Freeman, New York, 1993, p. 266.

29. E.D. Goddard and K.P. Ananthapadmanabhan, eds., *Interactions of Surfactants with Polymers and Proteins*, CRC, Boca Raton, FL, 1993.

30. E.S. Pagac, D.C. Prieve, and R.D. Tilton, *Langmuir* **14**(9), 2333 (1998).

31. E. Daniel and Z. Alexandrowicz, *Biopolymers* **1**, 473 (1963).

32. B. Lindman, M.-C. Puyal, N. Kamenka, R. Rymdén, and P. Stilbs, *J. Phys. Chem.* **88**, 5048 (1984).

33. R. Dorshow, J. Briggs, C.A. Bunton, and D.F. Nicoli, *J. Phys. Chem.* **86**, 2388 (1982).

34. K.M. Kale, E.L. Cussler, and D.F. Evans, *J. Phys. Chem.* **84**, 593 (1980).

35. S.G. Cutler, P. Meares, and D.G. Hall, *J. Chem. Soc. Farad. Trans. I* **74**, 1758 (1978).

36. D.F. Evans and H. Wennerström, *The Colloidal Domain: Where Physics, Chemistry, Biology and Technology Meet*, VCH, New York, 1994, pp. 141–142.

37. J.L. Parker, V.V. Yaminsky, and P.M. Claesson, *J. Phys. Chem.* **97**, 7706 (1993).

14 Total Internal Reflectance Fluorescence

MARIA M. SANTORE

Department of Chemical Engineering, Lehigh University, Bethlehem, Pennsylvania 18015

14.1 INTRODUCTION

The science of polymer interfaces is central to numerous technologies from controlled colloidal stability to lubrication, adhesion, bioadhesion and separations, and engineering plastics and composites. In these instances, a polymeric or inorganic material forms an interface with a polymer glass, melt, or solution (including proteins). Key to the success of a particular application is the extent to which, and mechanism by which, polymers adhere to the surface, the interfacial mobility, and the presence of secondary species. In probing such issues, a variety of experimental techniques have emerged to provide direct interfacial information. Indeed, great advances in our understanding of interfacial phenomena have been made possible through studies using ellipsometry, neutron reflectivity and scattering, hydrodynamic and electrokinetic measurements, and spectroscopic probes.

Ellipsometry is the most accessible of direct interfacial probes of surface excess.[1-4] This method measures the change in the polarization state of light reflected from a planar interface and, based on the refractive indices of the surface, second medium, and interfacial layer, the surface excess can be determined accurately. More sophisticated instruments such as nulling ellipsometers[5] also provide information about the layer thickness but typically require that the user postulate a model for the concentration profile of the adsorbed layer. Typically a step function is used. Neutron scattering[6] and reflectivity[7-9] methods elucidate the details of the concentration profiles of species adsorbed about an interface. The combination of scattering and reflectivity facilitates study of colloidal systems and planar interfaces, respectively. Neutron methods require, however, that the adsorbing species be available in deuterated form or that through the use of deuterated solvents sufficient contrast can be achieved. The methods are therefore

Colloid–Polymer Interactions: From Fundamentals to Practice, Edited by Raymond S. Farinato and Paul L. Dubin
ISBN 0-471-24316-7 © 1999 John Wiley & Sons, Inc.

restricted in their applicability to systems of interest. Also, data collection times are on the order of hours such that neutron methods can be used to investigate only the slowest of interfacial kinetics. Hydrodynamic methods such as dynamic light scattering (DLS)[10,11] and capillary flow[12,13] yield the hydrodynamic thickness of a layer adsorbed from solution onto a solid phase: This corresponds to the outermost portion of the adsorbed concentration profile. Dynamic light scattering and capillary flow methods are restricted to stable colloids or microcapillaries, respectively, and therefore cannot accommodate all substrates.

Spectroscopic methods such as attenuated total reflection infrared (ATR-IR)[14–16] and nuclear magnetic resonance (NMR)[17,18] have also provided great insight into adsorbed polymer layers and polymeric materials at interfaces, for example, in composites. When applied to polymers adsorbing from solution, ATR-IR can provide a measure of surface excess, and IR dichroism[19] can distinguish polymer segments oriented in the plane of the interface from those normal to the surface. For certain polymers available in deuterated form, ATR-IR can probe self-exchange rates and to the extent that well-designed flow geometries are implemented, extract interfacial mobility information. Nuclear magnetic resonance has also been implemented to provide information on surface excess and the conformation of adsorbed chains.[17,18]

In the last several years, polymer scientists have begun to exploit another spectroscopic-based technique previously employed in a large sector of the biotechnology community,[20–23] total internal reflectance fluorescence (TIRF). Like many of the techniques previously mentioned, TIRF can provide a measure of the surface excess of a fluorescently labeled polymer adsorbing from solution onto a solid surface. More important, however, is TIRF's capability to provide kinetic information about a polymer interface and its extreme sensitivity, which exceeds that of ellipsometry and neutron reflectivity. Kinetic resolution can be extremely fast, on the order of seconds, or extremely slow, on the order of days since the baseline stability of TIRF is less temperature sensitive than methods such as ellipsometry, which rely on refractive index.[24] For any polymer that is intrinsically fluorescent or can be tagged, TIRF can measure, in addition to the evolving surface excess, interfacial relaxations and mobility. In certain cases, depth profiling[25,26] can be accomplished to probe the interfacial concentration profile. In the biotech community, TIRF was employed primarily as a method for protein adsorption and interfacial relaxations. More recently we have exploited TIRF's unique ability to probe competitive equilibria and interfacial kinetics in multicomponent systems.[27,28]

14.2   BACKGROUND ON TOTAL INTERNAL REFLECTANCE FLUORESCENCE

14.2.1   Principles

Evanescent-wave-induced fluorescence, often called TIRF, has become a popular probe of polymer interfaces, though the method has not been commercialized. This is due largely to the need to tailor the optical details to the particular system of study. While the method is applicable to a variety of polymer interfaces, including the melt state of a polymer, our description of the technique will focus on a polymer solution in contact with a solid where adsorption occurs. The optics translate directly for any number of other scenarios.

In TIRF, a beam of excitation light is brought to the interface between two materials (e.g., a high refractive index waveguide contacting a lower refractive index polymer solution) where a polymer adheres or adsorbs. The light enters through the medium of higher refractive index and is totally internally reflected, as shown in Figure 14.1, to generate an evanescent wave (an exponentially decaying surface light wave) that serves as the excitation source for a fluorescence experiment. Though fluorescently tagged polymers may reside anywhere in the second medium, only those within reach of the evanescent wave are excited, with the fluorescence collected usually from behind the beam. With evanescent penetration depths on the order of 100 nm, the technique is surface sensitive and probes a length scale appropriate for macromolecular physics.

Total internal reflectance fluorescence is a relatively flexible method in that a wide variety of polymers, solvents, and substrates can be accommodated. The primary requirement for a system to be studied with TIRF is the optical clarity of the substrate. This is not, however, usually a problem as adsorption may be studied directly on the waveguide or the waveguide may be optically coupled to flats of higher refractive index and different surface chemistries. Also, spin-cast films and self-assembled monolayers may comprise the substrate. Any polymer that can be fluorescently tagged is amenable to study via TIRF. Any solvent can be accommodated, but the refractive index of the waveguide or substrate must exceed that of the polymer solution. Systems in which mild light scattering occurs can be accommodated to determine interfacial rates;[29] however, quantification of adsorbed amounts is compromised by scattering background.

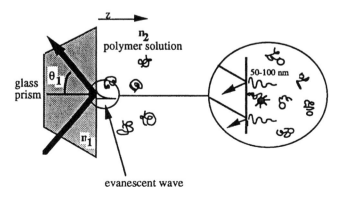

FIGURE 14.1 Physics of TIRF

In general, for a beam impinging on an interface, the relationship between the incident and transmission angles (both measured from the normal) is given by Snell's law:

$$n_1 \sin \theta_1 = n_2 \sin \theta_2 \qquad (14.1)$$

Here θ_1 and θ_2 are the angles of the incident and transmitted beams, shown in Figure 14.2. When the refractive index of the incident medium (n_1) exceeds that of the transmitting medium (n_2), as one increases the incident angle to more glancing conditions, the transmitted angle more rapidly approaches its maximum value of 90°. When this occurs, the intensity of the transmitted beam becomes vanishingly small such that all the energy impinging on the interface is reflected back into the incident medium. This is the condition of total internal reflection, which persists for values of θ_1 exceeding the critical angle, θ_c:

$$\theta_c = \sin^{-1}(n_2/n_1) \qquad (14.2)$$

Though all the incident energy is reflected back into the first medium, the electric field disturbance is not confined to medium 1. Instead, there persists an "evanescent" standing wave, propagating along the interface and whose intensity, I, decays normal to the interface (in the z direction) into medium 2.

$$I(z) = I_0 e^{-z/\Lambda} \qquad (14.3)$$

Here I_0 is the intensity at the interface, and Λ is the evanescent penetration depth, given by:

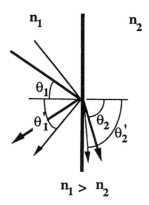

FIGURE 14.2   Application of Snell's law as total internal reflection is approached, going from θ_1 to θ_1'.

$$\Lambda = \frac{\lambda_0}{4\pi\sqrt{n_1^2\sin^2\theta_1 - n_2^2}} \tag{14.4}$$

where λ_0 is the wavelength of the incident light. While Λ is independent of the polarization of the incident beam, I_0 depends on polarization according to fundamental optical principles:[30]

$$I_{0,\parallel} = A_\parallel^2 \frac{4\cos^2\theta_1}{(n_2/n_1)^4\cos^2\theta_1 + \sin^2\theta_1 - (n_2/n_1)^2} \tag{14.5a}$$

$$I_{0,\perp} = A_\perp^2 \frac{4\cos^2\theta_1}{1 - (n_2/n_1)^2} \tag{14.5b}$$

Here, $A_\parallel$ and $A_\perp$ are the field amplitudes of the incident electromagnetic wave, parallel and perpendicular to the plane of incidence, respectively.

As indicated in Eq. (14.4), the penetration depth is a function of incident angle and also depends on the refractive indices of the two media comprising the interface. Values of Λ/λ_0 are presented in Figure 14.3 for common pairs of optical interfacial parameters useful in polymer interface studies. For a given pair of media, at the most glancing angle approaching 90°, the penetration depth approaches a limiting minimum value. Penetration depths near this minimum value are experimentally obtainable with minimal error over about 10 to 15° of range of the incident angle. As the incident angle is decreased toward θ_c, the penetration depth diverges to infinity at transmission. A significant amount of error is sustained when one attempts penetration depths significantly larger than the minimum value, because of limited precision in the incident angle and due to variations in the flatness of the materials forming the interface. For these reasons, it is important to choose a waveguide of appropriate refractive index to accommodate the polymer solution of interest. Chemical treatment

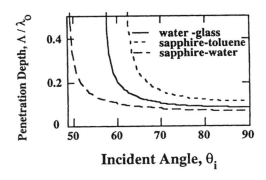

FIGURE 14.3 Penetration depths for different interfaces.

of the waveguide, or spin-cast films to control the substrate chemistry do not affect the evanescent penetration depth when the film's refractive index exceeds that of the waveguide. The reader may convince himself of this fact by repeated applications of Snell's law to a multilayered interface. The only refractive indices ultimately affecting the evanescent wave are those of the waveguide and polymer solution.

To the first order, changes in the fluorescence signal typically reflect changes in the number of fluorophores near the interface, for instance, during a polymer adsorption study where the species of interest is fluorescently labeled. Second-order changes in fluorescent signal may result from a restructuring of the adsorbed layer, either via movement of fluorophores within the evanescent field or an alteration in the environment surrounding the fluorophore, which alters its quantum yield. Scanning angle versions of TIRF change the penetration depth of the evanescent wave within a single experiment such that the fluorescence signal as a function of the penetration depth is proportional to the LaPlace transform of the interfacial concentration profile.

14.2.2   Experimental Setups

Figure 14.4 illustrates a typical experimental setup for TIRF. The light source may be a laser or a lamp passed through a monochrometer to obtain the desired wavelength. The laser version of TIRF is typically constructed on an optical table, though there are many groups employing total internal reflection cells built inside fluorescence spectrometers. For lasers on an optical bench, excitation light passes through a

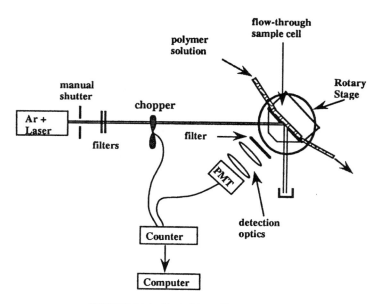

FIGURE 14.4   Typical laser TIRF apparatus.

chopper to facilitate subtraction of electronic noise. The beam is then directed into the waveguide and totally internally reflected at the interface of interest, proceeding to a trap. The emitted fluorescence passes through filters and/or a monochrometer to a detector, usually a photon-counting photomultiplier tube. In the case of a laser TIRF instrument on an optical table, narrow-cut high-pass filters prevent stray laser light from reaching the detector. Only the fluorescent light reaches the photomultiplier tube (PMT) and the signal is typically proportional to the total fluorescence. For TIRF cells inside a fluorescence spectrometer, a monochrometer is sufficient to distinguish between fluorescent and excitation light, and with sufficient fluorescent emissions, spectra can be resolved.

Laser-based TIRF instruments have the advantage of coherent excitation light that is well collimated such that the evanescent decay length is well-defined and such that interference patterns may be projected onto the interface for photobleaching studies of interfacial mobility. A laser is also a necessary excitation source if depth profiling is attempted. The TIRF cells in fluorescence spectrometers compromise a precisely controlled penetration depth (not a drawback for most kinetic and relaxation studies), but gain spectroscopic resolution that can provide information about the evolving interfacial environment, leading to molecular-scale interpretations of interfacial polymer conformations. Use of a fluorescence spectrometer also has the advantage that a wide range of fluorescent tags may be exploited. In contrast, the choice of a laser restricts one to fluorophores exciting at the lines of the particular laser.

14.2.3 Calibration

The translation of the fluorescence signal from a TIRF experiment to a quantitative measure of surface excess of polymer conformation is not as straightforward as absorbance measurements (e.g., IR) or ellipsometry because the amount of the fluorescent emissions counted by the detector depends on instrumental alignment and the collection efficiency. Furthermore, internal calibrations of TIRF must also address potential changes in quantum yield and artifacts such as scattering.

The amount of fluorescence in a TIRF experiment is expected to be the convolution of the evanescent excitation wave with the concentration profile of the fluorescently tagged species, $c(z)$:

$$F(\Lambda) = K(\Lambda) \int_0^\infty c(z) e^{-z/\Lambda} \, dz \tag{14.6}$$

The proportionality factor $K(\Lambda)$ relates the measured fluorescence to that emitted from the sample, and includes I_0, the detector efficiency, the efficiency of the collection optics, in addition to efficiency factors for the absorbance of the evanescent wave and the quantum yield of the fluorophore. Equation (14.6) also requires that the fluorophores in the evanescent zone are sufficiently dilute such that the evanescent wave is not significantly attenuated by absorption.

While it can be convenient to calibrate TIRF externally by comparison with methods such as optical reflectivity,[31] this option is not usually available. Therefore, a number of groups have proposed direct internal methods for obtaining $K(\Lambda)$. For instruments with scanning angle capability, Rondelez et al.[32] proposed a calibration with a nonadsorbing fluorescent dye that attains a uniform concentration, C_0, in the fluid side of the TIRF cell. Substitution of the uniform concentration profile for $c(z)$ into Eq. (14.6) yields

$$F_{cal}(\Lambda) = K(\Lambda)C_0\Lambda \qquad (14.7)$$

Equations (14.6) and (14.7) were then combined to eliminate $K(\Lambda)$. Angle scans were repeated for TIRF cells of different depths to quantify the effect of light scattering reaching fluorophores beyond the evanescent wave. This contribution to the fluorescence was eliminated by extrapolation to a cell depth of zero.

Lenhoff's group[33] proposed a second calibration procedure for instruments capable only of fixed-angle (fixed penetration depth) measurements. Here, a series of nonadsorbing dye solutions were used as calibration standards and their fluorescence in TIRF plotted as a function of concentration. At high fluorophore concentrations, the scattered light is attenuated by absorption (the inner filter effect) such that the incremental increase in fluorescence with concentration yields $K(\Lambda)$. The maximum contribution of scattered light reaching fluorophores beyond the evanescent wave can be determined by extrapolation of the high concentration slope (in Fig. 14.5) to zero concentration. This method requires that the integrated scattering intensity be much less than that of the evanescent wave.

We more recently proposed an internal calibration method for TIRF employing variations in both Λ and the concentration of a nonadsorbing fluorophore.[34] The combined method was found to accurately quantify the contribution of scattering to the TIRF signal in situ, without requiring repeated runs in different flow cells. Though our method is more tedious than Shibata and Lenhoff's approach,[33] it relaxes the constraint that the scattered intensity must be at least an order of magnitude less than the evanescent field.

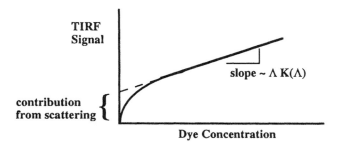

FIGURE 14.5   Calibration of TIRF at fixed angle.

14.3 EXPERIMENTAL

Adsorption experiments were conducted using either a home-built laser TIRF instrument, employing the 488-nm line of an argon ion laser and measuring the total fluorescence above 520 nm, or a TIRF compartment inside a Spex Fluorolog II Spectrofluorometer. In either case, adsorption was measured from gentle shearing flow, employing a controlled shear-flow cell modeled after that of Shibata and Lenhoff.[35] Controlled flow conditions allow the user to mediate the rate of chain arrival from bulk solution to the surface of interest. This allows quantification of intrinsic surface kinetics[35] when the mass-transfer rates are sufficiently rapid (see also Chapter 5).

The data employ a variety of water-soluble polymers, labeled with a fluorescent tag: fluorescein or coumarin. These labels have been attached at low levels (typically one tag per every 33,000 to 500,000 daltons of sample). Polyethyleneoxide (PEO) samples were narrow molecular weight standard materials from Polymer Labs. For PEO, we attached fluorescent labels onto a single end of each chain. Hydroxyethyl cellulose (Natrosol 250 GR, HEC) was a gift from Hercules and was randomly labeled with fluorescein at a density of one tag per every 80,000 to 100,000 daltons of backbone. Bovine serum albumin was obtained from Sigma and tagged with about one fluorescein per protein molecule. The recipes for the labeling chemistry and subsequent purification of product have been documented elsewhere.[36]

Substrates include chemically treated glass and spin-cast polystyrene (PS) film. Acid-etched microscope slides contain a silica outer layer on a bulk soda-lime slide.[31] The silica surface contains silanols that dissociate to give a surface potential near -200 mV when the bulk pH is near 7.[37] Spin-cast PS films are hydrophobic.

14.4 RESULTS

14.4.1 Simple Adsorption and Desorption

Though simple adsorption experiments do not exploit TIRF's unique capabilities for tracking interfacial populations and relaxation processes, adsorption kinetics measured via TIRF rival the resolution achieved with other methods, as illustrated in Figure 14.6 for PEO adsorption onto silica from aqueous solution.[38] In this particular case, the PEO is a narrow molecular weight standard, 33,000, end-tagged with a coumarin label. These data were taken in a fluorescence spectrometer with excitation at 405 nm and detection at 500 nm. Also in Figure 14.6 are reflectivity measurements[31] (based on refractive index), of the evolving surface excess for a repeat of the same experiment (see also Chapter 13). (The error in coverage for repeated runs of this type is on the order of the noise in the data, giving relatively high confidence in the absolute amount of coverage in the TIRF experiment.)

In Figure 14.6, the evolving adsorbed amount as measured by TIRF rises linearly with time, in accord with expectations for mass-transport-limited kinetics and

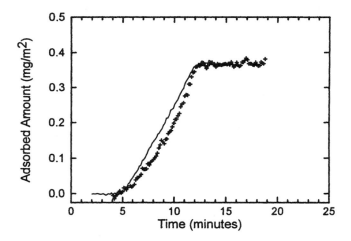

FIGURE 14.6   Comparison of PEO adsorption from a 5-ppm aqueous solution onto silica, measured by TIRF (—) and reflectivity (++).

high-affinity adsorption (see also Chapter 5). The TIRF measured adsorption kinetics are indeed more nearly linear (and in better agreement with expectations) than the reflectivity data, which are concave up as a result of changes in the adsorbed layer density as the coverage increases. In most instances,[31,39] reflectivity data are interpreted in terms of surface excess within 10% error. The TIRF yields results that are much more sensitive (below 0.2 mg/m²) and accurate with regard to relative interfacial mass. Indeed, the combination of the two methods facilitates calculations of the evolving layer thickness and density during adsorption.[38]

While a linear signal rise during adsorption is a feature of diffusion-limited kinetics and high-affinity adsorption for monodisperse polymer systems (see also Chapter 5), a second feature of high-affinity adsorption is the resistance of an adsorbed layer to removal during extended periods of flowing solvent. This is illustrated in Figure 14.7 for PEO adsorbed onto a PS film, PEO adsorbed onto silica, and HEC adsorbed onto silica. While the time scales in Figure 14.7 are on the order of several hours, we have measured the stability of the adsorbed layers to flowing solvent in each case for at least 24 h and found minimal changes in the adsorbed amount. (Such data do not, however, make for interesting figures.) While it was initially thought that such resistance to desorption was a result of a nonequilibrium state or a glassylike interface,[40] it has been recently shown that very slow desorption can indeed be quantitatively predicted by diffusion-controlled kinetics, as long as the surface is extremely high affinity[12] (see also Chapter 5). In this case, a nearly saturated surface can be in local equilibrium with nearby fluid elements containing less than a part per million of polymer. In the past year, our group has been the first to measure adsorption isotherms down to such extremely dilute conditions, substantiating the diffusion-limited explanation of the data in Figure 14.7.[31]

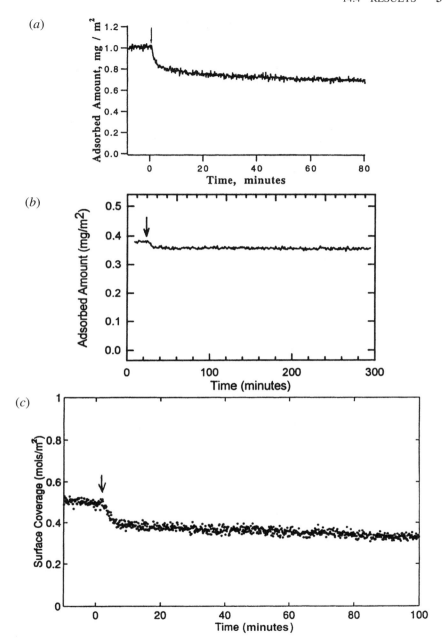

FIGURE 14.7 Minimal desorption experienced by adsorbed layers exposed to flowing solutions: (*a*) fluorescein-tagged PEO adsorbed on a polystyrene film in flowing phosphate buffer, (*b*) coumarin-tagged PEO on silica in flowing phosphate buffer, and (*c*) HEC on silica in flowing phosphate buffer. Arrow indicates replacement of adsorbate polymer solution with flowing solvent.

14.4.2 Interfacial Environment

While changes in fluorescent quantum yield signaling interfacial reconformations or confounding the calibration techniques above are often anticipated, we find in practice that significant changes in fluorescence per label rarely occur once the adsorbed layer is formed, at least for PEO adsorption.[34,38] This may be a result of our choice of labels with minimal affinity for the adsorbate surface. When the fluorescent tag is attached to the chain end and is not attracted to the surface, it remains in solution within the adsorbed layer, several nanometers away from the surface. At this location, the tag avoids the region densest in polymer concentration near the surface and samples a local polymer concentration on the order of 10%.[8] Here, the probability of direct fluorophore–quencher interaction is reduced, and the fluorophore tends not to selectively orient, having minimal influence on the absorbance.

Under some conditions, however, the environment within a few nanometers of the surface is drastically different from the bulk, influencing the fluorescence. For instance, near a highly charged silica surface, with a surface potential near -200mV,[37] the fluid nearest the surface will be richer in protons (more acidic) than the bulk solution, with the spatial decay of the acidity dependent on the overall ionic strength of the solution. This situation was explored for adsorbed layers of PEO containing fluorescein on the chain ends.[41] Fluorescein is a weak acid with a pK_a near 6.8. Only the protonated form is significantly fluorescent.[42] Because for classic homopolymer adsorption chain ends tend to reach into solution from entangled tails, and because fluorescein was found not to adsorb onto silica from aqueous solution, the role of the polymer was simply to tether the fluoresceins within a few nanometers of the interface, generating a sensor of local pH.[41]

Figure 14.8a illustrates the sensitivity of the tethered fluorescein emissions as a function of NaCl concentration and, independently, as a function of the concentration of buffering phosphate ions at different fixed ionic strengths. The dependence of the fluorescence on NaCl concentration is greatly influenced by the bulk pH as indicated by the large error bars. These result from day-to-day variations in dissolved CO_2. The fluorescent behavior of this interfacial sensor has been predicted, in good quantitative agreement with the data, by a Guoy–Chapman treatment of the electrostatic double layer giving a local proton concentration, plus acid–base equilibria for the fluorescein and phosphate buffering species.[41] Figure 14.8b illustrates the predicted influence of

---

FIGURE 14.8 (*a*) normalized fluorescence from an adsorbed fluorescein–PEO layer as a function of bulk solution ion composition and concentration from Ref. 42. The species that were varied were NaCl (□), phosphate (▲) in 0.001 M NaCl, and phosphate (■) in 0.1 M NaCl. (*b*) Predicted fraction of adsorbed fluorescein–PEO in dianion (fluorescent) form as a function of NaCl concentration, shown for fluorophores situated 1 (—) and 10 (. . .) nm from the surface. The model employs an acid–base equilibrium for fluorescein and the Guoy–Chapman treatment for the local potential and ion concentration from Ref. 42. Data from (*a*) for the layer fluorescence at varying NaCl concentration (▲) are shown for comparison. (*c*) Predicted equilibrium fraction of fluorescein in dianion (fluorescent) form as a function of total phosphate concentration, with a total local fluorescein concentration of 0.004 M from Ref. 42. The model includes an acid–base treatment of the fluorescein–phosphate interaction, and illustrates the effect of local potential.

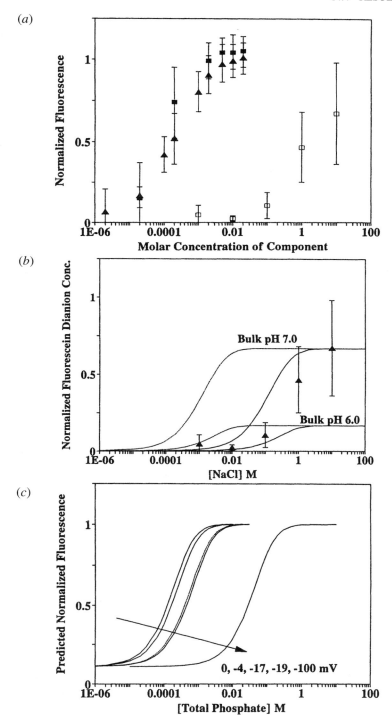

(a)

(b)

(c)

ionic strength, and the calculated effects of bulk pH variation between 6 and 7 bound the experimental results. Figure 14.8c illustrates the predicted influence of phosphate ions for fluorophores residing in planes of different local potential, from 0 to −100 mV, where the latter is dependent on the overall ionic strength. Comparison of Figures 14.8a and 14.8c demonstrate that the interfacial phosphate titration curves at 2 different NaCl concentrations (0.1 and 0.01 M) correspond to average local potentials near the fluorophores of 0 and −19 mV, which are reasonable values for fluorophores positioned 5 nm from the surface at these two ionic strengths. A particularly interesting conclusion from this data is that adsorption of PEO onto a silica surface has a minimal effect on the local potential from the underlying charge. This may be a result of PEOs hydrogen bonding with undissociated hydroxyls on the surface, and the low coverages levels, on the order of 0.4 mg/m^2.

14.4.3 Interfacial Dynamics

While surface titration experiments illustrate some unique capabilities of TIRF and provide new perspective into interfacial polymer physics, the feature most unique to TIRF is its ability to perform dynamic and kinetic measurements of specific populations of polymers adsorbed at an interface. Two types of studies will be discussed here: self-exchange rates for several polymer–substrate systems and the nature of molecular competition during polymer adsorption onto a bare surface, in this instance, driven by polydispersity.

It is generally observed, as shown in a previous section of this chapter, that polymers are fast to adsorb and relatively slow to desorb: They resist washing off a surface in flowing solvent. It has been observed, however, that adsorbed polymers can often readily be displaced by other molecules that are preferred on the surface.[43] Such exchange processes occur at experimentally accessible rates because the kinetic energy barrier for an exchange process is on the order of kT. The segments on incoming chains are thought to trade places with those of the previously adsorbed chains one segment at a time. In contrast, the energy barrier to pure desorption is much greater because even when equilibrium conditions favor complete desorption, all the segments from a single chain must desorb from a surface all at once, at an energetic cost on the order of 100 kT (for about 100 segment–surface contacts per high-molecular-weight polymer chain, with an energy of 1 kT/contact).

One particular kind of exchange process, self-exchange, has the potential to provide significant insight into the nature of adsorbed polymers. Self-exchange should be a strong function of polymer backbone chemistry, surface features, and the solvent, in addition to potential history effects. Therefore, the nature of self-exchange should highlight fundamental differences between various systems. Such issues are only recently beginning to be addressed, and therefore in this chapter we present only a qualitative introduction to the subject.

In self-exchange studies, a layer of tagged material is adsorbed to a surface and then allowed to age in contact with solvent for a specified time. Then it is challenged with the untagged analog of the same polymer. The latter must be identical to the originally adsorbed tagged material in all respects, including molecular weight or molecular weight distribution, since chain length differences can considerably affect adsorption equilibrium. The displacement of the tagged material by the untagged material will cause a fluorescence decay, the time scale and extent of which are of primary interest. (The complimentary experiment in which an untagged layer is preadsorbed, aged, and then challenged with tagged material should give the same results, and will not be presented in this brief overview.)

Figure 14.9 compares self-exchange studies for coumarin-tagged and native PEO on silica, fluorescein-tagged and native HEC on silica, and fluorescein-tagged and

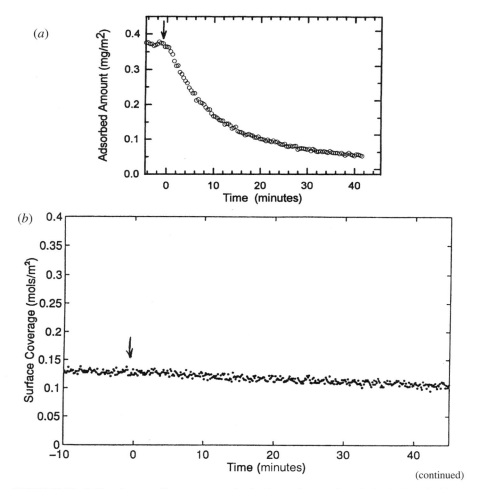

(continued)

FIGURE 14.9   Self-exchange studies where a preadsorbed layer of a tagged species is challenged with an untagged version of the same molecule: (*a*) adsorbed coumarin–PEO on silica challenged by a 5-ppm solution of native PEO, (*b*) fluorescein-tagged HEC on silica challenged by native HEC at 100 ppm, and (*c*) fluorescein-tagged bovine serum albumin on silica challenged by native albumin at 100 ppm. Arrow indicates time where flowing solvent was replaced by flowing solution of native species.

(c)

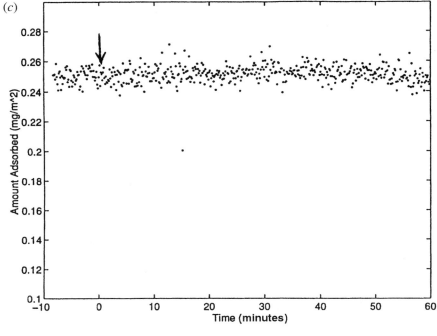

FIGURE 14.9 Continued

native bovine serum albumin on silica. In these studies, the labeled polymers have been preadsorbed and incubated for times on the order of 30 min. These adsorbed layers are exposed to an unlabeled version of the same polymer, at time zero. For PEO, there proceeds an obvious self-exchange process; however, hydroxyethyl cellulose and albumin do not undergo significant exchange over the same time period.

The results suggest that there are fundamental differences between the adsorption of PEO and the other polymers. While PEO and HEC are both uncharged water-soluble polymers capable of hydrogen bonding with silica, a key difference between these two chains is the backbone flexibility. It appears that chains with stiff backbones are generally slower to equilibrate interfacially. The reason for albumin's resistance to self-exchange is probably quite different from that of HEC. Albumin contains regions of hydrophobicity and positive and negative charge. When albumin relaxes on a surface, it unfolds to maximize the extent of electrostatic and hydrophobic interactions with specific points on the surface. Hence, in a qualitative sense, the results in Figure 14.9 argue that both backbone flexibility and polymer–surface interactions can be responsible for apparently irreversible adsorption.

A second aspect of interfacial dynamics occurs when two different species compete for the same surface. Such competition during adsorption involves the same fundamentals as the self-exchange above, but in addition to interfacial dynamics, is complicated by the kinetics of an evolving surface. Competition during adsorption is widespread, even in systems that are thought to be relatively simple, for example, one kind of homopolymer adsorbing onto a single surface. In most instances of homopolymer adsorption, the polymer sample is not monodisperse. In polydisperse samples, short and long chains of the same chemical makeup compete for a limited number of surface sites. For classical homopolymer adsorption involving a layer comprised of tails, loops, and trains, long chains are always preferred on the surface because the short chains remaining in solution maximize the translational entropy. At short times, however, short chains dominate the surface, simply because they diffuse to the surface from the bulk solution more rapidly than the long chains. Equilibration must therefore involve an exchange of long and short chains.

Figure 14.10 examines the issue of adsorption kinetics from a polydisperse solution by breaking the problem down to competition within a bimodal sample of short and

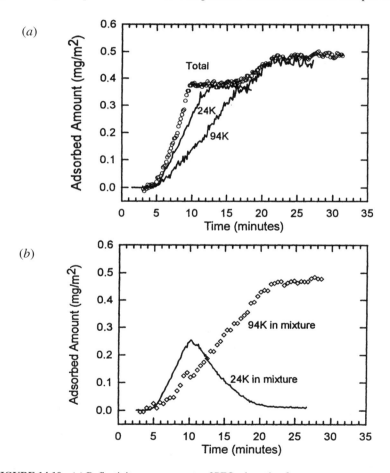

FIGURE 14.10 (*a*) Reflectivity measurements of PEO adsorption for one-component samples (2.5 ppm each) 20,000 and 94,000 PEO (——) and a 50–50 mixture of the two with a total concentration of 5 ppm (○). (*b*) TIRF measurements of 24,000 PEO (——) and 94,000 PEO (◇) populations during their adsorption from the binary mixture in part (*a*).

long chains. Here, the system of study is PEO, with molecular weights of 24,000 and 94,000. A series of experiments are presented: First the noncompetitive adsorption behavior of the two samples is determined, as illustrated by the lines. Comparing the two samples adsorbing noncompetitively, one finds, as expected, that the shorter chains exhibit the faster adsorption rate as a result of their greater free solution diffusivity. The longer chains show a slightly higher ultimate coverage. A 50–50 mixture of the two, with concentrations of each component corresponding to the pure species concentrations in the noncompetitive runs, shows an adsorption profile containing two steps. The steplike nature of the mixture run fits well with the runs for the individual species. The initial slope for the mixture run is the sum of the initial slopes for the individual runs, demonstrating that at short times there is no significant surface competition between the two species. When the total surface coverage of the mixture run reaches a coverage corresponding to saturation of the surface by short chains alone, the adsorption appears to level off momentarily. The adsorption process ultimately appears to continue again in a way that matches the tail end of the noncompetitive run for the long chains.

The TIRF experiments targeting individual surface populations of short and long chains elucidate the actual events in Figure 14.10*a*. In Figure 14.10*b* the evolution of surface populations, measured by TIRF, during the competitive run is presented. The results show that at short times the adsorption of each population proceeds independent of the other. In fact, the long chains adsorbing competitively behave identically to their noncompetitive adsorption in Figure 14.10*a*. The short chains, however, have a limited residence on the surface. At the time when the total surface coverage corresponds to saturation for the short chains, the short chains no longer adsorb, but are, in fact, displaced by the continued adsorption of the longer chains. A quantitative analysis of the rise and decay rates for the various populations reveals that the entire process proceeds at a mass-transport-limited rate.[27,28]

14.5   SUMMARY

This chapter provides a detailed description of total internal reflectance fluorescence (TIRF) in terms of its theory of operation, variations in instrumental capabilities, and calibration methods. Next, example data were presented to illustrate the application of TIRF in the field of polymer adsorption. Qualitative explanations of the data were provided where appropriate, with references to the complete works in the literature. These examples demonstrated TIRF's use in measuring polymer adsorption kinetics, desorption kinetics (or lack thereof), the interfacial environment, self-exchange, and exchange processes during competitive adsorption.

Acknowledgments   The work summarized here was made possible by Lehigh's Polymer Interfaces Center and grants from the National Science Foundation (CTS-9209290, 9310932), the Electric Power Research Institute (RP-801902), the Whitaker Foundation (RG 94-0355), the Exxon Educational Foundation, and the

Petroleum Research Foundation (23917-G7). Special thanks go to Z. Fu, E. Mubarekyan, and D. Lorenz for their help with figures.

REFERENCES

1. A. Takahashi, M. Kawaguchi, H. Hirota, and T. Kato, *Macromolecules* **13**, 884 (1980).
2. A. Takahashi and M. Kawaguchi, *Adv. Polym. Sci.* **46**, 1 (1982).
3. M. Kawamaguchi, K. Hayakawa, and A. Takahashi, *Macromolecules* **16**, 631 (1983).
4. J. Lee and G. Fuller, *Macromolecules* **17**, 375 (1984).
5. M. Malmsten and F. Tiberg, *Langmuir* **9**, 1098 (1993).
6. T. Cosgrove, T. Crowley, K. Ryan, and J. Webster, *Colloids Surfaces* **51**, 255 (1990).
7. T. Cosgrove, T. Heath, R. Richardson, and J. Phipps, *Macromolecules* **24**, 94 (1991).
8. E. Lee, R. Thomas, and R. Rennie, *Europhys. Lett.* **13(2)** 135 (1990).
9. S. Vaslin-Piana, F. Lafuma, and R. Audebert, *J. Colloid Interface Sci.* **119**, 476 (1990).
10. G. van der Beek, M.A. Cohen Stuart, and T. Cosgrove, *Langmuir* **7**, 327 (1991).
11. M.A. Cohen Stuart, F. Waajen, T. Cosgrove, B. Vincent, and T. Crowley, *Macromolecules* **17**, 1825 (1984).
12. J. Dijt, M.A. Cohen Stuart, and G. Fleer, *Macromolecules* **25**, 5416 (1992).
13. M.A. Cohen Stuart and H. Tamai, *Macromolecules* **21**, 1863 (1988).
14. H. Sakai, T. Fujimori, and Y. Imamura, *Bull. Chem. Soc. Jap.* **53**, 3457 (1980).
15. D. Kuzmenka and S. Granick, *Colloids and Surfaces*, **31**, 105 (1988).
16. P. Frantz and S. Granick, *Phys. Rev. Lett.* **66(7)**, 899 (1991).
17. K. Barnett, T. Cosgrove, B. Vincent, B. Sissons, and M.A. Cohen Stuart, *Macromolecules* **14**, 1018 (1981).
18. F. Blum, *Colloids Surfaces* **45**, 361 (1990).
19. E. Enriquez and S. Granick, *Colloids Surfaces A: Physchem. Eng. Aspects* **113**, 11 (1996).
20. B. Lock, Y. Cheng, and C. Robertson, *J. Colloid Interface Sci.* **91**, 104 (1983).
21. V. Hlady, D. Reinecke, and J. Andrade, *J. Colloid Interface Sci.* **111**, 555 (1986).
22. V. Hlady, J. Rickel, and J. Andrade, *Colloids Surfaces* **34**, 171 (1988).
23. T. Watts, H. Gaub, and H. McConnell, *Nature* **320(13)**, 179 (1986).
24. V. Rebar, Ph.D. Thesis, Lehigh University, 1995.
25. I. Caucheteux, H. Hervet, R. Jerome, and F. Rondelez, *J. Chem. Soc. Farad. Trans.* **86(9)**, 1369 (1990).
26. M. Santore and X. Liu (in preparation).
27. M. Santore and Z. Fu, *Macromolecules* **30**(26), 8516 (1997).
28. Z. Fu and M. Santore, *Langmuir* **14(15)**, 4300 (1998).
29. M. Yanagimachi, M. Toriumi, and H. Masuhara, *Chem. Matls.* **3(3)**, 413 (1991).
30. M. Born and E. Wolf, *Principles of Optics*, Pergamon, New York, 1983.
31. Z. Fu and M. Santore, *Colloids Surfaces A: Physiochem. Eng. Aspects* **135**(1–3), 63 (1998).
32. F. Rondelez, D. Ausserre, and H. Hervet, *Ann. Rev. Phys. Chem.* **38**, 317 (1987).

33. C. Shibata and A. Lenhoff, *J. Colloid Interface Sci.* **148**, 469 (1992).

34. V. Rebar and M. Santore, *Macromolecules* **29**, 6263 (1996).

35. C. Shibata and A. Lenhoff, *J. Colloid Interface Sci.* **148**, 485 (1992).

36. M. Kelly and M. Santore, *Colloids Surfaces A: Physiochem. Eng. Aspects* **6**, 199 (1995).

37. L. Bousse, N. DeRooij, and P. Bergvald, *Surf. Sci.* **135**, 479 (1983).

38. Z. Fu and M. Santore, *Langmuir* **13**, 5779 (1997).

39. J. Dijt, M.A. Cohen Stuart, J. Hofman, and G. Fleer, *Colloids Surfaces* **51**, 141 (1990).

40. P.G. deGennes, *Adv. Colloid Interface Sci.* **27**, 190 (1991).

41. V. Rebar and M. Santore, *J. Colloid Interface Sci.* **178**, 29 (1996).

42. J. Slavik, *Flourescent Probes in Cellular and Molecular Biology*, CRC, Ann Arbor, MI, 1994, p. 201.

43. E. Pfefferkorn, A. Carroy, and R. Varoqui, *J. Polymer Sci. (Phys.)* **23**, 1997 (1985).

15 Design and Applications of Oscillating Optical Tweezers for Direct Measurements of Colloidal Forces

H. DANIEL OU-YANG

Department of Physics, Lehigh University, Bethlehem, Pennsylvania 18015

15.1. INTRODUCTION

Optical tweezers, first developed by A. Ashkin et al. almost three decades ago,[1] have found wide use both in biological applications[2] and in colloid–polymer applications.[3-6] The technique became truly user friendly when Ashkin and co-workers demonstrated that it is possible to use a single beam with large gradient force to form an optical trap (also known as optical tweezers) for colloidal particles.[7] A review of optical tweezers applications that covers the development and many applications published between 1970 and 1995 can be found in an article by Kuo.[8]

In this chapter we demonstrate an approach that utilizes the phase-sensitive detection of the dynamic position of a single colloidal particle undergoing a forced oscillation driven by optical tweezers. By measuring the particle's position, we can calculate the dynamic forces on the particle.[9] This approach differs from that of conventional applications of optical tweezers where the optical trap is usually stationary. The advantage of using oscillating optical tweezers is that one can use it to measure frequency-dependent, dynamic properties of polymer–colloid systems at the colloidal level. The phase-sensitive lock-in method measures the phase shift and displacement of the oscillating particle relative to the oscillating optical tweezers. It can be shown that phase-sensitive measurements provide a greater sensitivity for dynamical measurements than most other techniques. Phase-sensitive measurements

Colloid–Polymer Interactions: From Fundamentals to Practice, Edited by Raymond S. Farinato and Paul L. Dubin
ISBN 0-471-24316-7 © 1999 John Wiley & Sons, Inc.

also have an advantage over direct displacement measurements in that they are less prone to the optical contrast variation that usually occurs in optical microscopy.

In what follows, we start with a brief description of the principle of optical tweezers trapping individual colloidal particles. Although there are several approaches to deal with the theory of optical trapping of particles, we choose the electrostatic approximation. It is probably the easiest to understand, and it is sufficient for someone who wishes to design a similar system.

We then review the equation of motion of a particle in a forced harmonic motion. For simplicity, we treat the medium as viscoelastic with constant viscosity and elasticity. Because we can explore high frequencies, the Stokes drag is modified to include frequency-dependent viscous drag and inertia terms; the latter includes not only the particle mass but also a part related to the frequency-dependent momentum transfer to the surrounding liquid. The equation of motion is solved to give the time-dependent displacement and phase shift of the particle's motion in the laboratory reference frame. When the displacement and phase shift are measured relative to the center of the optical trap, we can use expressions derived for the (moving) optical tweezers reference frame.

Following the theoretical treatment, an experimental setup with the necessary optical components is given. Three modes of detection are described, including direct imaging, the single-beam forward-scattering method, and the dual-beam forward-scattering method. The direct-imaging method is a laboratory frame measurement. It is easy to set up but fails at high frequencies. The forward-scattering method measures a particle's motion in the reference frame of the oscillating laser. It has an excellent high-frequency response but is insensitive at very low frequency. A dual-beam method, with one laser beam oscillating and the other stationary, measures a particle's motion in the laboratory frame and provides good response at both low and intermediate frequencies.

All three methods of detection are tested against the theoretical treatment. We ran these tests on individual colloidal particles in water, in water–glycerol mixtures, and in low-concentration polymer solutions. Most of the results are obtained using polystyrene latex spheres. Some results are from measurements of surfactant-stabilized oil droplets in water. The experimental results agree well with our theoretically calculated predictions.

The theory presented here for solutions with constant viscosity and elasticity can be extended readily to frequency-dependent viscoelastic polymer–colloid systems. One possible application is to study the microviscoelasticity of colloidal particles embedded in polymer gels.[10] Another potential application of this technique is to study the interactions between a pair of particles, each held by an optical tweezers.

15.2. PRINCIPLE OF OPTICAL TWEEZERS

By strongly focusing a laser beam, a very strong electric field is formed at the focal point, and a large electric field gradient is formed in both the axial (the laser propagation direction) and radial directions. A sufficiently steep field gradient can create a force on a colloidal particle large enough to counter Brownian motion, thus yielding a stable optical trap in all three dimensions. The balance of gradient and scattering forces in the axial direction causes the potential minimum for the trapped particle to be slightly downstream from the focal point of the lens. The achievement of axial stability, due to the availability of high numerical aperture (NA) objective lenses, makes the trap suitable for a wider range of applications. The magnitudes of the forces are generally quoted to be about 1 picoNewton per milliwatt (pN/mW) of power at the trap site. Because of their relatively noninvasive nature, laser tweezers are ideal for probing individual colloids and cells in their microscopic environments.

Several theoretical approaches have been proposed to explain the physics of laser trapping of dielectric particles. The simplest is an electrostatic model, such as the parallel-plate capacitor model: The potential energy of the system (capacitor plus dielectric) is lowered when a dielectric material is drawn into the capacitor. Optical tweezers can be explained in the same manner: Particles of higher (at optical frequencies) dielectric constant than the surroundings are drawn into the high electric field provided by the tightly focused laser beam. The fact that the laser electric field oscillates at high frequency does not matter because, as shown below, the potential energy of the particle in the field is proportional to the square of the field strength so that the particle can follow the direct current (DC) component potential whereas the second harmonic frequency is much too high for particles to follow.

However, the electrostatic model is not complete. The problem is that the electric field produced by a laser is not electrostatic—rather it is a traveling electromagnetic (EM) wave. A traveling EM field exerts radiation pressure on the surfaces on which it impinges. In other words, the laser beam tends to push the particle in the direction of laser propagation. Trapping stability depends on the competition between the scattering force in the direction of beam propagation, proportional to intensity, and the gradient force, proportional to the gradient of intensity. It is known that not only will lower dielectric constant particles be ejected from the trap, but so will hollow particles and highly reflective materials.[11] For a more precise description, the Mie or Rayleigh scattering models must be used for particles comparable to, or smaller than, the size of the optical trap. When a particle is much larger than the size of the trap, a geometrical optics model should be used to calculate the trapping force.[12]

The potential energy of a homogeneous, linear dielectric particle in an electric field E is given by[13]

$$U = \frac{1}{2} \int \mathbf{D} \cdot \mathbf{E}^* \, d^3x \qquad (15.1)$$

where $\mathbf{D}$ is the electric displacement vector equal to $\varepsilon\mathbf{E}$, where ε is the dielectric constant at optical frequencies. The difference in potential energy is

$$\Delta U = U_2 - U_1 = \frac{1}{2} \int [\mathbf{D}_2 \cdot \mathbf{E}_2^* - \mathbf{D}_1 \cdot \mathbf{E}_1^*] d^3x$$

$$= \frac{1}{2} \int [\mathbf{D}_1 \cdot \mathbf{E}_2^* - \mathbf{D}_2 \cdot \mathbf{E}_1^*] \, d^3x + \frac{1}{2} \int (\mathbf{E}_1 + \mathbf{E}_2^*) \cdot (\mathbf{D}_2 - \mathbf{D}_1^*) \, d^3x \quad (15.2)$$

where U_2 (U_1) is the energy when the particle is inside (outside) the trap. If the source charge density is assumed unchanged, the second integral on the right-hand side of Eq. (15.2) vanishes, in which case, Eq. (15.2) is reduced to[13]

$$\Delta U = \frac{1}{2} \int [\varepsilon_1 \mathbf{E}_1 \cdot \mathbf{E}_2^* - \varepsilon_2 \mathbf{E}_2 \cdot \mathbf{E}_1^*] d^3x \quad (15.3)$$

Assuming harmonic fields, we have

$$\mathbf{E} = \mathbf{E}_c \, e^{-i\omega t}$$

$$\mathbf{D} \cdot \mathbf{E}^* = \varepsilon|\mathbf{E}|^2 = \varepsilon|\mathbf{E}_c|^2 \quad (15.4)$$

Because we require $|\mathbf{E}|^2$ to be real, the amplitude $\mathbf{E}_c$ must be either purely real or purely imaginary. Thus

$$\mathbf{E}_1 \cdot \mathbf{E}_2^* = \mathbf{E}_1^* \cdot \mathbf{E}_2$$

$$\Delta U = \frac{1}{2} \int (\varepsilon_1 - \varepsilon_2)\mathbf{E}_1^* \cdot \mathbf{E}_2 \, d^3x \quad (15.5)$$

It can be shown that for a dielectric sphere in an external field,[13]

$$\mathbf{E}_2 = \left(\frac{3\varepsilon_1}{\varepsilon_2 + 2\varepsilon_1} \right) \mathbf{E}_1 \quad (15.6)$$

And thus,

$$\Delta U = \frac{3\varepsilon_1}{2} \left(\frac{3\varepsilon_2 - \varepsilon_1}{\varepsilon_2 + 2\varepsilon_1} \right) \int_{V_2} |\mathbf{E}_1|^2 \, d^3x \quad (15.7)$$

where V_2 is the volume of the particle. Assuming that the radius of the particle is the same size as or smaller than the trap spot size, and that the electric field is approximately uniform over the particle volume, we can rewrite Eq. (15.7):

$$\Delta U = -\frac{3n_1^2\varepsilon_0}{2} \left(\frac{n_2^2 - n_1^2}{n_2^2 + 2n_1^2} \right) |\mathbf{E}_1|^2 \, V_2 \quad (15.8)$$

where n is the index of refraction, ε_0 is the dielectric permitivity of free space. Here, we have used $\varepsilon = K \varepsilon_0$, and $K \approx n^2$. Thus, when $n_2 > n_1$, $U_2 < U_1$ and the energy is lower if the particle is in the trap. For example, at optical frequencies, the index of refraction for water is $n_{\text{water}} = 1.33$, and for polystyrene spheres, $n_{\text{ps}} = 1.57$; micron-sized polystyrene spheres in water can be trapped easily.

Since we will need only to calculate the trapping forces in the radial direction, we will not discuss the rather complicated calculation of the axial trapping force. All we need for the present application is a large enough NA objective so that the particle is stable axially. In what follows, we present an electrostatic argument for the radial component of the gradient force.

In order to estimate the radial trapping force, we rewrite the potential energy in terms of the intensity via the time averaged Poynting vector:

$$I = <\mathbf{S}> = \frac{1}{2} \frac{1}{\mu_1 n_1 c} |\mathbf{E}_1|^2 = \frac{c\varepsilon_0 n_1}{2} |\mathbf{E}_1|^2$$

$$I = I_0 e^{-r^2/R^2} \tag{15.9}$$

where c is the speed of light in vacuum, μ is the magnetic permeability, and r is the radial position measured from the center of the trap. A Gaussian intensity profile with a $1/e$ width of R at the center of the trap has been assumed for the radial direction, and the potential energy measured relative to a particle at infinity is

$$U = \frac{-3V_2 n_1}{c} \left(\frac{n_2^2 - n_1^2}{n_2^2 + 2n_1^2} \right) I_0 e^{-r^2/R^2} \tag{15.10}$$

Now the force F can be found for a particle with a radial displacement from the center of the trap:

$$\mathbf{F} = -\nabla U = -\frac{6r V_2 n_1}{cR^2} I_0 \left(\frac{n_2^2 - n_1^2}{n_2^2 + 2n_1^2} \right) e^{-r^2/R^2} \hat{r}$$

$$\equiv -k_{\text{ot}} r e^{-r^2/R^2} \hat{r} \tag{15.11}$$

For small displacements the force is a Hooke's law force with spring constant k_{ot} that is, $F = k_{\text{ot}} r$. For a polystyrene sphere ($n_2 = 1.57$, radius $a = 0.5$ μm) in water ($n_1 = 1.33$) and assumed power at the trap ($R = 0.5$ μm) of 1 mW the spring constant is approximately 8 mdyn/cm. At a displacement of $r \sim 0.25R$ the force is approximately 1 pN. A comparison of the linear approximation with the force given by Eq. (15.11) is shown in Figure 15.1. It is calculated that the linear approximation is good to within 4% when the particle is displaced about $0.2\,R$ from the center of the trap or within 10% when particle is displaced about $0.32\,R$.

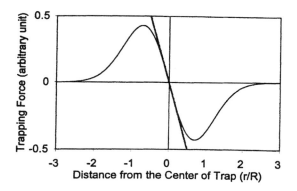

FIGURE 15.1   Comparison of the Hooke's law force (linear approximation) with the force given by Eq. (15.11).

15.3.   EQUATION OF MOTION FOR A COLLOIDAL PARTICLE IN FORCED OSCILLATION IN A VISCOELASTIC MEDIUM

An individual particle in suspension can be forced into an oscillatory motion by optical tweezers that are steered by a piezoelectric transducer (PZT) controlled mirror. A single particle of radius a, set in motion (say, along the x axis with velocity v) experiences the following forces: (1) a springlike force $- k_{ot}\, x$, exerted by the optical tweezers; (2) the viscous force $-6\pi\eta a v$, due to the Stokes drag in solution; and (3) an elastic restoring force $-kx$, where k is the effective spring constant of the solution of viscosity η. In a viscoelastic medium, both the viscosity and elasticity depend on frequency; for simplicity, we assume both are constant. However, because we consider oscillations over a broad oscillation frequency range, we need to include the frequency-dependent hydrodynamic modification terms to the Stokes drag.[14] Figure 15.2 illustrates all the forces experienced by a particle in a viscoelastic medium.

The equation of motion for the particle is

$$m^*\ddot{x} + 6\pi\eta^* a\dot{x} + (k_{ot} + k)x = k_{ot}A\,\cos(\omega t) \tag{15.12}$$

where m^* is the effective mass of the particle, a is the radius of the colloidal particle, η^* is the effective viscosity, k is the elastic modulus of the solution, and A is the amplitude of the tweezers' oscillation of frequency ω. It is important to note that the effective mass m^* includes not only the bare mass of the particle m_0 but also the inertia of the liquid around the particle when the particle is set into an oscillating motion. According to Landau and Lifshitz,[14] the effective mass m^* can be expressed as

$$m^* = m_0 + \frac{2\pi}{3}\,a^3\rho_s + 3\pi a^2\,\sqrt{\frac{2\eta_s\rho_s}{\omega}} \tag{15.13}$$

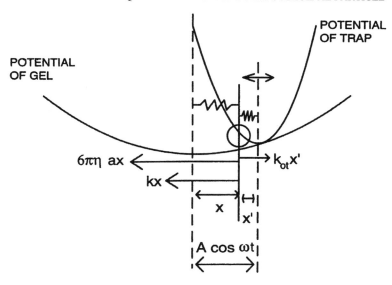

FIGURE 15.2 Forces on a single particle set in forced oscillation in a viscoelastic medium.

and the effective viscosity η^* can be expressed as

$$\eta^* = \eta_s\left(1 + \frac{\sqrt{a^2\rho_2\omega}}{2\eta_2}\right) \qquad (15.14)$$

where ρ_s and η_s are the solution density and viscosity, respectively. Equation (15.12) has a steady-state solution:

$$x(t) = D(\omega)\cos(\omega t - \delta(\omega)) \qquad (15.15)$$

where the amplitude and the phase shift of the response are

$$D(\omega) = \frac{k_{ot}A}{\sqrt{(k_{ot} + k - m^*\omega^2)^2 + m^{*2}\beta^2\omega^2}} \qquad \delta(\omega) = \tan^{-1}\frac{m^*\beta\omega}{k_{ot} + k - m^*\omega^2} \qquad (15.16)$$

where

$$\beta = \frac{6\pi\eta^* a}{m^*} \qquad (15.17)$$

In Figure 15.3a, we show the calculated phase shift versus $\log(\omega)$ for a 1.0-μm-diameter polystyrene particle in water (density = 1 g/cm^3, η_s = 0.01 poise). The spring

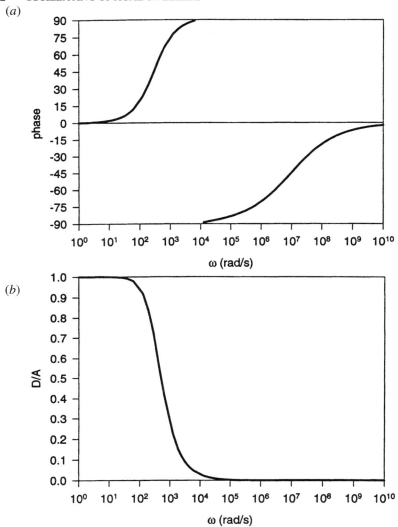

FIGURE 15.3 (*a*) Phase shift vs. log(ω) for a 1.0-μm diameter polystyrene particle in water (density = 1 g/cm^3, η_s = 0.01 poise). The spring constant k_{ot} is 3 mdyn/cm. (*b*) Relative displacement D/A vs. log(ω) for a 1.0-μm polystyrene particle (density 1.05 g/cm^3) in water (density = 1 g/cm^3, η_s = 0.01 poise). The spring constant k_{ot} is 3 mdyn/cm.

constant k_{ot} is 3 mdyn/cm. In Figure 15.3*b*, we show the calculated D/A versus log(ω) in the same system. We can see that for most systems of interest the system is highly overdamped, and both $D(\omega)$ and $\delta(\omega)$ vary appreciably over the range $0 < \omega < 10^4$ rad/s. This allows for measurement of the viscoelastic properties of the polymer solution in the frequency range of $0 < \omega < 10^4$ rad/s (shear rate range up to 10^3 to 10^5 s^{-1}). It is therefore feasible to determine the spring constant of the trap from the phase

[$\delta(\omega)$] and displacement [$D(\omega)$] measurements made on polystyrene particles in water, as water is not elastic in our frequency range (<7 kHz) and the viscosity of water is a known function of temperature.

In the next section, we describe the experimental setup and the procedures to calibrate the spring constant k_{ot} of the optical tweezers.

15.4. EXPERIMENTAL SETUP

Here we introduce a setup that allows us to measure the motion of a colloidal particle undergoing forced oscillation, and we describe the phase lock-in detection technique. A schematic of the experimental setup for both single-beam and dual-beam optical tweezers is shown in Figure 15.4. In this diagram, we include all three methods of

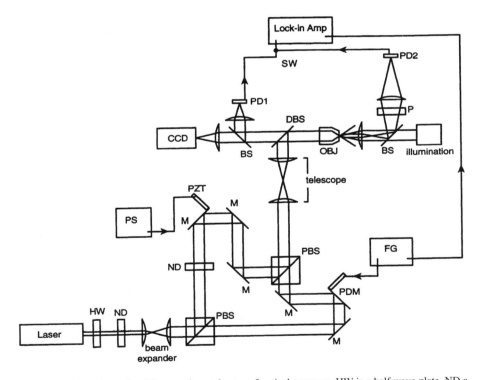

FIGURE 15.4 Schematic of the experimental setup of optical tweezers. HW is a half wave plate, ND a neutral density filter, PBS a polarizing beam splitter, M a mirror, PS a power supply for piezoelectric-driver (PZT), PDM a piezoelectric-driven mirror, FG function generater, BS a beam splitter, DBS a dichroic beam splitter, which reflects the green laser light and allows the illuminating light to pass. The OBJ is the high NA objective lens, PD1 and PD2 are splitted photodiode detectors, P a polarizer and SW a switch to allow signal from PD1 or PD2 to go to the lock-in amplifier. The sample chamber and the trapped particle, both not shown, are located directly to the right of OBJ.

detection: direct imaging, single-beam forward scattering, and dual-beam forward scattering.

As shown in Figure 15.4, our optical tweezers are built on an Olympus IX-70 inverted microscope with an Olympus Plan-Apo 100X (N.A. 0.5–1.35) oil-immersion objective lens. The laser is a Spectra-Physics Millennia Nd:YVO$_4$ laser at a wavelength of 532 nm (frequency-doubled, maximum power 5.5 W). The laser beam is steered by a PZT-driven mirror (Physik Instrumente, P830-40). A sinusoidal signal created by a Stanford Research Systems frequency synthesizer (SRS DS-345) is fed in a piezoelectric driver (Physik Instrumente, P863) to drive the steering mirror. A CCD (MTI CCD72) camera is used to generate a video image of the trapped particle for viewing and optical alignment. The actual measurements of the particle motions are made on the signal detected by a split photodiode detector (Hamamatsu S4204). The output electrical current signal from the split photo diode detector is fed into a lock-in amplifier (SRS 830). The reference signal is taken from the frequency synthesizer (SRS DS-345) that is used to drive the mirror. The lock-in amplifier measures the magnitude and the phase shift of the signal from the photodiode detector relative to that of the driving signals. A beam split from the beam reflected off the steering mirror can also be detected by a separate photodiode detector for determination of the frequency dependence of the steering mirror's displacement and phase shift. Note that the phase shift described in Eq. (15.16) is the relative phase shift of the particle motion to the mirror motion. The displacement described in Eq. (15.16) is the relative motion of the measured particle's displacement to the laboratory reference frame. The amplitude A of the motion of the optical tweezers is, to a good approximation, linearly proportional to the amplitude of the motion of the mirror.

15.5. EXPERIMENTAL RESULTS AND DISCUSSIONS

15.5.1. Direct Imaging Method

By imaging the trapped particle onto a split photodiode detector (PD1 in Fig. 15.4), we are actually creating a shadow of the particle in a bright background. When the particle is set into motion by the oscillating tweezers, the light distribution, with a dark spot near the center of the split photodiode detector, changes spatially with time. The lock-in amplifier analyzes the photo currents measured by the split photodiode PD1 and provides a measure of the magnitude and phase of the particle's motion. The lock-in amplification technique suppresses random noises caused by the particle's Brownian motion and other sources of noise caused by fluctuation of the bright background.

We tested the direct imaging method by measuring the phase shift of a 1.1-μm-diameter polystyrene latex sphere in water, from which we obtained the measured optical tweezers' spring constant k_{ot}. Using the same spring constant, we are able to determine the viscosity of dilute polymer solutions in water and to compare our data with direct viscosity measurements. The polymer used was 100,000 g/mol polyethyleneoxide (PEO) with both ends capped with a 16-carbon alkyl group

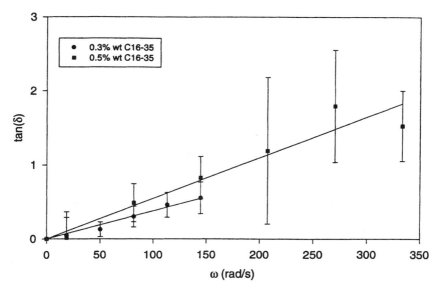

FIGURE 15.5   Plot of tan δ vs. angular frequency ω for C16-100 PEO in water solution.

(C16-100 PEO). Figure 15.5 shows a plot of tan δ versus angular frequency ω. The linear dependence in the plot is expected from Eq. (15.16). With a known spring constant, we can determine the solution viscosities of the polymer solutions. Figure 15.6 shows a comparison of the data obtained by this method and that obtained with

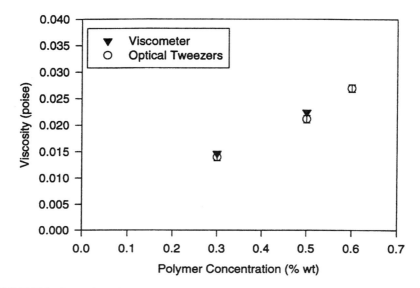

FIGURE 15.6   Comparison of viscosities measured by the optical tweezers and by a capillary viscometer.

a capillary viscometer; both methods are at shear rates less than 100 s $^{-1}$. The direct imaging method works well for low frequencies, but the bright background noise sets an intrinsic limit on the range of frequency response. High-frequency response can be accomplished by forward-scattering methods, as shown below.

15.5.2. Forward-Scattering Method

The motion of the trapped particle relative to the optical tweezers affects the forward-scattering pattern created by the laser. We can use the time variation of the forward-scattering intensity distribution at a split photodiode (PD2 in Fig. 15.4) to measure the desired particle motion in the trap.[15] The concept of the detection is given in Figure 15.7, where a split photodiode is positioned at the back focal plane (BFP′) of the condenser. As shown in Figure 15.7, the trapped particle can be treated as a lens that collimates the laser beam and projects the forward scattering into a tight spot on the split photodiode. When the particle is at the center of the optical trap, as shown in (15.7a), the forward-scattering light is projected to the center of the split photodiode, yielding a zero photo current. When the particle is away from the center of the trap, as shown in (15.7b), the forward-scattered light is projected off-center of the photodiode, yielding a photo current proportional to the displacement of the particle from the center of the trap. When the trapped particle undergoes oscillation in the direction perpendicular to the laser beam, the forward-scattering spot executes a similar motion at the position of the split photodiode. It should be noted, however, that the signal so obtained is a measure of the motion of the particle relative to the trapping beam center—a moving reference frame.

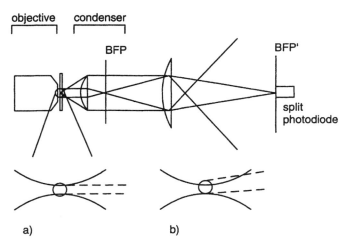

FIGURE 15.7 Working principle of forward-scattering detection method. When the particle is at the center of the optical trap, as shown in (a), the forward-scattered light is projected to the center of the split photodiode. When the particle is away from the center of the trap, as shown in (b), the forward-scattered light is projected off-center of the photodiode.

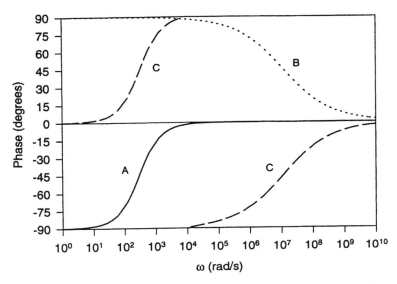

FIGURE 15.8 Plot of calculated phase shift in the moving trap reference frame δ vs. log (ω) for a 1.0-μm polystyrene particle (density 1.05 g/cm^3) in water (density = 1 g/cm^3, η_s = 0.01 poise). The spring constant k_{ot} is 3 mdyn/cm. The curve A is the moving frame phase shift δ', the curve B is the first term, and curve C is the second term on the right side of Eq. (15.20).

The solution given in Eq. (15.16) is for the motion of a particle relative to the laboratory reference frame. Transformation between the two reference frames is straightforward. It can be shown that the measured phase shift in the trap frame is related to the phase shift in the laboratory frame by:

$$\delta'(\omega) = \tan^{-1}\frac{m^*\beta\omega}{m^*\omega^2 - k} + \tan^{-1}\frac{m^*\beta\omega}{(k + k_{ot}) - m^*\omega^2} \qquad (15.18)$$

Note that the second term on the right side of the above equation is exactly the phase shift in the laboratory frame. Figure 15.8 shows each term in Eq. (15.18) as a function of log(ω). The curve A is the moving frame phase shift δ', the curve B is the first term on the right side of the equation, and curve C is the second term.

The measured displacement in the trap frame, normalized to the amplitude of the trap motion, is

$$\frac{D'(\omega)}{A} = \sqrt{\frac{(m^*\beta\omega)^2 + (k - m^*\omega^2)^2}{(m^*\beta\omega)^2 + (k_{ot} + k - m^*\omega^2)^2}} \qquad (15.19)$$

Figure 15.9 shows the relative displacement $D'(\omega)/A$, both as a function of log(ω).

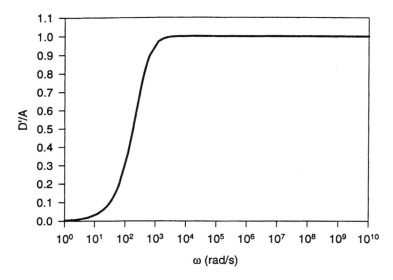

FIGURE 15.9   Plot of normalized displacement in the moving trap reference frame D'/A vs. log (ω) for a 1.0-μm polystyrene particle (density 1.05 g/cm^3) in water (density = 1 g/cm^3, η_s = 0.01 poise). The spring constant k_{ot} is 3 mdyn/cm.

In doing the forward-scattering measurements, we obtain directly the moving reference frame data for both phase shift δ' and displacement D'/A. To compare experimental phase shift δ' to Eq. (15.18) we can fit the equation to the data by iteration, where the spring constant k_{ot} is the only fitting parameter ($k = 0$ in water). The data were fit to the phase shift up to 90° (at 4000 rad/s), beyond which the data were not reliable. An error of a few degree near 90° phase shift can cause large errors in the fit. We believe the errors are caused by misalignment of the optical tweezers. Once the parameters are known, one can readily transform the data back to the laboratory reference frame. Figure 15.10 shows the laboratory frame phase shift δ versus log(ω) for a 1.1-μm diameter polystyrene latex sphere in water at the room temperature of about 23°C. (At the estimated laser power of less than 1 mW at the optical trap, we do not expect the trap temperature to be different from the ambient temperature.) The solid line is a fit to Eq. (15.18) with a laser power of 15 mW (reference power) measured just before the laser beam enters the microscope. Depending on the optical elements inside the microscope, only 2 to 5 % of the reference power reaches the focal point, where the optical tweezers are located; a precise percentage of the transmitted power is difficult to determine.

Figure 15.11 shows the trap frame displacement normalized to the amplitude of the trap motion. The condition of the experiments is the same as that shown in Figure 15.10. The displacement and phase shift data were taken simultaneously by the lock-in amplifier.

According to Eq. (15.11), the spring constant should be linearly proportional to the laser power at the optical trap. We carried out a calibration of the spring constant for 1.1-μm-diameter polystyrene particles and for similar sized

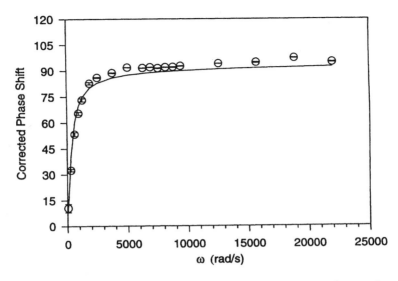

FIGURE 15.10 Laboratory frame phase shift δ vs. log (ω) for a 1.1-μm diameter polystyrene latex sphere in water at the room temperature of about 23°C.

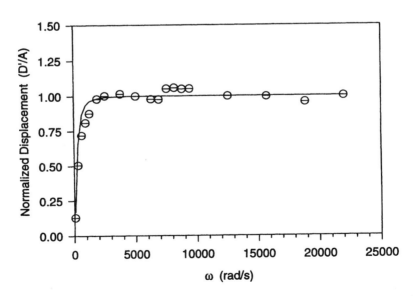

FIGURE 15.11 Trap frame displacement normalized to the amplitude of the trap motion. The condition of the experiments is the same as that shown in Figure 15.10. The displacement and phase shift data were taken simultaneously by the lock-in amplifier.

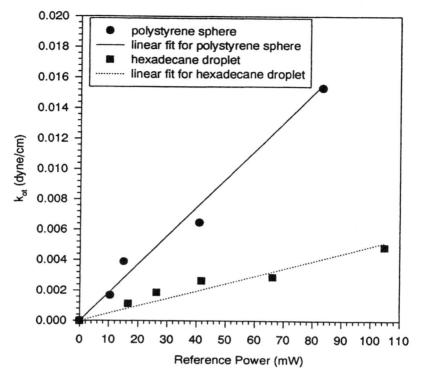

FIGURE 15.12 Plot of the measured spring constant vs. reference laser power. The linear relation between the spring constant and the laser power is predicted in Eq. 15.11.

surfactant-stabilized (TritonX-100) hexadecane oil droplets, both in pure water. As shown in Figure 15.12, reasonable linear fits are obtained for both the hard particles and oil droplets. The solid line in Figure 15.12 is not a best-fit line; instead it is calculated from Eq. (15.11) with all the known parameters for polystyrene particle, and with an assumption that $a = R$.

The dashed line in Figure 15.12 is a best fit to the oil droplet data. To calculate the spring constant for oil droplets in the trap is more difficult. The main difficulty is that the Stokes drag coefficient of the oil droplet may depend on the effects of surfactant molecules on the oil–water interface. The results shown here are based on the drag coefficient of a moving liquid particle in another liquid, with the assumption that the Marangoni effect is negligible. The drag coefficient is given by[16]

$$F_{drag} = -4\pi\eta a \left[\frac{1 + (3\eta_{oil}/2\eta_s)}{1 + (\eta_{oil}/\eta_s)} \right] V_\infty \qquad (15.20)$$

where η_{oil} (= 3.032 cp) is the viscosity of the oil droplet and V_∞ is the fluid velocity far from the droplet.

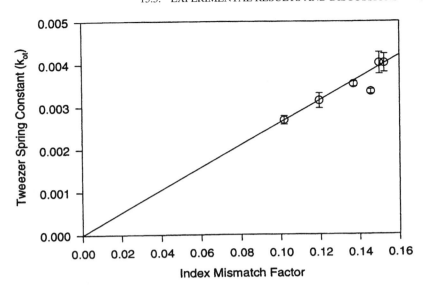

FIGURE 15.13 Measured spring constant vs. refractive index mismatch factor at a constant reference laser power of 15 mW. The data shown are for samples of weight percent of glycerol in water of 0, 5, 10, 20, 40 and 60. Because the refractive index of the glycerol (1.474) is higher than that of water, the highest glycerol content sample gives the lowest spring constant.

According to Eq. (15.11) the spring constant k_{ot} should be linearly proportional to the refractive index mismatch between the particle and the solution, that is,

$$k_{ot} \propto n_1 \frac{n_2^2 - n_1^2}{n_2^2 + 2n_1^2} \qquad (15.21)$$

We carried out experiments in mixtures of water and glycerol to test this relationship. We chose water–glycerol mixtures because they cover a broad range of viscosity and a broad range in the refractive index mismatch. We took the viscosity of water–glycerol from the values given in the *CRC Handbook of Chemistry and Physics*[17] and calculated the refractive index mismatch by linear superposition of the refractive indices of water and glycerol according to the mass of each components. Figure 15.13 shows an excellent linear relationship between the measured spring constant k_{ot} and the calculated values of the refractive index mismatch, as is predicted in Eq. (15.21).

15.5.3. Forward-Scattering Method with Two Laser Beams

It is possible to measure the phase shift in the laboratory frame and still obtain broad frequency response by a forward-scattering method. This requires two laser beams aligned colinearly at the optical trap. The first laser beam, linearly polarized, forms the stationary optical tweezers. The second laser beam, also linearly polarized but in a direction perpendicular to the first beam, forms the oscillating optical tweezers. Both optical tweezers act on the trapped particle and affect the motion of the particle. We let only the stationary laser beam reach the split photodiode detector by placing a polarizer between the optical tweezers and the detector to block the oscillating beam. The equation of motion for the particle is similar to that of Eq. (15.12), except that the effective spring constant k_{ot} on the left-hand side of the equation is replaced by the combined effective spring constant $(k_{ot1} + k_{ot2})$. With this approach, a reasonable response at high frequencies can still be achieved.

Figure 15.14 shows the data obtained by the two methods. In the dual-beam measurements the probe beam, forming the stationary optical trap, is more than an order of magnitude weaker in power than the oscillating beam. The oscillating beam here is at the same reference power as the single-beam experiments, thus we expect the spring constants measured by the two methods should be comparable. Indeed, as shown in Figure 15.14, the two sets of data gave a similar spring constant. Both sets of data in the figure are plotted (corrected) in the laboratory frame. The k_{ot} for the single-beam (trap frame) method is 4.130 mdyn/cm whereas the k_{ot} for the dual beam (laboratory frame) is 4.167 mdyn/cm. The major deviation of the data from best-fit lines that occurs at 60 Hz ($2\pi \cdot 60$ rad/s) is caused by the phase shift of the line frequency notch filter in the lock-in amplifier.

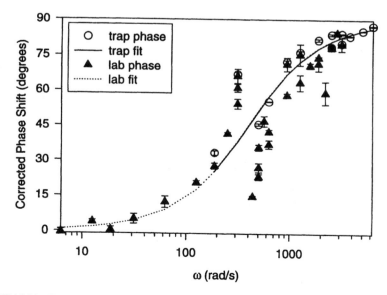

FIGURE 15.14 Comparison of the single laser beam and dual laser beam methods for measurement of the spring constant for a 1.1-μm polystyrene sphere in water at 23 °C and at 15 mW reference laser power

15.6. POTENTIAL APPLICATIONS

15.6.1. Microviscoelasticity

The equation of motion developed in Section 15.3 assumes constant solvent viscosity and elasticity. However, theory presented there can be readily extended so that the frequency-dependent microviscoelastic properties of polymer gels can be investigated. In this case, the solvent viscosity η_s and the medium elasticity k should be replaced by corresponding frequency-dependent functions. One can easily relate the frequency-dependent viscosity and elasticity to the storage and loss moduli $G'(\omega)$ and $G''(\omega)$. In general, one will need to measure both the phase and displacement response function to solve for $G'(\omega)$ and $G''(\omega)$.

Although the theory for macroscopic viscoelastic properties of associative polymer solutions is well established,[18,19] relatively little is known of the role of the adsorption of these polymers in the presence of the colloids to the rheological properties of the solution. In particular, the contrast between the case that polymers strongly adsorb and the cases that the polymers weakly adsorb or do not adsorb to the particles.[20] We suspect that the microviscoelasticity, the viscoelasticity measured within nanometer range of the particle surface, will be very different for each of the above cases. Thus, probing microviscoelasticity can provide valuable insight to the polymer–colloid interactions at the microscopic level, a key to a better understanding of the overall polymer–colloid solution rheology.[21]

15.6.2. Colloidal Forces Between Two Particles

The oscillating optical tweezers technique can also be used for probing dynamic interactions between two polymer-coated colloidal particles. In this case, the force on one of the particles is measured when a second particle, held by a separate optical tweezers, approaches.[22]

The most straightforward application might be probing hydrodynamic interactions between a pair of colloidal spheres. Theories for calculation of hydrodynamic interactions have been developed for decades,[23,24] however, few direct measurements of these interactions exist.[25] Hydrodynamic interactions occur when two particles move relative to each other. These motions are either from Brownian motion or from the forced oscillation discussed in this chapter. One ongoing experiment in our laboratory is to measure the Stokes drag of one oscillating particle as a second particle is approaching. The hydrodynamic force between particles moving in the longitudinal or shear directions is measured as a function of interparticle distance. Results of this study will be published elsewhere.

The applications to the two-particle interactions can be further extended for particles coated with polymers. Problems such as compression or shear forces between polymer brushes,[26–28] or bridging of polymers between two colloids[29,30] are just some possibilities of this kind of applications.

15.7. CONCLUSION

In this chapter, we demonstrate that it is possible to construct oscillating optical tweezers for dynamical measurements of forces on a single colloidal particle. The theoretical treatment includes the basic principle of optical tweezers and the solutions for the equations of motion of a particle driven into harmonic oscillation for both the laboratory reference frame and the optical trap reference frame.

A working experimental setup is provided to carry out tests using all three methods of detection: the imaging method, the single-beam forward-scattering method, and the dual-beam forward-scattering method. The imaging method gives reasonable results but suffers from limited frequency response. The single-beam forward-scattering method yields the best frequency response up to about 1 kHz, beyond which the results are found to be unreliable. We believe the errors at high frequencies are caused by misalignment of the optical tweezers. A phase error of a couple of degrees near the 90° phase shift can cause a large error in the data fitting. The dual-beam forward-scattering method compares well with the single-beam method and yields better data at low frequencies but cannot quite match the latter at the high frequencies.

Comparisons between experiments and theory are made for both solid polystyrene latex particles and for surfactant-stabilized hexadecane oil droplets. Water, dilute polymer in water solutions, and water–glycerol mixtures are used as solution media. Good agreements are made between the experiments and theory. Potential applications for microviscoelastic force measurements and possibilities to extend the applications of this technique to two-particle interactions are briefly mentioned.

Acknowledgments   The author would like to acknowledge his graduate students, Luke E. Dewalt, Samantha J. Parmley, Joshua C. Daghlian, and Lawrence A. Hough and his two undergraduate students, Megan T. Valentine and Christopher J. Killian, all of whom performed the experiments. LED was supported by a Grant from the Polymer Interfaces Center, an NSF-IUCRC, at Lehigh University. SJP and JCD were supported by Graduate Fellowships from the U.S. Department of Education. LAH and CJK were supported by a grant from the National Science Foundation (SGER CTS-9805887). MTV was supported by a grant from NSF-REU Program at Lehigh University and a grant from the Lehigh University Forum Undergraduate Research.

REFERENCES

1. A. Ashkin, J. Dziedzic, J. Bjorkholm and S. Chu. *Opt. Lett.* **11**, 288 (1986).

2. S.M. Block. *Nature* **360,** 493 (1992). S.C. Kuo and M.P. Sheetz. *Trends Cell Biol.* **2**, 116 (1992). K. Svoboda and S.M. Block. *Annu. Rev. Biophys. Biomol. Struct.* **23**, 247 (1994). K. Svoboda, C.F. Schmidt, B.J. Schnapp and S.M. Block. *Nature* **365**, 721 (1993). K. Svoboda and S.M. Block. *Cell* **77**, (1994) 773.

3. J.C. Crocker and D.G. Grier. *Phys. Rev. Lett.* **73**, 352 (1994).

4. A. Ashkin. *Phys. Rev. Lett.* **24** 156 (1970). A. Ashkin and J.M. Dziedzic. *Science* **187**, 1073 (1975). J.P. Barton, D.R. Alexander and S.A. Schaub. *J. Appl. Phys.* **66**, 4594 (1989). W.H. Wright, G.J. Sonek and M.W. Berns. *Appl. Phys. Lett.* **63**, 715 (1993).

5. L.P. Ghislain, N.A. Switz and W.W. Webb. *Rev. Sci. Inst.* **65**, 2762 (1994).

6. R.M. Simmons, J.T. Finer, S. Chu and J.A. Spudich. *Biophys. J..* **70**, 1813 (1996).

7. A. Ashkin, J.M. Dziedzic, J.E. Bjorkholm and S. Chu. *Opt. Lett.* **11**, 288 (1986).

8. S.C. Kuo, *J. Microscopy Soc. Am.* **1**, 65 (1995).

9. M.T. Valentine, L.E. Dewalt and H.D. Ou-Yang. *J. Phys: Condens. Matt.* **8**, 9477 (1996).

10. M.T. Valentine and H.D. Ou-Yang. 71st Colloid and Surface Science Symposium Abstracts, No. 143 (1997).

11. A. Ashkin, *Science* **210** (4474), 1081 (1980).

12. A. Ashkin. *J. Biophys.* **61**, 569 (1992).

13. J.D. Jackson, *Classical Electrodynamics* 2nd ed, Sect 4.7, Wiley, New York, 1975.

14. L.D. Landau and E.M. Lifshitz. *Fluid Mechanics* Pergaman, Oxford, 1982.

16. A.Y. Rednikov, Y.S. Ryazantsev and M.G. Velarde. *Phys. Fluids* **6**, 451 (1994).

17. D.R. Lide, *CRC Handbook of Chemistry and Physics,* 73rd ed., CRC Press, Boca Raton, FL, 1992–93.

18. F. Tanaka and S.F. Edwards. *J. Non-Newtonian Fluid Mech.* **43** 247 (1992).

19. R.D. Jenkins. Ph.D Thesis, Lehigh University, 1991.

20. Y.W. Inn and S.Q. Wang. *Phys. Rev. Lett.* **76** 467 (1996).

21. M.T. Valentine, L.E. Dewalt and H.D. Ou-Yang. *Res. Prog. Rep.* **12**, 68 (1997).

22. H.D. Ou-Yang. 71st Colloid and Surface Science Symposium Abstracts, No. 268 (1997).

23. G.K. Batchelor. *J. Fluid Mech.* **74**(1), 1 (1976).

24. S. Kim and S.J. Karrila. *Microhydrodynamics*: *Principles and Selected Applications* Butterworth-Heinemann, Boston, 1991.

25. J.C. Crocker. *J. Chem. Phys.* **106**, 2837 (1997).

26. J. Klein, D. Perahia and S. Warburg. *Nature* **352**, 143 (1991).

27. G.S. Grest. *Phys. Rev. Lett.* **76**, 4979 (1996).

28. P.S. Doyle, E.S.G. Shaqfeh and A.P. Gast. *Phys. Rev. Lett.* **78**, 1182 (1997).

29. L.E. Dewalt, Z Gao and H.D. Ou-Yang. *ACS Adv. Chem.*, **248**, 395 (1996).

30. A. Johner and J.F. Joanny, *J. Chem. Phys.* **98**, 1647 (1993).

INDEX

Acrylamide polymerization, 31
2-Acrylamido-2-methylpropane-
    sulfonate, 37
Acrylamidopropyltrimethylammonium
    chloride, 35
Acryloxyethyltrimethylammonium
    chloride, 34, 308, 311
Acryoyloxyethyltrimethylammonium
    chloride, *see* Acryloxyethyl-
    trimethylammonium chloride
Adhesion, 60
    adhesion, AFM tip, 264
    adhesion, depletion-induced, 149, 154
    adhesion, particle, 101–102, 109
    adhesion, selective, 102
    adhesion strength, 109–111
Adsorbed amount (Γ), 129, 194, 201,
    291, 341–342, 344, 347
Adsorbed layer thickness, 194, 198, 201
Adsorbed polymers, 27, 210–211, 274–
    275, 288, 309, 352, 354, 355, 360
Adsorption
    adsorption, calcium, 232, 235
    adsorption, competitive, 139, 141
    adsorption, kinetic-limited, 348
    adsorption kinetics, 109, 115, 128,
        133, 136, 237, 331, 332, 336,
        346, 349, 350, 357, 359, 373, 381

adsorption, mass-transfer limited,
    129, 136, 373
adsorption, multicomponent, 139,
    142, 351
adsorption, multilayer, 111, 117, 238
adsorption, particle, 102–103
adsorption, patchwise, 344
adsorption, polyelectrolyte, 15, 175,
    179, 181–182, 275, 309, 355,
    360–361
adsorption, polymer, 58, 106, 128,
    193
adsorption, protein, 226, 346,
    348–350
adsorption, random sequential, 103
adsorption, sequential, 360–361
adsorption, surfactant, 351, 352,
    354–361
adsorption, transport-limited, *see also*
    Adsorption, mass-transfer
    limited, 136, 348
AETAC, *see* Acryloxyethyltrimethyl-
    ammonium chloride
AFM, *see* Atomic force microscope
Aggregation
    aggregation, diffusion limited, 19, 148
    aggregation, reaction limited, 19, 148
    aggregation, reversible, 149

Alpha particles (radiation), 232
Alum, 17, 84
Alumina, 97, 216, 217, 269, 280
Aluminum hydroxide, 84
AMPS, *see* 2-Acrylamido-2-methyl-
    propanesulfonate
Anchor blocks, 297
Anionic PAM, *see* Polyacrylamide,
    anionic
Anionic polyacrylamide, *see*
    Polyacrylamide, anionic
APTAC, *see* Acrylamidopropyl-
    trimethylammonium chloride
Association, 318
Atomic force microscope, 109, 153,
    154, 156, 159, 253
  atomic force microscope cantilevers,
    257
  atomic force microscope, imaging,
    254
  atomic force microscope, surface
    force measurement, 257
Avidin, 280

Bentonite, 57
Biocompatibility, 280
Beta particles (radiation), 229–232
Biotin, 280
Bjerrum length, 177–178
Block copolymer, *see also* Diblock
    copolymer, Triblock copolymer,
    203, 297, 339, 352–354
Blocks
  blocks, anchor, 297
  blocks, buoy, 297
Bound fraction, 194, 201
Bovine serum albumin, 181, 373, 380
Bridging, 16, 57, 84, 92, 106, 110,
    118–120, 147, 165–166, 194, 308,
    312, 314, 317, 324, 403
  bridging flocculation, 57, 194
  bridging force, 275
Brownstock washing, 64
Buoy blocks, 297

Brewster angle, 332
Brewster angle microscopy, 332
BSA, *see* Bovine serum albumin

Capillary suction time, 21, 26
Carboxymethylcellulose, 170
Calcium carbonate, 14, 61, 97
Cellulose, 51, 53, 288, 305
  cellulose, carboxymethyl, 170
  cellulose, ethyl(hydroxyethyl),
    299–305
  cellulose, hemi-, 309
  cellulose, hydroxyethyl, 163, 165,
    373, 380
  cellulose, methoxy, 226
  cellulose, trimethylsilyl, 305–306
Cetyltrimethylammonium bromide,
    261, 277, 354–361
Chain expansion, 179
Charge neutralization, 6, 13, 106, 309
Charge patch, 17, 84, 88, 309
Charge reversal, 119, 271, 309
Chemical pulping, 53
Chitosan, 307
Chromium (III) oxide (Cr_2O_3) particles,
    102, 112–118
Cr_2O_3 particle layer on glass, 102, 112,
    114–118
Clarification, 22–24, 41
Clay, *see also* Bentonite, Montmorrilo-
    nite, 17
Cleveland method (AFM), 260
Cloud point, 299
CMA, *see* Acryoyloxyethyltrimethyl-
    ammonium chloride
CMC, *see* Carboxymethylcellulose
Coacervate, 166
Coadsorption, 351–352, 354–361
  coadsorption, polymer-surfactant,
    351–352, 354–361
  coadsorption kinetics, 357–359
Coalescence, 7
Coagulation, 7, 269
Cobalt aluminate, 102

Collision efficiency, 18, 20
Collision rate, 18
Colloid
 colloid destabilization, 6
 colloid–polymer interactions, 6
 colloid stability, 112, 193, 288, 331
 colloid titration, *see* Polyelectrolyte
  titration
Complexes
 complexes, hydrogen-bonded, 74
 complexes, polyelectrolyte, 52, 58,
  61–63, 91, 97, 181, 183
 complexes, polymer–surfactant, 97
Compressible surfaces, 261
Concentration fluctuations, 198
Conditioning, 15, 44
Conformational entropy, 148, 188, 296
Constant compliance region (AFM),
 257, 261–262
Contrast match, 197
Contact opportunity, 19
Core-shell particles, 163
Coumarin, 373
Critical association concentration (cac),
 318
Cr_2O_3, *see* Chromium (III) oxide
CST, *see* Capillary suction time
CTAB, *see* Cetyltrimethylammonium
 bromide

DADMAC, *see* Diallyldimethylammonium chloride, *see also*
 Poly(diallyldimethylammonium
 chloride)
DCS, *see* Dissolved colloidal substances
DDJ, *see* Dynamic Drainage Jar
Debye length (parameter), 12, 178, 269
Depletion
 depletion attraction, 148, 153, 164,
  165
 depletion flocculation, 147, 148, 161,
  164–166, 168, 170, 193
 depletion force, 8, 17, 152, 154, 156,
  161, 164

depletion-induced adhesion, 149
depletion-induced phase separation,
 148, 163–166, 170
depletion interaction, 6, 148, 149,
 150, 153, 162, 278
depletion layer, 149, 278
depletion layer relaxation, 150
depletion layer thickness, 148, 150,
 158, 161, 162, 278
depletion (pair) potential, 148,
 150–152, 166–168
depletion, polyelectrolyte, 156
depletion, polymer, 149
depletion stabilization, 151, 161, 167
Deposition, particle, 288
Derjaguin approximation, 260, 292, 299
Derjaguin–Verwey–Landau–Overbeek
 theory, *see* DLVO theory
Desorption, 127, 331, 332, 355–356,
 360, 361, 378
 desorption kinetics, 137–139, 226,
  242, 331, 332, 355–356, 361
 desorption, polymer, 127
 desorption rate, 226, 242, 299
Destabilization, colloid, 6
Dewatering, 3, 4, 22, 25, 26, 42
Dewatering testing, 40–42
Diallyldimethylammonium chloride,
 35–37
Diblock copolymer, 297
 diblock copolymer, PBO-b-PEO, 297
 diblock copolymer, PEO-b-PS, 203,
  242, 339
 diblock copolymer, PM4VP-b-PS,
  243, 244
 diblock copolymer, PS-b-PIP, 163
 diblock copolymer, PS-b-VP,
  214–216
 diblock copolymer, PS-b-2VP, 227
 diblock copolymer, PSil-b-VPO, 243
Differential cross-section, 196
Diffusion, 128–132
Diffusion limited aggregation, 19, 148
Diffusion limited kinetics, 373–374

Dipalmitoylphosphatidylcholine, 343, 353
Dissolved colloidal substances, 52
DLA, *see* Diffusion limited aggregation
DLVO theory, 9, 269
Double layer, 9, 10, 12
    double layer force, 268, 311
    double layer interaction potential, 12
    double layer repulsion, 103–104, 114
DPPC, *see* Dipalmitoylphosphatidyl-choline
Drag coefficient, 400
Dry strength (agents), 56, 57
Dual addition systems, *see* Dual polymer systems
Dual polymer systems, 83–86
Dual systems, *see* Dual polymer systems
Dynamic Drainage Jar, 59

EHEC, *see* Ethyl(hydroxyethyl) cellulose
Electrical double layer, *see* Double layer
Electroacoustic sonic amplitude technique, 112
Electrokinetic methods, 13
Electron spin resonance, 212
Electrophoretic mobility, 313, 320
Electrostatic patch, *see* Charge patch
Electrostatic potential, 12, 178, 269
Electrosteric repulsions, 22, 69, 75, 312
Electrosteric stabilization, 53
Ellipsometry, 365
Emulsions, 203, 280
Entropy, conformational (configurational), 148, 188, 296
ESA, *see* Electroacoustic sonic amplitude technique
ESR, *see* Electron spin resonance
Ethyl(hydroxyethyl) cellulose, 299–305
Equilibrium, restricted, 300
Erosion control, 5
Evanescent wave, 366, 367
Excluded volume, 179, 293
Expansion, chain, 179

FECO, *see* Fringes of equal chromatic order
Fiber optic flocculation sensor, 87
Filtration, 25, 26, 41
Floc
    floc breakage, 19, 21
    floc density, 21
    floc formation, 6, 17
    floc strength, 16, 21, 22, 57, 60, 96
    floc structure, 20–21
Flocculation, 7–8, 288
    flocculation, bridging, 57, 194
    flocculation, depletion, 147, 148, 161, 164–166, 168, 170, 193
    flocculation, patch-wise, 88, 309
    flocculation, rate, 18
Flotation, 22, 24, 41
Fluorescein, 373
FOFS, *see* Fiber optic flocculation sensor
Force
    force, bridging, 275, 312–313
    force, depletion, 8, 17
    force, double layer, 268, 311
    force, electrosteric, 275
    force, hydration, 272–273
    force, hydrophobic, 273
    force, oscillatory, 272, 279
    force, steric, 8, 17, 274, 300
    force, structural, 272
    force, surface, 253, 288
    force, trapping, 389
    force, van der Waals, 312
Form factor, 197
Fractal dimension, 21
Fringes of equal chromatic order, 290
Froths, 280

Gamma particles (radiation), 232
Glass, 102, 110, 112, 114, 118–120, 203, 275, 273, 281, 373
    glass, Vycor, 203
Glycerol, 401
Gold, 226–227, 266, 267, 274

Gouy–Chapman model, 12
Graphite, 266, 267, 272

Hamaker constant (coefficient), 11, 265
HEC, *see* Hydroxyethylcellulose
Hemicellulose, 309
Hexadecyltrimethylammonium
    bromide, *see* Cetyltrimethyl-
    ammonium bromide
Hindered settling, 25
Hit-and-stick, 103, 108
Hydration, 272, 295
Hydration force, 273
Hydration, secondary, 273
Hydrodynamic interactions, 403
Hydrogen-bonded complexes, 74
Hydrogen bonds, 295–296
Hydrolyzed PAM, *see* Polyacrylamide,
    hydrolyzed
Hydrophobic interaction, 273, 297
Hydrophobically modified latex, 163
Hydrophobically modified silica, 154,
    162, 163–164, 213, 273, 278
Hydroxyapatite, 217
Hydroxyethylcellulose, 163, 165, 373,
    380

IEP, *see* Isoelectric point
Imaging, atomic force microscope, 254
Interparticle structure factor, 197
Ion exchange, 58, 234, 235
Iron oxide, 102
Isoelectric point, 13, 14, 267, 269
Isotopes, 209, 232

Jamming limit, 104, 120
Jar test, 18, 41, *see also* Dynamic
    Drainage Jar

Kinetics, 338

kinetics, adsorption, 109, 115, 128,
    133, 136, 237, 331, 332, 336,
    346, 349, 350, 357, 359, 373, 381
kinetics, coadsorption, 357, 359
kinetics, desorption, 137–139, 331,
    332, 361
kinetics, diffusion limited, 373–374
kinetics, polymerization, 31
Kraft process, 53, 63
Kuhn model, 177

Labeled polymers, 233–234
Langmuir adsorption model, 132
Langmuir–Blodgett, 266, 275, 290, 306
Larmor frequency, 208
Lattice mean field theories, 293
Layer thickness, *see* Adsorbed layer
    thickness
Lévêque equation, 131, 349
Lifshitz theory, 11, 267
Lignin, 51, 61
Lipid, 154
Lipid monolayer, 339, 343, 353
London–van der Waals interactions, 10,
    11, *see also* Van der Waals
    interactions
Loops, 211
Lysozyme, 350

MAPTAC, *see* Methacrylamidopropyl-
    trimethylammonium chloride
MASIF, 292
Mass-transfer limited adsorption, 129,
    136, 373
Mass transport, 128
Mean field theory, 133–134, 153, 194,
    293
Mechanical pulping, 52
Methacrylamidopropyltrimethyl-
    ammonium chloride (MAPTAC),
    35, 311, 312
Methoxycellulose, 226

Mica, 154, 156, 242, 266, 267, 271, 272, 273
Microparticle, 57, 84, 96
Micro-viscoelasticity, 403
Mixing, 19
Molecular modelling, 74
Molecular weight
   molecular weight effect on bridging, 119
   molecular weight effect on phosphor haze, 118–119
Monolayer
   monolayer, insoluble, 353
   monolayer, lipid, 339, 343, 353
Monte Carlo simulation, 181–183, 186–188, 194, 312, 317
Montmorrilonite, 84, 97
Multilayer adsorption, 111, 117, 238

NaPSS, *see* Poly(sodium styrene sulfonate)
Neutron scattering, *see* Small angle neutron scattering
Neutron scattering cross-section, 195
NMR, *see* Nuclear magnetic resonance
Normalized parallel reflectivity, 338
Nuclear magnetic resonance, 207–210

Octamethylcyclotetrasiloxane, 272
OMCTS, *see* Octamethylcyclo-tetrasiloxane
Optical model (reflectometry), 343
Optical tweezers, 385–386
Orthokinetic, 18

PAA, *see* Poly(acrylic acid)
PAC, *see* Polyaluminum chloride
Pake pattern, 218
Palladium, 102
Paper making, 51, 55
Particle
   particle adhesion, 102, 109–110
   particle adsorption, 102–103

particle deposition, 288
particle layers, thick, 120
particle layers, thin, 112, 114, 120
Patch-wise flocculation, 309
Patchy surface, 270, 273, 344
PCMA, *see* Poly(acryoyloxyethyl-trimethyl- ammonium chloride)
PDADMAC, *see* Poly(diallyldimethyl-ammonium chloride)
PDMS, *see* Poly(dimethylsiloxane)
Peclet number, 131
PEG, *see* Poly(ethylene glycol)
PEI, *see* Poly(ethyleneimine)
Pentaoxyethylene dodecyl ether, 293
PEO, *see* Poly(ethylene oxide)
Phase separation, 6, 148, 149, 162–166, 170, 293, 296
   phase separation, depletion-induced, 148, 163–166, 170
Phosphors, 102, 117–120
PIPA, *see* Poly(isopropylacrylate)
PMMA, *see* Poly(methylmethacrylate)
Poisson–Boltzmann distribution (equation), 268–269
Polyacrylamide, 27, 31, 32, 226
   polyacrylamide, anionic, *see also* Poly(acrylamide-co-acrylate), Poly(acrylamide-co-sodium 2-acrylamido-2-methylpropanesulfonate), Polyacrylamide, hydrolyzed, 37, 84–85
   polyacrylamide, cationic, *see also* Poly(acrylamide-co-acryoyloxy-ethyl trimethylammonium chloride), Polyacrylamide, Mannich, Poly(AMD-AETAC), 34–35
   poly(acrylamide-co-acrylate), 17, 33, 37–38
   poly(acrylamide-co-acryoyloxyethyl trimethylammonium chloride), 34
   poly(acrylamide-co-sodium 2-acryl-amido-2-methylpropane-sulfonate), 38

polyacrylamide, hydrolyzed, 33, 84
polyacrylamide, Mannich, 32–33
polyacrylamide modification, 32
polyacrylamide, sulfomethylated, 33
Polyacrylate, 14, 29
Poly(acrylic acid), 108, 159–161,
  165–166, 170, 275
Poly(acryloxyethyltrimethylammonium
  chloride), *see* Poly(acryoyloxyethyltrimethylammonium chloride)
Poly(acryoyloxyethyltrimethylammonium chloride), 311, 319
Polyaluminum chloride, 17
Poly(AMD-AETAC), *see* Poly(acrylamide-co-acryoyloxyethyl
  trimethylammonium chloride)
Polyamidoamine, 84–85
Polyamine, 37
Poly(butanyl viologen), 245
Poly(butyleneoxide-*b*-ethyleneoxide),
  297
Poly(diallyldimethylammonium
  chloride), 35–37, 61–63, 88–96
poly(diallyldimethylammonium
  chloride–co-acrylamide), 37, 114
poly(diallyldimethylammonium
  chloride)-lignin complex, 61–63
Poly(dimethylsiloxane), 154–155, 162,
  168, 217, 278
Polydispersity, 168, 378, 381
Polyelectrolyte, 27, 275, 288, 309, 352,
  354, 355, 360
  polyelectrolyte adsorption, 15, 175,
    179, 181–182, 309, 355,
    360–361
  polyelectrolyte chemistry, 27
  polyelectrolyte complexes, 52, 58,
    61–63, 91, 97, 181, 183
  polyelectrolyte depletion, 156
  polyelectrolyte effect, 27
  polyelectrolyte rearrangement, 15,
    315
  polyelectrolyte–surfactant mixtures,
    288, 318–325
  polyelectrolyte synthesis, 30
  polyelectrolyte titration, 61
Poly(ethyleneglycol), 154–155, 274
Poly(ethyleneimine), 60, 84–85, 97
Poly(ethyleneoxide), 27, 73–79,
  134–139, 197, 203, 213, 274–275,
  294–297, 373–374, 378–380, 382
  poly(ethyleneoxide) clusters, 77, 296
  poly(ethyleneoxide)-cofactor
    complexes, 73, 74–75, 77
  poly(ethyleneoxide), C_{16} end capped,
    394
Poly(ethylene oxide-b-styrene), *see*
  Diblock copolymer, PEO-b-PS
Poly-L-glutamic acid, 217
Polyimides, 217
Poly(isopropylacrylate), 217
Polylysine, 217, 354–361
Polymer(s)
  polymer, adsorbed, 27, 210–211,
    274–275, 288, 309, 352, 354,
    355, 360
  polymer adsorption, 58, 106, 127, 193
  polymer bridging, *see* Bridging
  polymer depletion, 149
  polymer desorption, 127
  polymer-enhanced brownstock
    washing, 63–66
  polymer, labeled, 233, 234
  polymer, nonionic, 274, 293
  polymer–polymer systems, *see* Dual
    polymer systems
  polymer stabilized sol, 102
  polymer–surfactant complexes, 97,
    351
  polymer synthesis, 30
Poly(methylacrylate), 214, 218
Poly(methylmethacrylate), 164, 213
Poly(*N*-methyl-2-vinylpyridine), 237,
  238
Poly(propylene), 273
Poly(*p*-vinyl phenol), *see* Poly(vinyl
  phenol)

Poly(siloxane-*b*-vinylpyridinoxide), *see* Diblock copolymer, PSil-b-VPO
Poly(sodium styrenesulfonate), 156–159
Polystyrene, 197, 213, 226, 227, 242, 373
Poly(styrene-*b*-methyl-4-vinylpyridine), *see* Diblock copolymer, PS-b-4VMP
Poly(styrene-*b*-vinylpyridine), *see* Diblock copolymer, PS-b-VP
Poly(styrene-*b*-2-vinylpyridine), *see* Diblock copolymer, PS-b-2VP
Poly(styrene-*b*-4-vinylpyridine), *see* Diblock copolymer, PS-b-4VP
Polystyrene latex, 60, 61, 75, 165, 170, 201, 213, 271, 394, 398, 400
Poly(styrene-*ran*-*t*-butylstyrene), 239–242
Poly(styrenesulfonate), 213, 239, 245, 278
Poly(styrenesulfonate-*b*-*t*-butylstyrene), *see* Diblock copolymer, PSS-b-tbS
Polystyrene terminally grafted silica, 199
Poly(tetrafluoroethylene), 266–267
Poly(vinyl acetate), 218
Poly(vinyl alcohol), 27, 119, 213, 217
Poly(vinylimidazole), 106, 109, 111
Poly(vinylimidazole), methylated, 106, 109, 111
Poly(vinylmethylether), 165
Poly(vinyl phenol), 74
Poly(vinylpyridine), 179
Poly(2-vinylpyridine), 275
Poly(vinylpyrrolidone), 118–119, 213
Poly(vinylpyrrolidone-co-vinyl acetate), 216
Porod law, 199
Porous media flow, 26–26
Porous solids, 203
Precipitation, 97
Primary solids, 4
Protein
    protein adsorption, 226, 346, 348–350
protein, purple membrane, 280
protein rejection, 295
protein separation, 181
PS-b-PIP, *see* Diblock copolymer, PS-b-PIP
PTFE, *see* Poly(tetrafluoroethylene)
Pulping, chemical, 53
Pulping, mechanical, 52
Purple membranes, 280
PVA, *see* Poly(vinyl alcohol)
PVAc, *see* Poly(vinyl acetate)
PVI, *see* Poly(vinylimidazole)
PVME, *see* Poly(vinylmethylether)
PVP, *see* Poly(vinylpyrolidone)
P2VP, *see* Poly(2-vinylpyridine)
P4VP, *see* Poly(4-vinylpyridine)

Radiochemical methods, 226
Random sequential adsorption, 103
Reaction limited aggregation, 3, 148
Rearrangement, polyelectrolyte, 315
Red-ox polymerization, 30
Reflectometry, 104, 105, 108, 134–135, 139, 143, 331–333, 336, 341, 343, 349, 351–353, 361
    reflectometry, dynamic scanning angle, 338
    reflectometry, scanning angle, 338, 341, 343, 353
Refractive index, 291, 342
Refractive index increment, 342
Retention aids, 57
Reversible aggregation, 149
Rheology, suspension, 17, 113
RLA, *see* Reaction limited aggregation
Root mean square thickness, 201
RSA, *see* Random sequential adsorption
Rutile, 267, 271

Salt effects, 116, 237, 267, 313–315
SANS, *see* Small angle neutron scattering
Scaling theory, 158, 194, 293

Scattering length, 195
Scattering vector, 196
Scheutjens–Fleer theory, 134–136, 139, 165, 194
Scheutjens–Fleer–Vincent theory, 158
Schopper–Riegler tester, 87
Scintillation, 228–229
Scintillator, liquid, 228
Scintillator, plastic, 228–229
SCMFT, *see* Self-consistent mean field theory
SDS, *see* Sodium dodecyl sulfate
Second moment, 201
Secondary solids, 4
Secondary treatment, 4
Sedimentation (of suspensions), 24–25, 113, 114
Self-consistent mean field theory, *see* Mean field theory
Self exchange, 128, 139–142, 378–380
Settling, *see* Sedimentation (of suspensions)
SFA, *see* Surface forces apparatus
SF theory, *see* Scheutjens–Fleer theory
SFV theory, *see* Scheutjens–Fleer–Vincent theory
Shape factor, 196
Shear plane, 13, 110, 271
Shear rate, effective, 18
Silica, 57, 84, 96–97, 106, 108, 109, 111, 118–119, 134, 136, 139, 154, 156, 158, 159, 161, 164, 165, 170, 213–214, 217–222, 266, 268, 270–273, 275–277, 281, 308, 373, 379–380
silica, hydrophobically modified, 154, 162, 163–165, 168, 199, 213, 273, 278, 305
silica, polystyrene terminally grafted, 199
Silicon nitride, 266, 269, 271, 274, 275
Single particle optical sizing, 18
Small angle neutron scattering, 166–167, 193, 194

Snell's law, 368
Sodium dodecyl sulfate, 279, 282, 319–324
sodium dodecyl sulfate micelles, 279
Soil conditioning, 5
Sol, polymer stabilized, 102
Solid–liquid separation, 3, 22
Solvency, 288, 296
Specific resistance to filtration, 26
SPOS, *see* Single particle optical sizing
Spring constant, 257, 389, 390, 392, 393, 394, 398
SRF, *see* Specific resistance to filtration
Stability
    stability, colloid, 112, 193, 288, 331
    stability factor, 10
    stability, suspension, 112–113, 114–116, 119
Stabilization
    stabilization, depletion, 151, 161, 167
    stabilization, electro-steric, 53
    stabilization, enhanced steric, 288
    stabilization, steric, 297
    stabilization, suspension, 114–115
Stagnation point flow, 60, 105, 108, 109, 115
Starch, 57
Steric
    steric force, 8, 17, 274, 300
    steric interaction, 194
    steric stabilization, 297
    steric stabilization, enhanced, 288
    steric stabilizer, 297
Streptavidin, 280
Stretched exponential, 185
Structural forces, 272
Surface
    surface charge, 268–269, 271
    surface charge density, 12
    surface, compressible, 261
    surface excess, *see* Adsorbed amount
    surface force measurement, 253, 257
    surface forces, 253, 288

Surface (*Continued*)
  surface forces apparatus, 154, 156, 288
  surface, patchy, 270, 273, 344
  surface potential, 12, 13, 268, 269
  surface potential, constant, 271
  surface-to-volume ratios, 297
Surfactant, 354
  surfactant adsorption, 351, 352, 354–361
  surfactant aggregate structures, 275
  surfactant–polyelectrolyte mixtures, 288, 318–325
  surfactant–polymer coadsorption, 351–352, 354–361
Suspension
  suspension rheology, 113
  suspension stability, 112–115
SVP, *see* Diblock copolymer, PS-b-VP
Swelling ratio, 179
Synperonics™, *see* Triblock copolymer, PEO-PPO-PEO

Tails, 211
Teflon™, *see* Poly(tetrafluoroethylene)
Testing methods, 40–42, 59, 87, 113
Theory
  theory, DLVO, 9, 268–269
  theory, lattice mean field, 293
  theory, Lifshitz, 11
  theory, mean field, 134, 154, 194, 293
  theory, scaling, 158, 194, 293
  theory, Scheutjens–Fleer, 165, 194
  theory, Scheutjens–Fleer–Vincent, 158
  theory, self-consistent mean field, 134
Theta solvent, 293
Thickening, 22, 38
TIRF, *see* Total internal reflectance fluorescence
TIRM, *see* Total internal reflection microscopy
Titania, 271

Total internal reflectance fluorescence, 134, 365, 366
Total internal reflection microscopy, 158
Trains, 211
Transport limited adsorption, *see* Adsorption, mass-transfer limited
Trapping force, 389
Triblock copolymer, PEO-PPO-PEO, 163
Trimethylsilyl cellulose, 305–306
Trisodium citrate, 274
Turbidity, 320

Van der Waals force, *see also* London–van der Waals interactions, 312
Van der Waals interactions, *see also* London–van der Waals interactions, 265
Veolcity gradient, *see* Shear rate, effective
Viscoelasticity, 386
Vitamin H, *see* Biotin
Volume fraction profile, 194
VPS, *see* Diblock copolymer, PS-b-VP

Waste activated sludge, 20, 40
Wastewater treatment, 4, 38, 39
Wet strength (agents), 57
Wood fibers, 53–54

Y_2O_3, *see* Yttrium oxide, yttria
Y_2O_2S, 119
Yttrium oxide, yttria (Y_2O_3), 102

Zero point of charge, *see* Isoelectric point
Zeta potential, 13, 271
Zinc sulphide (ZnS), 118, 267, 272
Zirconia, 271, 275
ZnS, *see* Zinc sulphide